D0212569

The
Physical Chemistry of
MEMBRANES

of related interest

Lipids in plants and microbes
J. L. Harwood and N. J. Russell

The Physical Chemistry of MEMBRANES

An Introduction to the Structure and Dynamics of Biological Membranes

Brian L. Silver

TECHNION—ISRAEL INSTITUTE OF TECHNOLOGY
HAIFA, ISRAEL

ALLEN & UNWIN
Boston · London · Sydney

The SOLOMON PRESS
New York

Permission requests should be addressed to: **Publishers Creative Services Inc., 89–31 161 Street, Suite 611, Jamaica, New York 11432, USA**

First published in 1985

This book is a joint publication of **Allen & Unwin** and **The Solomon Press** (a division of Publishers Creative Services Inc.)

It is distributed by Allen & Unwin:

Allen & Unwin Inc., Fifty Cross Street, Winchester, Mass. 01890, USA

George Allen & Unwin (Publishers) Ltd,
40 Museum Street, London WC1A 1LU, UK

George Allen & Unwin (Publishers) Ltd,
PO Box 18, Park Lane, Hemel Hempstead, Herts HP2 4TE, UK

George Allen & Unwin Australia Pty Ltd, 8 Napier Street, North Sydney, NSW 2060, Australia

The publishers and author thank those individuals and organizations who have given permission for the use of their copyright material. Any omissions or errors in crediting any copyright holders are unintentional and will be rectified at the earliest opportunity.

British Library Cataloguing in Publication Data

Silver, Brian
The physical chemistry of membranes : an introduction to the structure and dynamics of biological membranes.
1. Membranes (Biology) 2. Biological chemistry
I. Title
574.8'75 QH601

ISBN 0-04-574028-3

Library of Congress Cataloging in Publication Data

Silver, Brian L.
The physical chemistry of membranes.

Bibliography: p.
Includes index.
1. Cell Membranes. 2. Bilayer lipid membranes.
3. Cytochemistry. 4. Chemistry, Physical organic.
I. Title. [DNLM: 1. Cell Membrane. 2. Chemistry, Physical. QH 601 S587p]
QH601.S48 1985 547.87'5 84-24407
ISBN 0-04-574028-3 (alk. paper)

Book design by Raymond Solomon
Set by Mathematical Composition Setters Ltd, Salisbury, UK
Printed and bound in Great Britain by
Anchor Brendon Ltd, Tiptree, Essex

For Yoni

One day,
in the middle of the night . . .

Preface

This book is an account of what physical chemistry has to say about the structural, electrical and transport properties of biological membranes and their simplest model—the lipid bilayer. The accent throughout is on basic ideas. In contrast to the essentially descriptive approach characteristic of texts on membrane biochemistry, our underlying themes are the role of force and entropy in maintaining membrane organization, in determining the electric fields and ionic environment of membranes, and in regulating the passage of molecules and ions across membranes. Although experimental findings will always be the touchstone against which theory will be tried, no attempt is made to present an exhaustive survey of experimental data. On the other hand, there is discussion of the nature and limitations of the results obtainable by the major laboratory techniques. The treatment is at the level of an advanced undergraduate course or an introductory survey suitable for postgraduate students carrying out research in biochemistry, biophysics, or physiology. The mathematical demands on the reader are trivial. The few forbidding equations appearing in Chapter 7 are soon whittled away to simple practical expressions. Although the current–voltage characteristics of nerves are traditionally the province of biophysics rather than physical chemistry, certain aspects relevant to the electrical activity of nerves are nevertheless included in this text, namely, membrane and diffusion potentials and conductivity fluctuations.

Where rival theories exist, conflicting convictions have been presented, but not necessarily accorded equal approbation. The author has a viewpoint.

The overwhelming majority of original research work referred to in the book was published after 1970. In general older papers are evoked when they fall within the categories of unavoidable classics or seminal studies. In the early papers, the student will meet angstroms, kilocalories, and other relics of the past. I have retained these units in quoting experimental results rather than destroy their antique charm by conversion to SI units.

Part of this book was written while I enjoyed the hospitality of the Department of Chemistry of the Imperial College of Science and Technology, London. In particular I wish to thank Dr. John Gibson. Professor J. Barber of the Department of Pure and Applied Biology contributed greatly to my knowledge of matters photosynthetical.

The manuscript was typed by an international bevy of impressively accurate typists, the bulk of the work being shared by Norma Jacobs, Charlotte Diament, and Doreen Walsh in Haifa and Nadine Green in London.

Raymond Solomon of Publishers Creative Services was patient and helpful far beyond the call of duty as deadline after deadline bit the dust.

Table of Contents

List of Figures

Figure

Figure

Figure

Figure

Figure

Figure

Figure

Figure

Figure

Figure

List of Tables

Introduction

Our image of the cell membrane has progressed, in this century, from a structureless, almost functionless packaging for the cytoplasm to a complex organ responsible for many of the most fundamental processes characterizing the living cell. Up to the late 1950s, the spectacular advances were made on the biochemical front. Membrane-bound proteins were found to be capable, among other things, of effecting the active transport of ions and molecules, to be implicated in immunological reactions, to be the site of the initial steps in photosynthesis, and to be involved in oxidative phosphorylation. Although our understanding of protein function has deepened immensely over the past decade, the molecular structure of membrane proteins remains on the whole obscure. By contrast it was early suspected that the proteins swim in a ubiquitous sea—a bilayer of lipid molecules.

The building blocks of the bilayer were long ago identified as amphiphilic molecules; that is, molecules containing both hydrophobic and hydrophilic regions. Standard chemical structure determination has established their primary structure, showing them to be, in the main, phospholipids with long hydrocarbon chains attached to polar head groups. By the mid-1930s it was already fashionable to depict membranes as flexible bilayers impregnated with lumps of protein. This book takes up the story from that point but not from that date. It was not until the 1950s that real progress was made in understanding the behavior of membranes in terms of their structure. The upsurge in membrane research and the torrent of papers now devoted to membranes are very largely a result of technical innovations in the laboratory. The separation and characterization of complex mixtures has undergone a revolution, and molecular structure and dynamics are under attack from a range of powerful techniques, such as nuclear magnetic resonance, electron spin resonance, fluorescence spectroscopy using lasers, and high-sensitivity calorimetry. Physical chemistry has driven deep into membrane science and has enabled us to discuss the stability of the lipid bilayer in terms of intermolecular forces, to analyze its electric field in terms of classic electrochemistry, to explain intermembrane forces with the aid of the modern theory of dispersion forces, to account for the transport of ions in terms of diffusion equations or activated rate theory, and to understand the time-dependent electrical conductivity of nerve membranes in terms of the theory of stochastic processes. The following chapters contain a general view of the physical chemistry of membranes. We keep well away from biochemical aspects since these have been the subject of several monographs. Our concern will be with structure, forces, thermodynamics, and kinetics. We will see how the traditional concepts of physical chemistry are rapidly laying a rational foundation for the observed physical properties and biological functions of membranes.

Our starting point is the crystal structure of pure phospholipids, which will be seen to have strong similarities to the arrangement and molecular conformation of the bilayer portion of natural membranes. Anhydrous lipids undergo conformational changes on heating that are understandable mainly in terms of the disordering or "melting" of all-trans hydrocarbon chains.

Since natural membranes exist in an aqueous mileu, it is necessary to examine some of the many phases adopted by hydrated phospholipids. Hydrated phospholipids can be induced to adopt a rich variety of phases by a change in temperature and variation in water content. Hydrated phospholipids at physiological temperatures exist in a number of forms, including bilayers, micelles, and the closed spherical structures known as vesicles or liposomes. It will be shown that by using simple considerations of molecular shape, it is possible to predict the particular phase formed by a given phospholipid as a function of temperature, pH, and the nature of the cations in the ambient solution.

In biological membranes the phospholipids have melted chains and the chain conformations are continually changing. Diffraction and magnetic resonance methods have provided a means of quantitatively specifying the average conformational state of the individual chain segments in melted chains. The state of the chains in bilayers is strongly affected by the addition of proteins and the interactions between lipids and other membrane components are a major area of experimental and theoretical investigation. There is rapidly mounting evidence that the properties of lipids are very relevant to the functioning of membrane proteins and to the interactions between cells. The complex phase diagrams of hydrated lipids and the vital biological importance of protein–lipid interactions have prompted theoretical attacks on the structure and dynamic properties of bilayers and membranes. A great range of techniques has been employed, from the use of cardboard disks to represent molecules to quantum mechanical calculations. Any realistic model of a molecular ensemble needs to take account of geometry and forces. In Chapter 7 we consider the forces that maintain the integrity of bilayers, control their ionic environment, and determine the forces between membranes. We will find it necessary to enter the modern theory of van der Waals' forces and to accept the existence of huge repulsive forces between adjacent hydrated bilayers. In selecting for perusal theoretical studies on bilayers or membranes, I have concentrated on classical statistical thermodynamics, on calculations based on the topology of simple geometrical models, and on the recent use of molecular dynamics. Inevitably quantum mechanics has been mobilized by some investigators; however, I barely mention these efforts as they are so few and as yet are based on oversimplified models. Every theory outlined in the text suffers from its own approximations and enjoys its own success. Among the areas that theory has illuminated to date are the nature of the lipid chain-melting transition, the average conformational state of the melted chains, and the effect of proteins on these chains.

An understanding of the static properties of membranes is an essential prelude to the study of the movement of molecules within the membrane. The lateral and rotational diffusion of proteins and lipids in the plane of the membrane has been observed via a number of techniques, and although the quantitative prediction of diffusion rates is still beyond us, satisfying qualitative explanations are beginning to emerge for the effects of protein and cholesterol concentration on lipid diffusion. The lateral diffusion of proteins and other membrane components is implicated in a number of important phenomena, such as the intramembranal transfer of electrons in photosynthesis and the redistribution of proteins on the cell surface in some immunological reactions. The diffusion of one or more membrane components may be a rate-determining step in these processes. In Chapter 10 we present a survey of the principle observations and theories concerning lateral and rotational diffusion in membranes.

The movement of molecules through membranes has not surprisingly attracted perhaps more experimental and theoretical attention than any other aspect of membrane science. Nineteenth century theories of the transport of matter through liquids were based on phenomenological diffusion equations, such as Fick's first law, which assumed the medium to be a continuum and defined a diffusion constant the magnitude of which was taken to be an indication of the frictional resistance to molecular motion through the medium. It was natural to use a similar formalism when the problem of ion transport through membranes was first attacked. The Nernst-Planck equation, the central result of the approach, takes into account both simple diffusion and the effect on ion transport of electric potential gradients across the membrane. The first major application of the theory to real systems was the famous Goldman-Hodgkin-Katz (GHK) expression, which accounted successfully for the relationship between the extra-membranal concentrations of ions and the potential difference across the membrane. In Chapter 12 the GHK equation is derived and its failures are also hinted at. The apparent inapplicability of diffusional equations to a growing number of experimental systems prompted an increased appeal to activated rate theory in which ions passing through membranes are presumed to move over a hilly potential energy surface. In Chapter 13 the basic formalism of the rate theory is described and developed. One need not know how to swim to cross a river; one alternative is to row across. Molecular boats, capable of ferrying ions and molecules across membranes, have been isolated and characterized. In Chapter 14 there is a discussion of the kinetics of transport by carriers in which it is shown that a number of at first-sight puzzling phenomena are explicable in terms of simple equations and ideas.

Thermodynamics appears throughout our story, often implicitly, but explicitly in Chapters 8 and 15. Natural membranes are nonideal mixtures but contain too many components to be suitable subjects for theory, and thermodynamic analysis of membrane models and bilayers has concentrated on binary mixtures. The treatment for mixtures of lipids is followed through using standard equilibrium thermodynamics. The effect of proteins on the free energy of lipid bilayers has been analyzed in terms of the effect of rigid impurities on the average conformational state of the lipid chains. The results have been used to account for the nature of the thermal transition in bilayers and also to suggest the presence of lipid-mediated forces between protein molecules. In the 1950s irreversible thermodynamics, which deals with systems not at equilibrium, was directed at biophysics. The two main targets were the "passive" transport of water and the "active" transport of ions and molecules through membranes. The theory promised somewhat more than it delivered. In Chapters 15 and 16 we take a look at the methodology of irreversible thermodynamics, some of its successes, and one or two misapplications. The reader will again be conscious of the gap between our primitive knowledge of membrane protein structure and our ability to predict or account for the general characteristics of the dynamic membrane processes presided over by protein molecules.

Macroscopic equilibrium does not preclude microscopic fluctuation. In membranes one of the clearest symptoms of fluctuations is the time-varying electrical conductivity of membranes, which is often a consequence of the random swinging to and fro of the molecular "gates" at the entrances to the channels through which ions carry charge. In Chapter 17 enough of the elementary theory of random processes is presented to allow the reader to appreciate the way in which kinetic information can be extracted from experimental data. Basically the same theory is shown to be relevant to the interpretation of certain spectroscopic observations on lipid chains.

The theories and laboratory methods of physical chemistry have reached beyond the problems presented by model systems intended to replicate some of the properties of real membranes. The membranes of cells are increasingly the subject of experimental probing and the structural and dynamic features of simple models have been demonstrated to be replicated in living systems. The physical chemistry of membranes is coming of age.

1 The Solid-State Structure of Lipids

It has been nearly 100 years since Overton[1] found that the rates at which uncharged molecules pass through the membranes of plant cells are roughly correlated with the solubility of the same molecules in fats. He concluded that membranes are composed largely of lipids. In Figure 1 there are four representations of one type of lipid molecule — polar head group upmost with non-polar tails trailing downwards. Such molecules, containing both strongly hydrophobic and hydrophilic groups, are termed amphiphilic. Figure 2 shows the chemical structure of the main varieties of lipid found in the membranes of living cells. Depending on the origin of the cell, lipids contribute from ~20% to ~80% of the weight of the membrane[2]. In the next few chapters we will be concerned almost exclusively with phospholipids, the properties of which are uncannily suited to building the thin, flexible, yet stable, barrier that separates the cell contents from the rest of the universe.

Over 50 years ago Gorter and Grendel[3] proposed that membrane lipids were organized into a sheet, two molecules thick, with polar head groups facing outwards as shown in Figure 3. In 1935 Danielli and Davson[4], on slender experimental evidence, correctly suggested that proteins were major structural components of membranes. Since, until recently, not much was known about membrane proteins, it was, and still

Figure 1. Four representations of a typical phospholipid molecule:

A. Chemical structure.

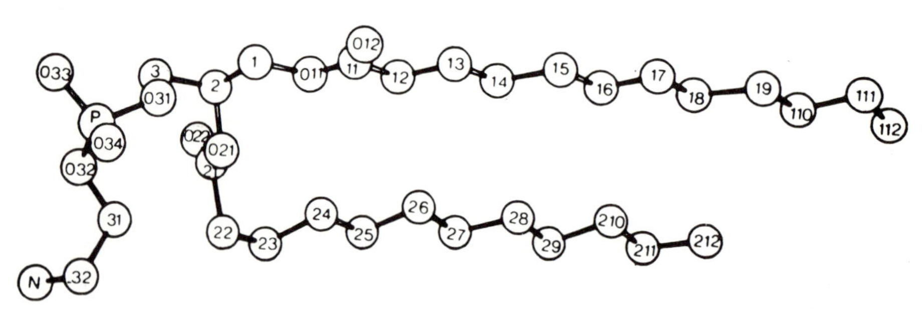

B. Conformation in the crystalline state—for an explanation of the numbering system see text. [From Hitchcock, P. B., et al., *Proc. Nat. Acad. Sci.* 71:3036 (1974).]

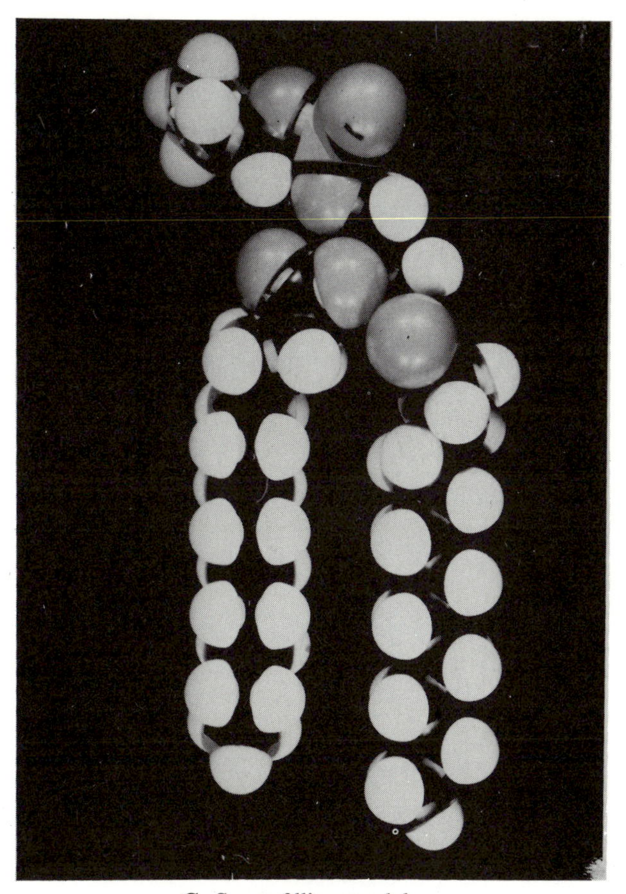

C. Space-filling model.

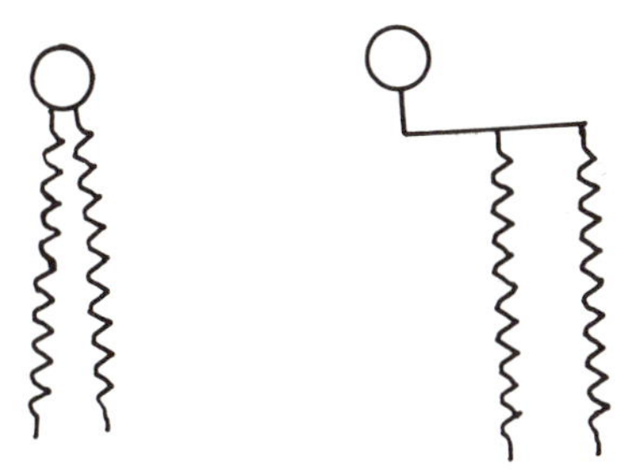

D. Diagrammatic sketch.

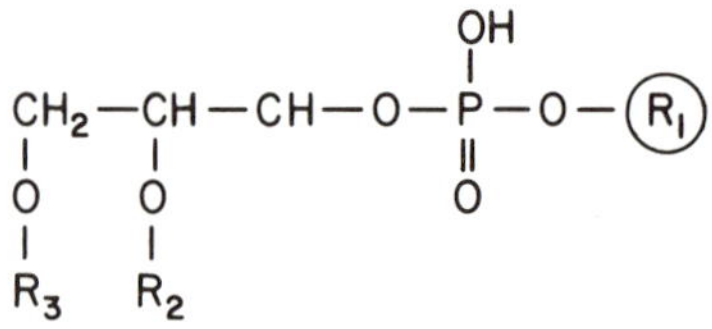

$$
\begin{array}{l}
\qquad\qquad\qquad\qquad\quad OH \\
\qquad\qquad\qquad\qquad\quad | \\
CH_2-CH-CH-O-P-O-(R_1) \\
\,| \qquad\; | \qquad\qquad\qquad \| \\
O \qquad O \qquad\qquad\quad\;\; O \\
\,| \qquad\; | \\
R_3 \quad\; R_2
\end{array}
$$

R_1	Name	Some common derivatives
$-H$	Phosphatidic Acid	$R_1 = R_3 =$ Palmitic Acid. DPPA (Dipalmitoylphosphatidic Acid
$-CH_2-CH_2-NH_2$	Phosphatidylethanolamine	$R_2 = R_3 =$ Palmitic Acid. DPPE
		$R_2 = R_3 =$ Lauric Acid. DLPE
$-CH_2-CH_2-\overset{+}{N}(CH_3)_3$	Phosphatidylcholine	$R_2 = R_3 =$ Palmitic Acid. DPPC (Lecitin)
		$R_2 = R_3 =$ Lauric Acid. DLPC
		$R_2 = R_3 =$ Myristic Acid. DMPC
$-CH_2-CH(NH_2)-C(=O)OH$	Phosphatidylserine	$R_2 = R_3 =$ Palmitic Acid DPPS
$-CH_2-CH(OH)-CH_2-OH$	Phosphatidylglycerol	$R_2 = R_3 =$ Palmitic Acid DPPG
$-\overline{CH-(CHOH)_4-CH}-OH$	Phosphatidylinisitol	

Figure 2A. Common head groups of naturally occurring phospholipids.

I. Saturated $CH_3(CH_2)_n COOH$

n	Common Name	Shorthand nomenclatures		
10	Lauric Acid	12:0	$C_{12:0}$	C_{12}
12	Myristic Acid	14:0	$C_{14:0}$	C_{14}
14	Palmitic Acid	16:0	$C_{16:0}$	C_{16}
16	Stearic Acid	18:0	$C_{18:0}$	C_{18}

II. Monounsaturated. $CH_3(CH_2)_n CH{=}CH(CH)_m COOH$

m	n	Common Name	Shorthand nomenclatures
7	5	cis−9−16:1	Palmitoleic Acid
7	7	cis−9−18:1	Oleic Acid
9	5	cis−11−18:1	cis Vaccenic Acid

Figure 2B. Some fatty acids commonly found in natural membranes.

Cholesterol

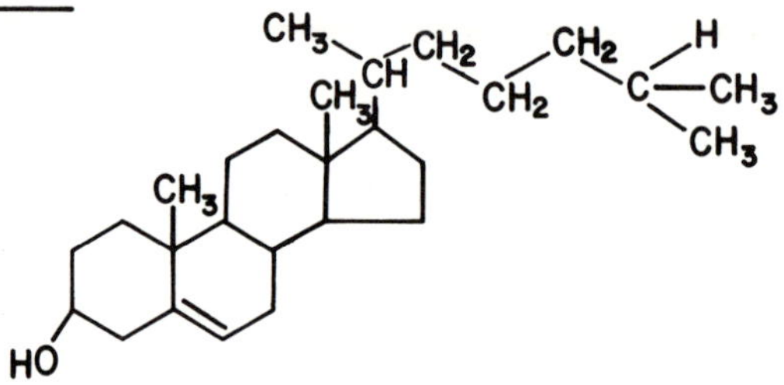

Sphingomyelin

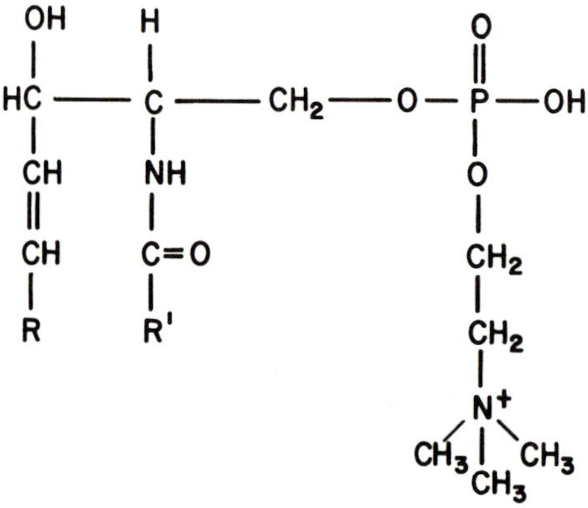

Sphingosines

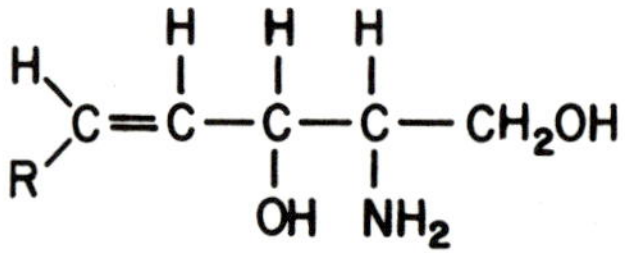

Cardiolipin

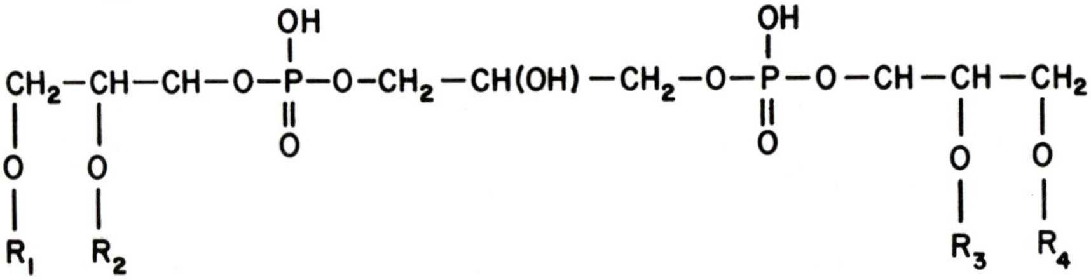

Figure 2C. Some other lipids found in cell membranes.

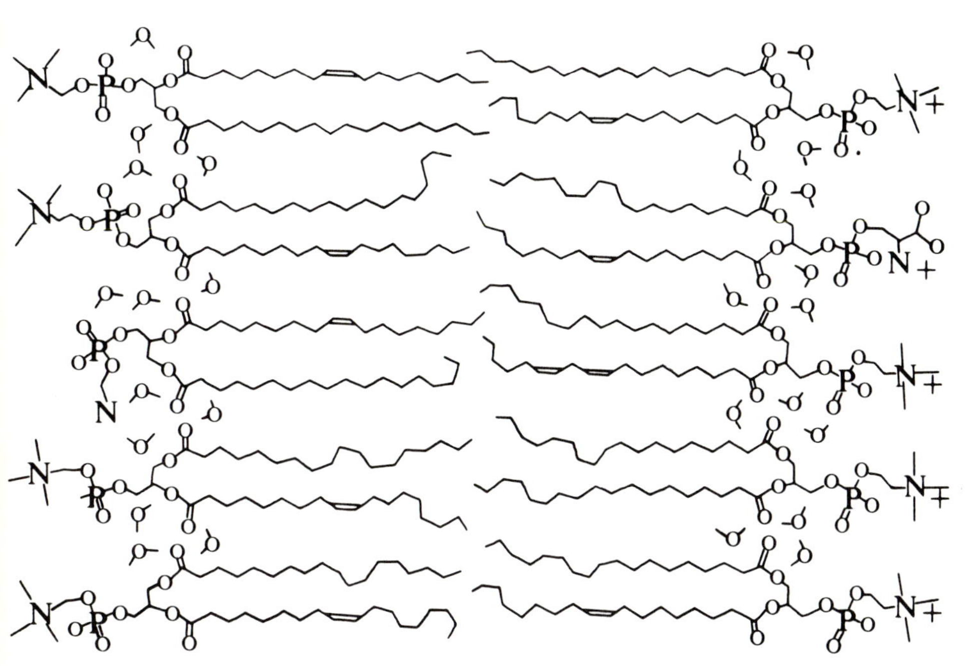

Figure 3. A representation of the structure of a lipid bilayer. Compare Figure 4. [From Griffith, O. H., et al., *J. Mem. Biol.*, 15:159(1974).]

is, customary to symbolize them by assorted potatolike lumps, some of which are covered with funguslike growth indicating the carbohydrate side chains of glycoproteins (Figure 4). Qualitative sketches of this kind were being made before there was any direct supportive visual evidence of the sort provided by the subsequent development of the electron microscope. The present status of biological membranes as "visible" objects is illustrated in Figure 5. The electron microscope (EM) reveals blurred railway tracks that, it is hoped, reflect something of the structure of the membranes before the staining and fixing procedure. The bilayer hypothesis gains support from many such photographs, which provide massive evidence for the existence of blurred railway tracks around both cells and cell organelles, such as mitochondria. The EM photograph of a freeze-fractured membrane (Figure 5), reveals particles that probably are protein. The role of proteins in the structure of membranes and their interactions with the lipids surrounding them are a major research challenge.

Our inability directly to observe the detailed molecular structure of membranes need not deter us since, as we shall see, physical techniques have provided a wealth of quantitative knowledge, particularly of lipids,

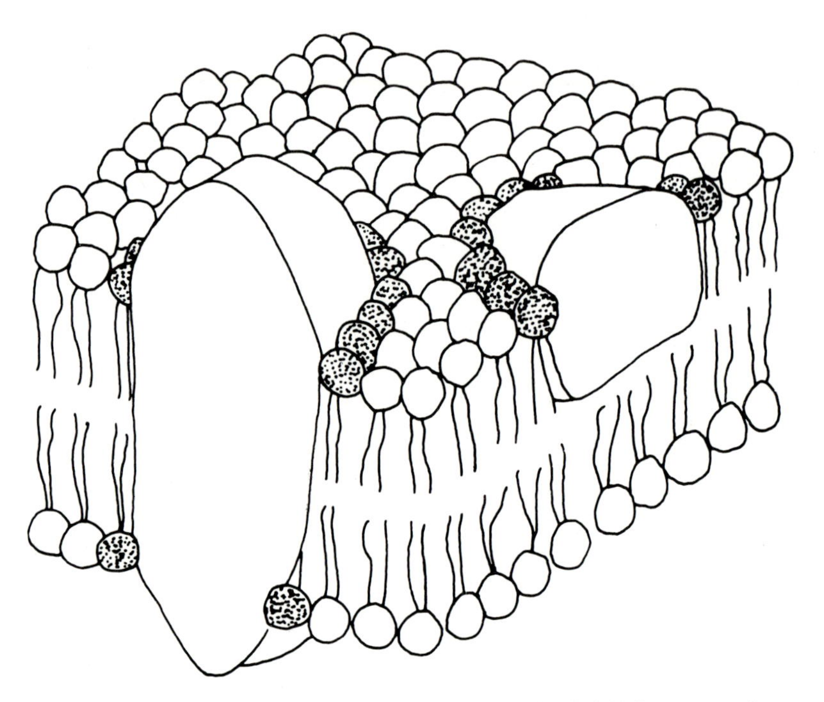

Figure 4. Illustrating a simple qualitative model of a lipid bilayer containing proteins. [From Singer, S. J., *Ann. N.Y. Acad. Sci.,* 195:21 (1972).]

and in this and the following chapters we will look at those basic properties that are relevant to an understanding of the role and behavior of lipids in biologic membranes. As a first step we revue the crystal structure of five lipids. To forestall scepticism as to the connection between the solid state and the living cell, we note here that the conformations of lipids in crystals and membranes have a great deal in common.

Figure 1b shows the shape of a single molecule of 1,2 dilauryl-DL-phosphatidylethanolamine (DLPE) as determined[5] by an X-ray crystallographic study of the phospholipid crystallized in acetic acid. The atoms labeled 1, 2, and 3 are the carbon atoms of the glycerol backbone that is common to many lipids (compare Figure 1A). The numbering system for the other atoms is illustrated by the following examples:

0 3 1 indicates the first oxygen atom on the chain originating on carbon atom 3.

0 3 2 is the second oxygen atom (carbonyl oxygen) on the same chain.

2 1 0 is the tenth carbon atom on the chain originating on carbon 2.

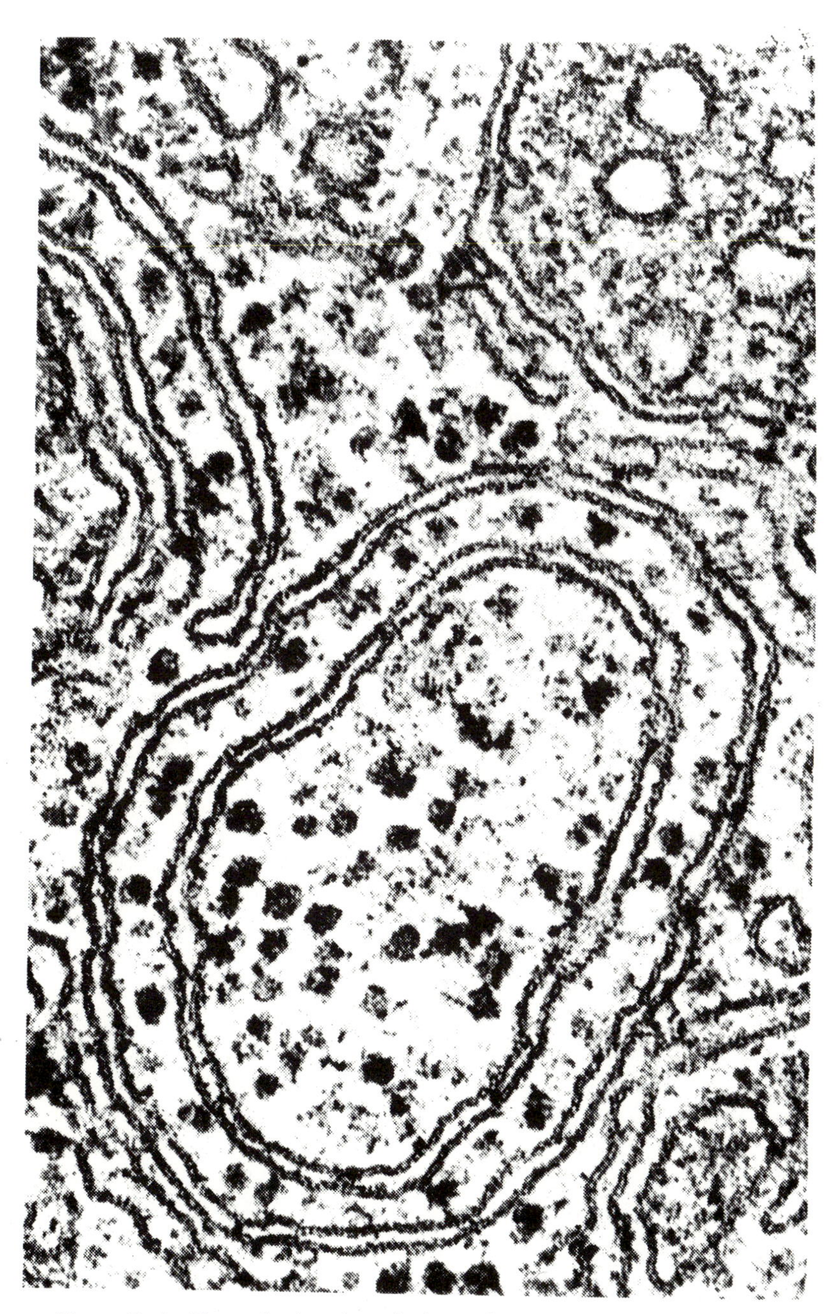

Figure 5. A. The endoplasmic reticulum of the rat adrenal gland. [From Hall, J. L., and Baker, D. A., *'Cell Membranes and Ion Transport'*, Longman, London, 1977.]

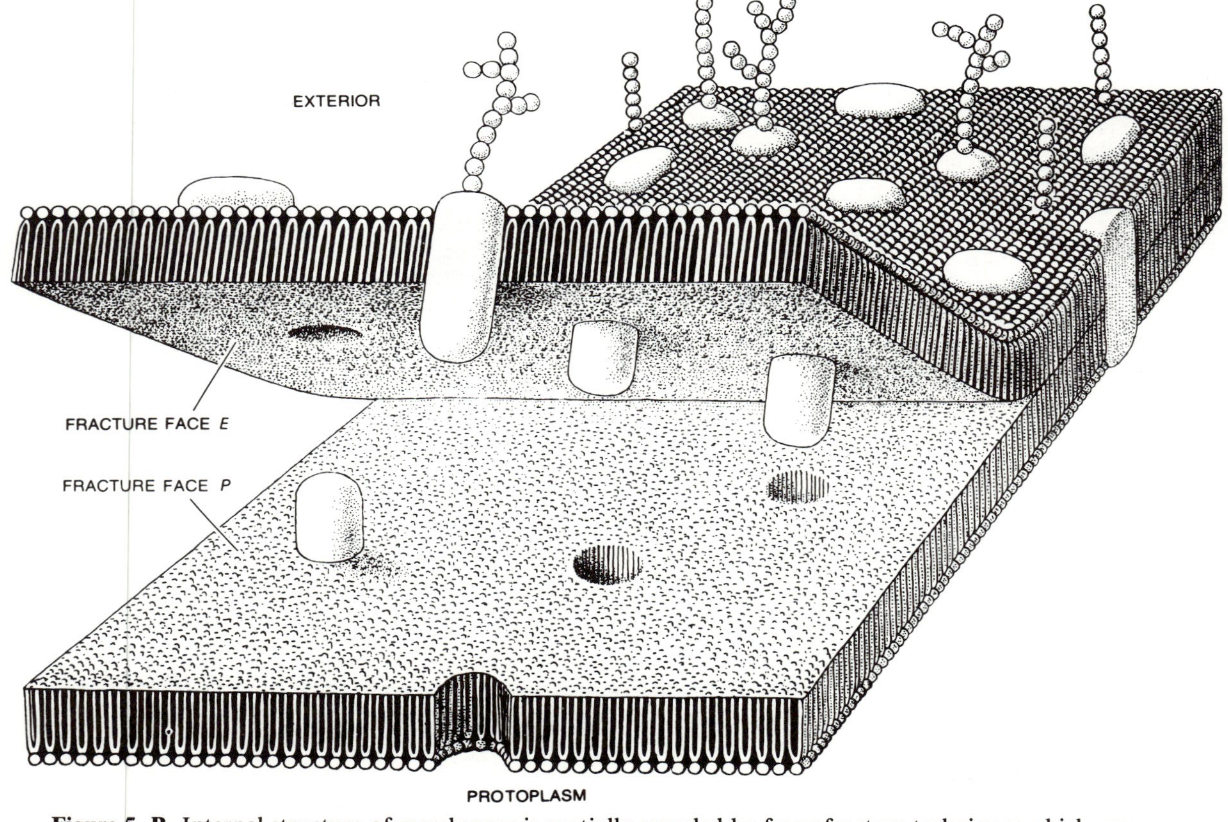

Figure 5. B. Internal structure of membranes is partially revealed by freezefracture techniques which can separate the two layers of the bilayer as shown in the diagram. [From Satir, B., *Sci. Am.* Oct. (1975).]

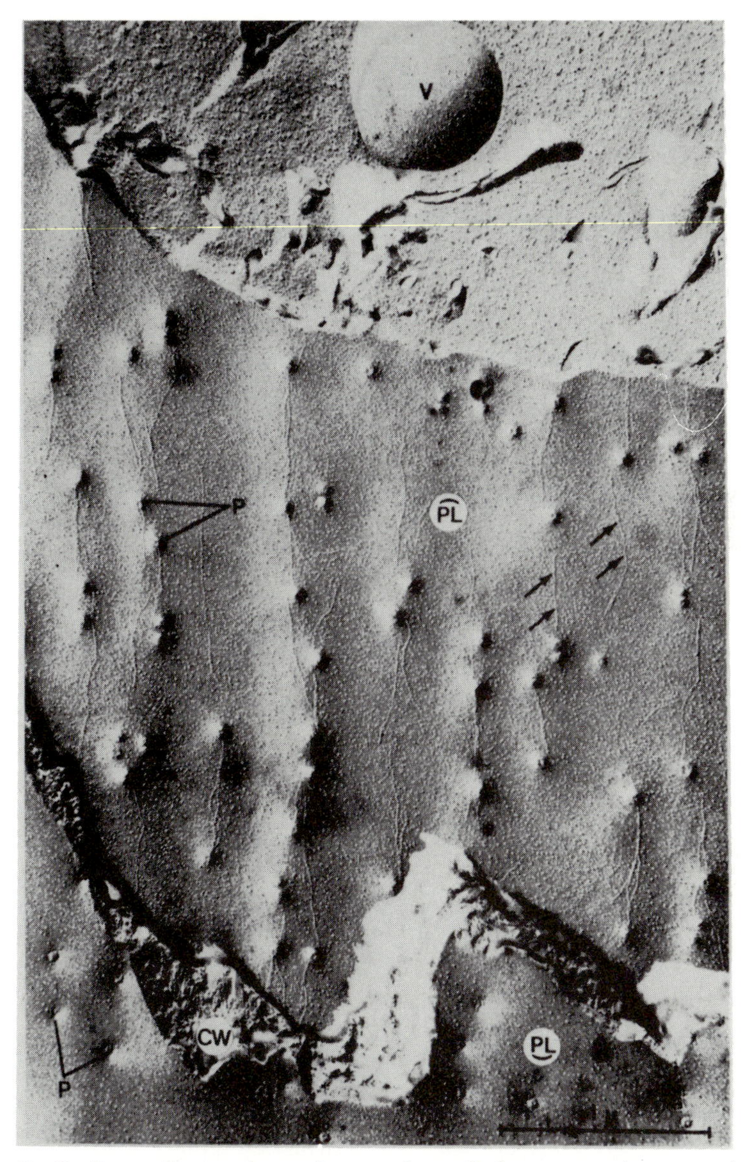

Figure 5. C. Freeze-fractured membrane of a cell from an onion root tip. The uppermost area in the photograph is the cell cytoplasm with a vacuole (V) evident. The center area shows the fracture face of the cell membrane (plasmalemma, PL). A face associated with an adjoining cell appears at the bottom of the picture. The fracture has also ruptured the cell wall (CW). Plasmodesmata (P), fine threads connecting the cytoplasm of adjacent cells, have been broken off. Isolated membrane particles are scattered over the fractured faces but also appear to be organized in rows (arrows). (From Branton, D., and Deamer, D. W., 'Membrane Structure', Springer-Verlag, New York, 1972.)

Three major features of the molecule depicted in Figure 1b are:

1. The two hydrocarbon chains are in the all-trans conformation and their axes are parallel.
2. The phosphoethanolamine head group is aligned roughly at right angles to the axes of the hydrocarbon chains. Specifically, the phosphorus → nitrogen vector is inclined at 15 degrees to the chains.
3. The initial part of the β-chain, on carbon 2, is at right angles to the axes of the chains, bending by 90 degrees at carbon 22 to bring the β-chain into line with the γ-chain on carbon 1. The β-chain thus appears to be shorter than the γ-chain.

If we put this molecule into the crystal, we find that it is one of four in each unit cell (Figure 6). The molecules are arranged in bilayers separated by sheets of acetic acid molecules. Within each bilayer the phospholipids are stacked with their polar head groups outward, the glycerol backbone roughly parallel to the bilayer normal. The hydrocarbon chains lie feet to feet, echoing the general description proposed for biologic membranes. The chains are perpendicular to the bilayer surface and the apparent discrepancy in their length is neatly taken into account by the interdigitation (locking of fingers) of the terminal methyl groups. The bilayer is 47.7 Å thick, approximately 30 Å being taken up by the hydrocarbon core.

We have been looking at the projection onto the crystallographic *a–b* plane. We now view the head groups along the normal to the bilayer surface (Figure 7).

The outstanding structural feature is the linking of the phosphate groups to the ammonium groups by short hydrogen bonds of lengths of 2.74 and 2.86 Å so as to give infinite ribbons separated by 7.8 Å but bound together by ethanolamine bridges. Maintaining our viewpoint but delving beneath the head groups, we find that the four hydrocarbon chains of the two molecules at our end of the unit cell are arranged as shown in Figure 8. If we add a few surrounding cells, we find that the packing is quasi-hexagonal. This specific type of packing, which has been labeled[6] HSI, is a centred orthorhombic subcell arrangement. Figure 8 is actually based on an electron diffraction study of anhydrous dipalmitoylphosphatidylethanolamine.[7] The range of wavelengths used in electron diffraction is usually between 0.01 and 0.05 Å, which should be compared with 0.7 to 2.3 Å for X rays; electron diffraction allows much higher resolution in the determination of atomic positions. Apropos of these positions it is of interest, and of later import, that for the hydrocarbon carbon atoms the Debye thermal factor, a measure of the degree of vibrational freedom of atoms in the crystal, increases going

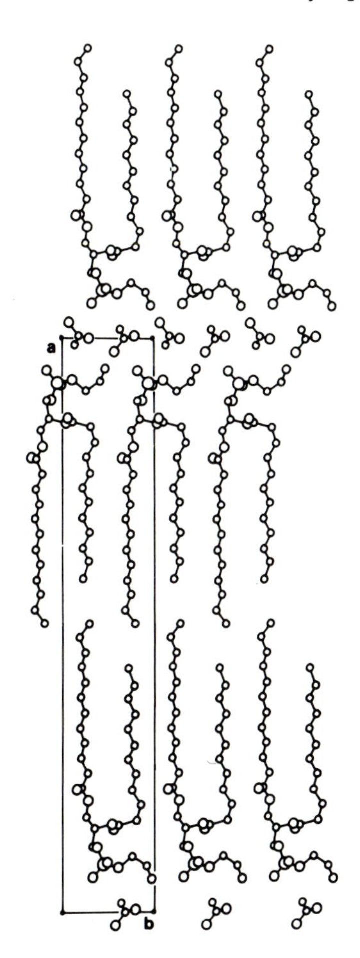

Figure 6. Crystal structure of dilaurylphosphatidylethanolamine (DLPE). [From Hitchcock, P.B., et. al., *Proc. Nat Acad. Sci.* 71:3036 (1974).]

along the chain, indicating that the rigidity of the crystal lattice is greatest near the tightly bound head groups and least at the center of the hydrophobic core of the bilayer (Figure 9).

Dividing a given area of bilayer by the number of contained head groups, we find that each lipid molecule takes up 38.6 $Å^2$ at the bilayer surface. The average cross-sectional area occupied by each of the two hydrocarbon chains is 19.3 $Å^2$.

It is instructive to compare the crystal structure of DLPE with that of another major component of biologic membranes, 2,3, dimyristoyl-D-glycero-1-phosphorylcholine (DMPC), the dihydrate of which has been studied by Pearson and Pascher[8]. We first concentrate on the similarities. Both substances give monoclinic crystals with four molecules per unit cell and a layered structure in which the head groups are roughly

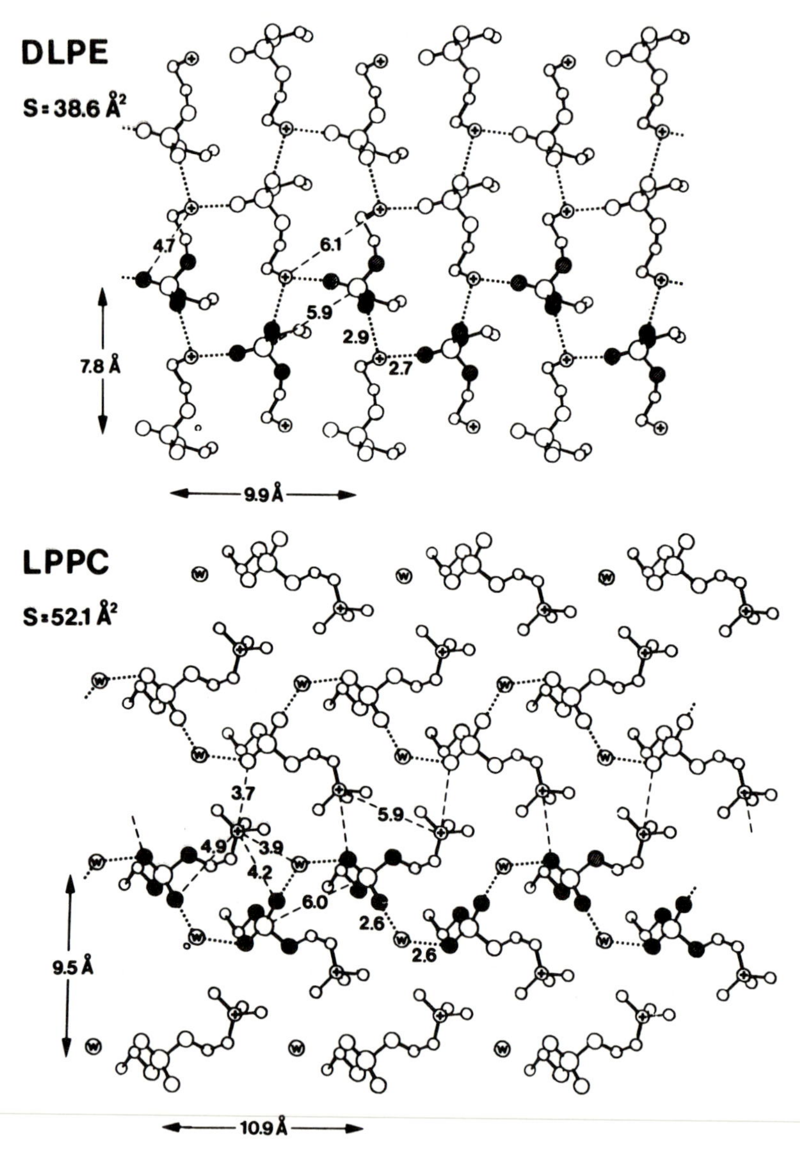

Figure 7. The head group packing of crystalline DLPE. [From Hauser, H., et al., *Biochim. Biophys. Acta* 650:21 (1981).]

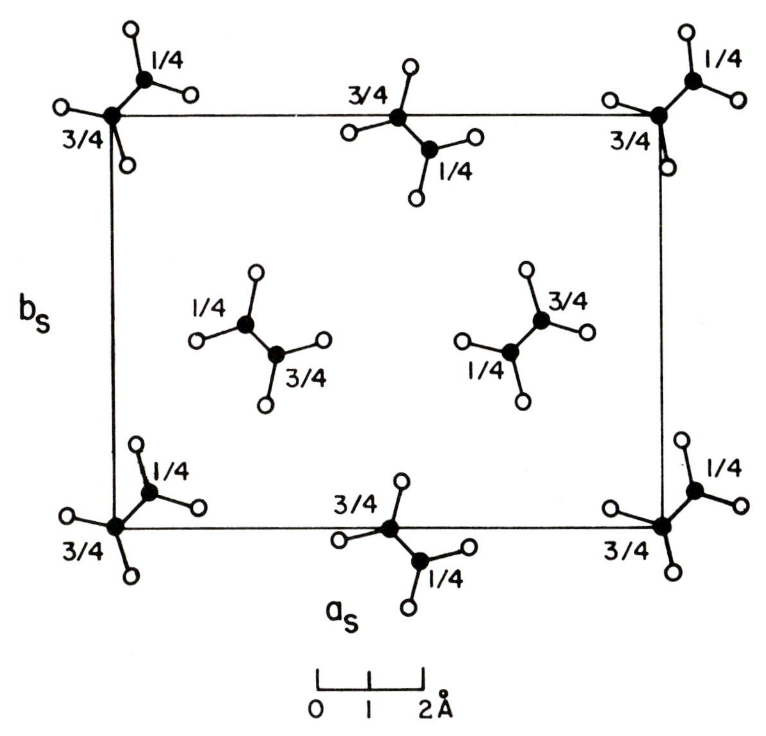

Figure 8. Illustrating the HSI mode of chain packing. (From Abrahamsson, S, et al., *J. Chem. Fats. Lipids* 16:125 (1978).]

parallel to the bilayer surface. In both cases one of the all-trans hydrocarbon chains is initially bent and the resulting seeming inequality in length is accommodated by interdigitation. The differences between the two structures are best appreciated from Figures 10A and 10B, the former shows that in DMPC the hydrocarbon chains are tilted away from the perpendicular to the bilayer normal, the experimental angle being 12 degrees, and that, in contrast to the head groups of DLPE, which are strictly coplanar, neighboring head groups of DMPC are staggered by 2.5 Å with respect to the layer normal. Furthermore, there are two conformationally distinct molecules in the unit cell, as shown more clearly in Figure 11. The head groups of the two conformers have approximately mirror-image symmetry. From Figure 10B we see that phosphate groups and water molecules are apparently linked into infinite chains by hydrogen bonds. The strong P-N hydrogen bonds in the DLPE crystal are precluded in DMPC by the choline methyl groups and are

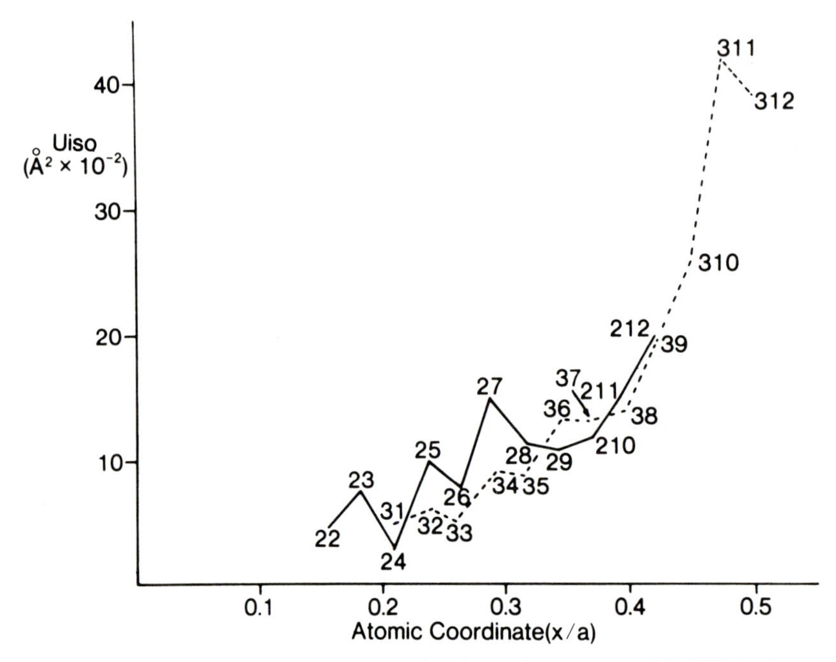

Figure 9. Debye thermal factor u_{iso}, for the carbon atoms of DLPE in the crystalline state as a function of position along the *a*-axis, which is normal to the bilayer. Note the increase in motional freedom in going down the chain. [Hitchcock, P. B., et al., *Proc. Nat. Acad, Sci.* 71:3036 (1974).]

replaced by a long-range (3.7 Å) electrostatic interaction between the phosphate oxygens and the choline nitrogen atoms. Thus while the phosphate ribbons in DLPE are separated by 7.8 Å, those in DMPC are 8.9Å apart. The head groups of DMPC form a looser, more open network than that found in DLPE.

The positioning of water molecules in membranes will be of later interest. We note here that the water molecules in the DMPC crystal are, not unexpectedly, associated with the polar head groups. One type of water molecule appears to be partially occupied with bridging the interface between neighboring bilayers (Figure 10B). The other, according to Pearson and Pascher,[8] is strongly hydrogen-bonded to the nonester phosphate oxygen. This conclusion has been challenged[9]. Addition of water to anhydrous DMPC leaves the frequency of the PO_2^- antisymmetric mode unchanged.

If the molecules we have been discussing are approximated as objects of constant cross section packed at an angle ϕ to the bilayer normal, then the area S taken up by a molecule at the bilayer surface is given by

$$S \cos \phi = n\Sigma \tag{1.1}$$

where n is the number of hydrocarbon chains and Σ the cross-sectional area of each chain. This holds reasonably well for DLPE and DMPC, and is an equation that we use later.

We have looked at DLPE and DMPC because they are major components of membranes, both representing the class of double-chained lipids. The important four-chain phospholipid cardiolipin has not yet been crystallized, but a single-chain lipid has,[10] and the molecular packing of 3-lauroylpropane-diol-1-phosphorylcholine monohydrate (LPPC) is shown in Figure 12. The unit cell contains four molecules — two pairs of conformational enantiomers. The main structural features are:

1. The hydrocarbon chain is all-trans and inclined at 41 degrees to the bilayer normal.
2. The chains of molecules on opposite sides of the layer are interdigitated along their entire lengths.
3. The head groups are very nearly parallel to the bilayer surface, to which the P → N dipole is inclined at only 7 degrees. The molecular cross section S at the surface is correspondingly large, 52.1 Å^2.

The interdigitation is clearly partially a result of the packing requirements for a molecule with a head group cross section considerably larger than the cross-sectional area of its single hydrocarbon chain. The molecule solves the problem by managing to pack two chains behind each head group.

The three amphiphiles that we have discussed are all commonly classified as phospholipids. Of the other lipid components of membranes, the crystal structure of cholesterol and of one sphingolipid have been determined. Cholesterol (Figure 13) is a nonpolar molecule with a relatively rigid ring system linked to a short hydrocarbon chain. It dissolves well in the hydrocarbon core of lipid bilayers but does not form bilayers by itself under any circumstances. The hydroxyl group is the only hydrophilic point in the molecule, and we therefore expect the orientation of cholesterol in lipid bilayers to be such that the -OH group will be in or near the head group lattice.

The sphingolipids (Figure 2) have received less attention from physical chemists than the structurally simple phospholipids despite their apparent physiologic importance. Cerebroside [β-D-galactosyl-N-(2-D-hydroxyoctadecanol)-D-dihydrosphingosine] crystallized from ethanol

Figure 10. A. Crystal structure of DMPC projected onto the crystallographic *a-c* plane. [From Pearson, R. H. and Pascher, I., *Nature* 281:499 (1979).]

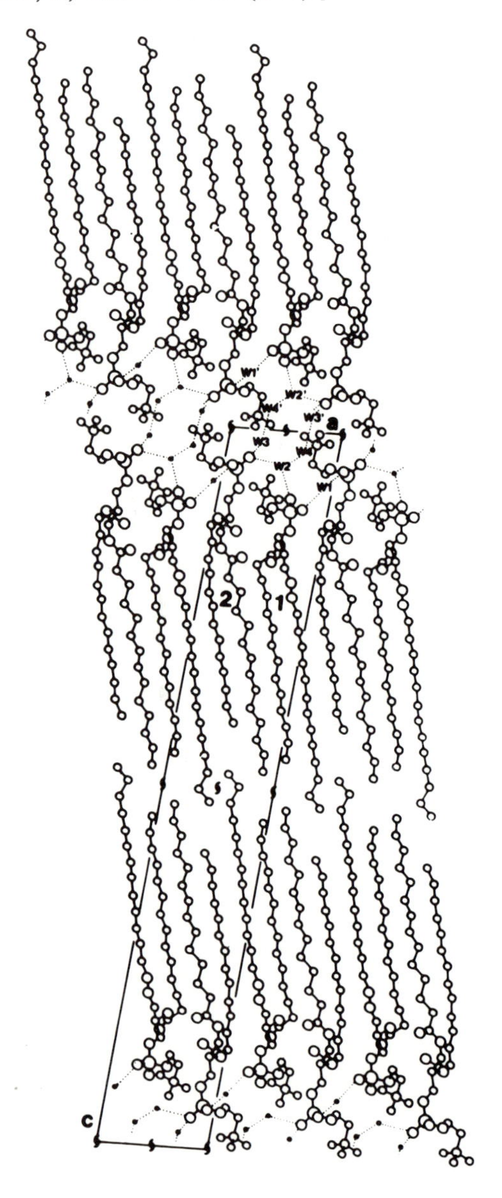

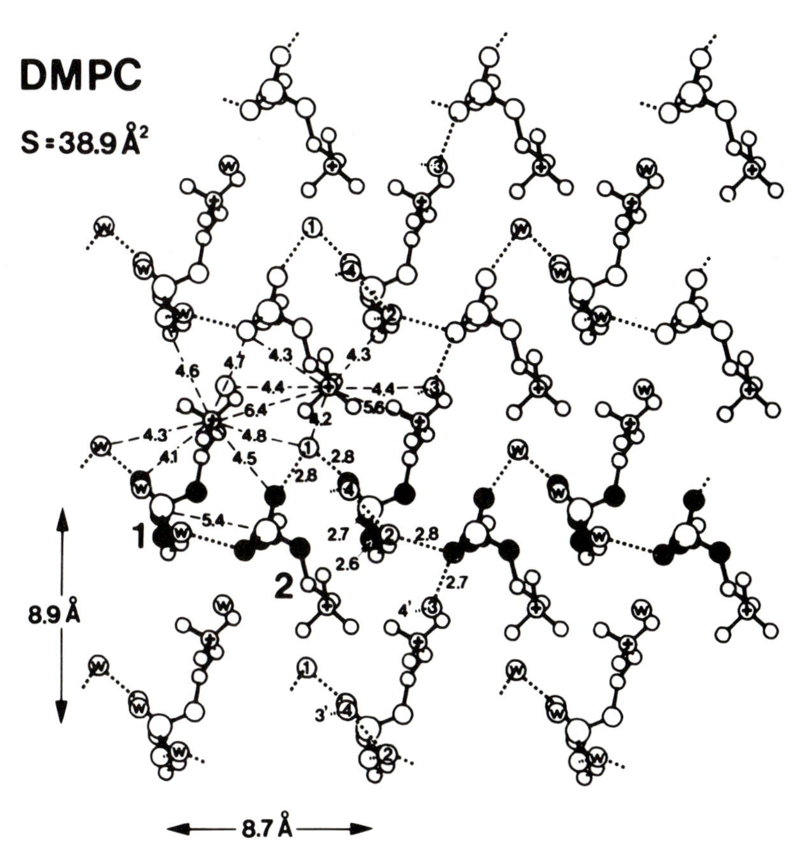

Figure 10. B. The headgroup packing of DMPC viewed perpendicularly to the bilayer surface, that is to the *a-b* plane.
[From Hauser, H., et al. *Biochim. Biophys. Acta* 650:21 (1981).]

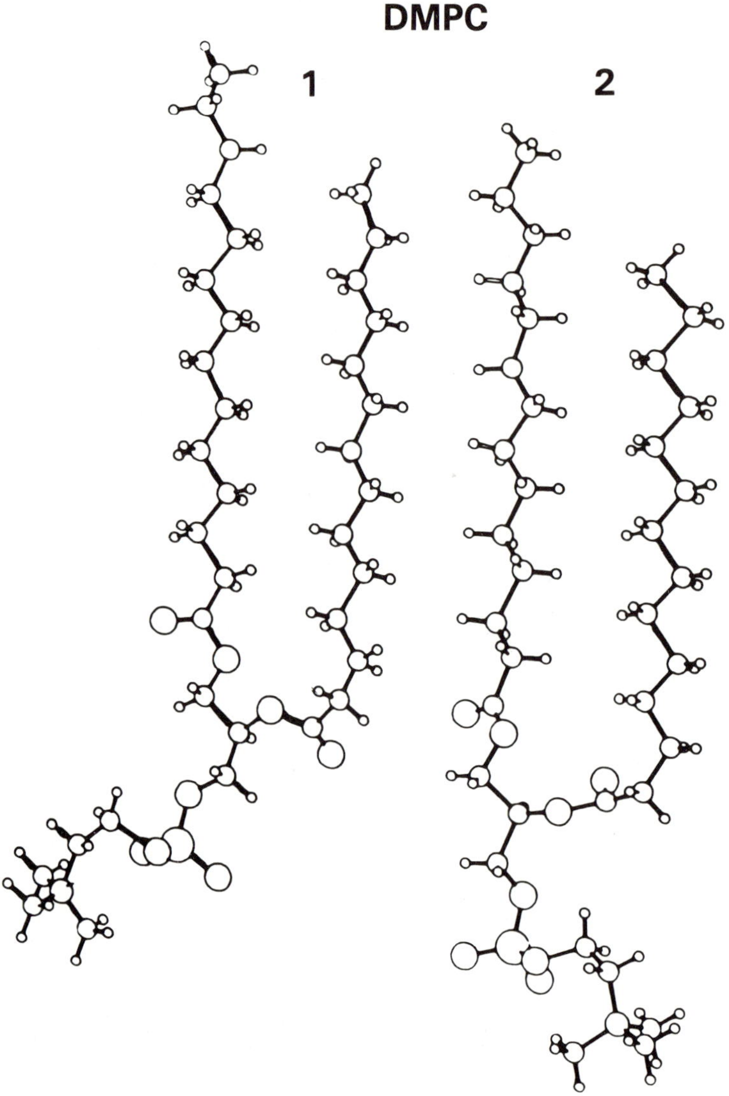

DMPC

1

2

Figure 11. Two conformers of DMPC in the crystalline state. [From Hauser, H., et al., *Biochim. Biophys. Acta* 650:21 (1981).]

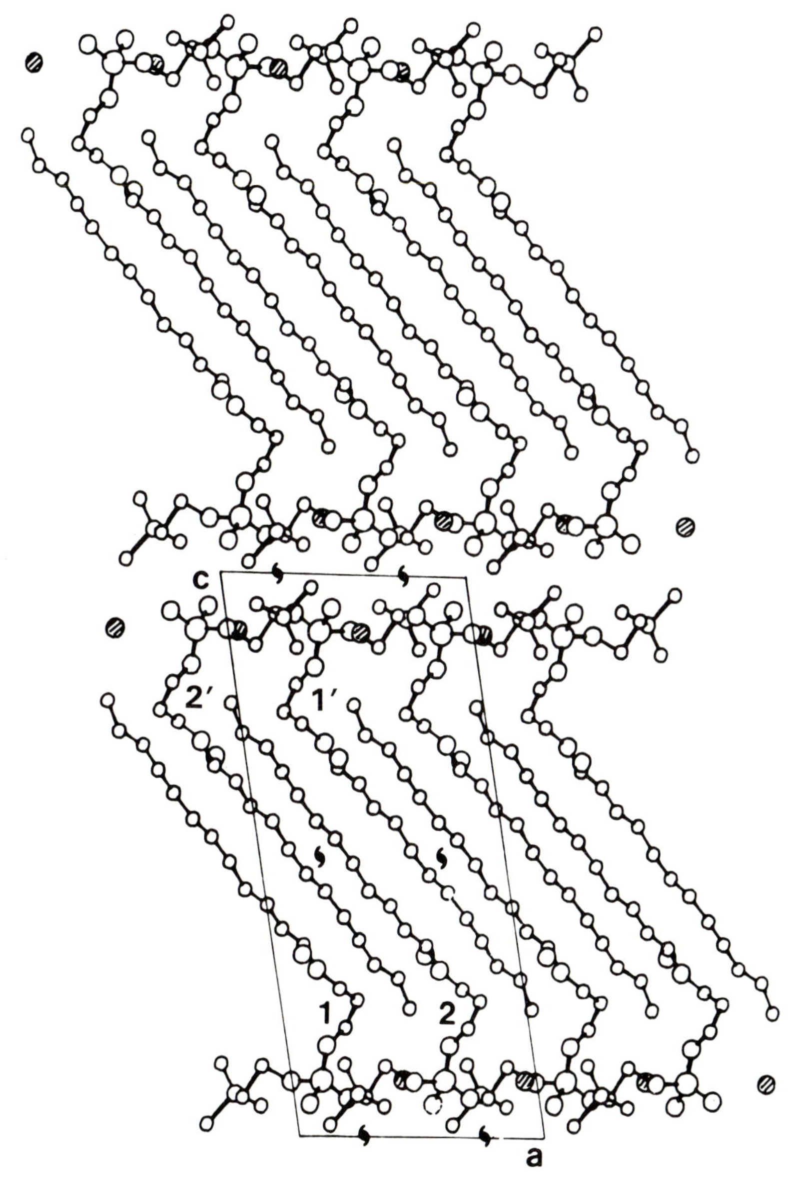

Figure 12. Crystal structure of LPPC (3-lauroylpropanediol-l-phosphatidylcholine monohydrate) projected onto the *a-c* plane. [From Hauser, H., et al., *J. Mol. Biol.* 137:249 (1980).]

Figure 13. Shape of the cholesterol molecule. In solution or in liquid crystal state of a bilayer, the hydrocarbon chain will flail around and effectively occupy a greater volume relative to the ring system than that shown here. See Figure 35. [From Rogers, J., et al., *Biochim. Biophys. Acta* 552:23 (1979).]

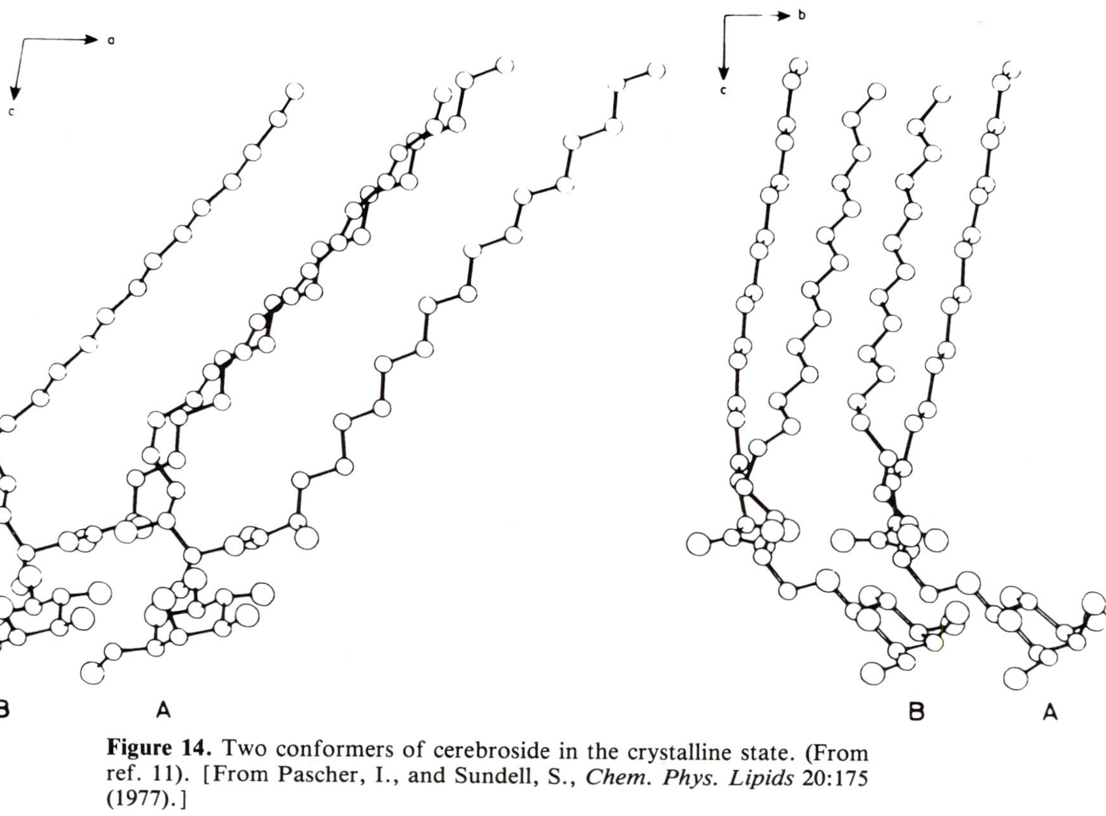

Figure 14. Two conformers of cerebroside in the crystalline state. (From ref. 11). [From Pascher, I., and Sundell, S., *Chem. Phys. Lipids* 20:175 (1977).]

has been subjected to X-ray crystallography[11] and displays a bilayered structure with a chain tilt of 41 degrees to the bilayer normal. The sugar rings lie with their planes almost parallel to the surface and their hydroxyl groups are involved in a complex hydrogen-bond network that includes the nitrogen atom of the amide group. The conformation of the two stereochemically unrelated molecules in the bilayer are shown in Figure 14.

The principal conclusions of this chapter are that in the crystalline state the long-chain lipids are stacked in bilayers, their chains all-trans, and their head groups folded approximately parallel to the bilayer surface and linked, chiefly by hydrogen bonds, into a firm network.

While it might appear that the phospholipids were designed by divine providence to self-assemble into bilayers, it should be pointed out that such behavior is displayed by many molecules of no obvious biologic significance, in particular, a wide range of liquid crystals. The structural feature common to such molecules is their elongated form. In the presence of water, those that are amphiphilic form lamellar phases, that is, hydrated bilayers, known in the trade as lyotropic smectic liquid crystals.[12]

2 Phase Transitions of Anhydrous Lipids

The phases adopted by phospholipids are rich in variety. In this chapter we look at the phase transitions of anhydrous phospholipids. The experimental results to be reviewed were obtained on polycrystalline solids since single crystals of anhydrous phospholipids have not yet been prepared, but we will see in this and the next chapter that the solid-state structure reviewed in Chapter 1 forms a partial basis for a qualitative understanding of the phase transitions of both anhydrous and hydrated lipids.

First-order phase transitions such as the melting of pure substances are accompanied by the absorption of energy over a small temperature range. Differential thermal analysis (DTA) relies on this fact. In DTA we supply heat to two samples, one a reference material and the other the compound of interest. The temperature is raised by electrical heating and the temperature difference ΔT between the samples is recorded. ΔT remains at zero until there is a phase transition in the compound, a process requiring the absorption of latent energy. The compound's temperature remains approximately constant while the temperature of the reference keeps rising so that during phase changes ΔT is nonzero, returning to zero at the completion of the event. Most solids show a single peak in the plot of ΔT against temperature — at the melting point. The DTA plot of 1,2,-dimyristoylphosphatidylethanolamine shown in Figure 15 is typical of many phospholipids and indicates that *three* phase transitions occur, one at ~200°C, close to the capillary melting point, This peak is very small in comparision with that at ~115°C, which indicates that the phase change at the lower temperature requires more energy per mole than the melting process. The main features of the DTA plot can be explained by assuming that the phospholipid sample, although not a single crystal, has the same microstructure as the crystals described in the previous chapter. The absorption of thermal energy results in an increase in vibrational amplitudes of all the atoms, the first major result of which is that the all-trans conformation of the hydrocarbon chains is broken down by rotational isomerization at a temperature well below that required to disrupt the ionic network of the head groups. The melting process thus involves two stages, with the second being independent of the energy requirements or transition temperature of the first. We will, in this chapter, present evidence for the above model, and in later chapters discuss in considerable detail the rocking and rolling of the liberated hydrocarbon chains above their private melting points. The transition from the all-trans to other, short-lived, chain conformations has been given many names: order to disorder, solid to liquid, all-trans

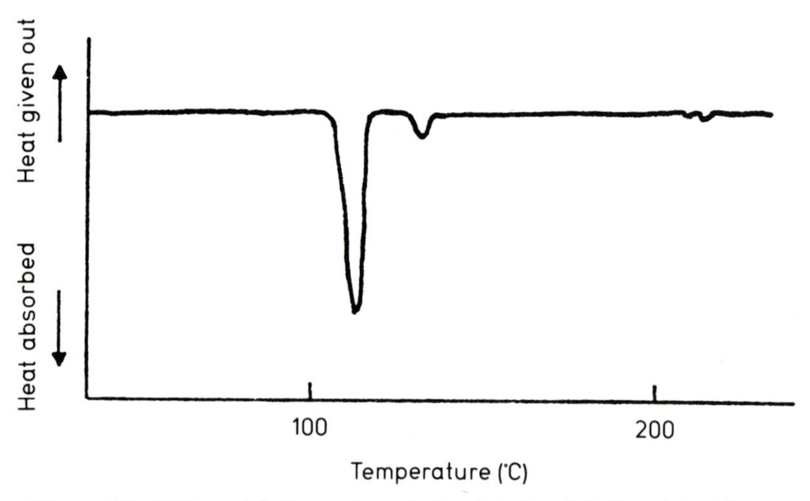

Figure 15. Differential thermal analysis plot for 1,2-dimyristoylphosphatidylethanolamine. Note the small absorbtion of heat at about 20°C above T_t and the very small change associated with complete melting above 200°C. [From Chapman, D., and Collins, D. T., *Nature* 206:189 (1965).]

to rotationally disordered, gel to liquid crystal. The last hints that the melted hydrocarbon chains have something of the nature of liquid crystals in which molecules have a measure of motional freedom reminiscent of the liquid state but a degree of alignment that results in anisotropy of the bulk properties of the sample.

The transition temperatures T_t of a number of phospholipids are listed in Table 1, together with the corresponding melting points and some data on related compounds. A number of generalizations and rationalizations can be made.

1. The melting points of the phosphatidylcholines, phosphatidylethanolamines, and sphingomyelins are relatively independent of hydrocarbon chain length. This is to be expected since at the melting point it is effectively only the ionic lattice that melts, the hydrocarbon chains are already "liquid". On the other hand, for the phosphatidic acids, which do not interact strongly via their head groups, the melting process consists almost entirely of the disruption of the contacts between the hydrocarbon chains, and since the forces between the chains increase with length, the melting point also increases.
2. In contrast to the melting points, the values of T_t for the high-melting-point amphiphiles depend significantly on the length of their hydrocarbon chains (Figure 16). This is understandable since the melting of the chains is similar to the melting of the phosphatidic acids discussed above, that is, it represents the triumph of the thermal energy over the van der Waals forces and steric repulsion between chains. Strong support for the occurrence of premelting of the hydrocarbon chains comes from many studies of physical properties of phospholipids and we now summarize some of the available evidence.

The infrared spectrum of 1,2-dilauroyl-DL-phosphatidylethanolamine at various temperatures is shown in Figure 17. The sharp lines at $-186°C$ are characteristic of solid-state spectra. As the temperature is raised, the line widths increase until, at 120°C, a temperature some 80 degrees below the melting point, a sudden broadening occurs resulting in an ill-resolved spectrum closely resembling that of the phospholipid dissolved in an organic solvent. The interpretation is obvious. The proton NMR spectrum[13] of phospholipids at liquid nitrogen temperatures consists of a broad line with a width of ~ 15 gauss. Such large line widths are typical of the NMR spectra of solids. The line width decreases gradually with rising temperature, a phenomenon explained by a moderate increase in the freedom of molecular motion and consistent with the temperature dependence of the line widths of the IR spectra below T_t. At $\sim 120°C$ the line width for

Table 1
The Transition Temperature T_t and the Melting Points for Some Amphiphiles

	T_t (°C)	M. pt. (°C)	Charge State at pH 7
Dimyristoylphosphatidic acid, DMPA	51		−1
Dipalmitoylphosphatidic acid, DPPA	67		−1
Dilauroylphosphatidyl glycerol, DLPG	4		−1
Dimyristoylphosphatidyl glycerol, DMPG	23		−1
Dipalmitoylphosphatidyl glycerol, DPPG	41		−1
Distearoylphosphatidyl glycerol, DSPG	55		−1
Dioleoylphosphatidyl glycerol, DOPG	−18		−1
Dimyristoylphosphatidyl ethanolamine, DMPE	49.5	~198	Negative, partially titrated
Dipalmitoylphosphatidyl ethanolamine, DPPE	60		
Dimyristoylphosphatidyl serine, DMPS	38		−1
Dipalmitoylphosphatidyl serine, DPPS	51		−1
Dilauroylphosphatidyl choline, DLPC	−1.8		0
Dimyristoylphosphatidyl choline, DMPC	23		0
Dipalmitoylphosphatidyl choline, DPPC	41	~230	0
Distearoylphosphatidyl choline, DSPC	55		0
Dioleoylphosphatidyl choline, DOPC	−22		0

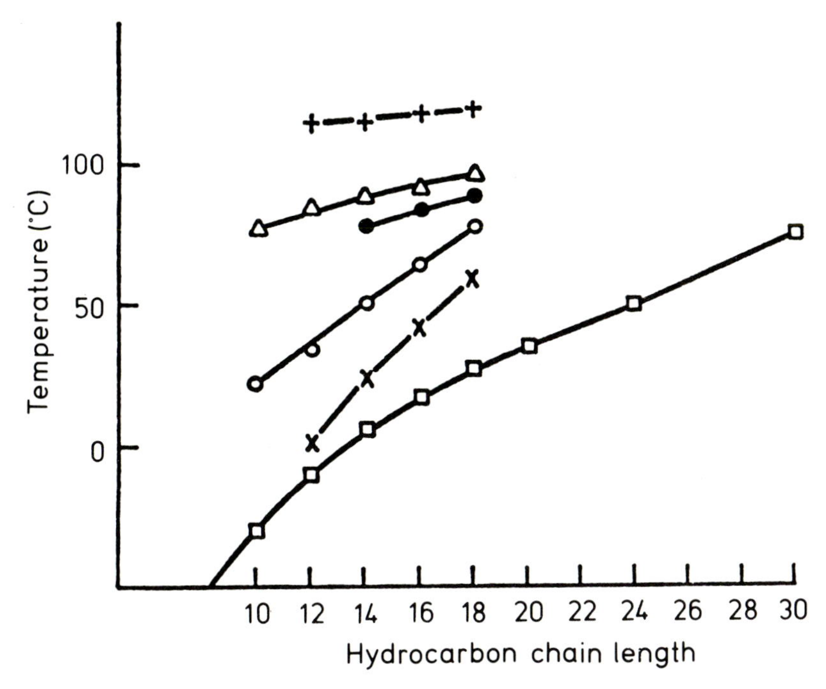

Figure 16. Chain-melting transition temperatures T_t for some amphiphiles as a function of chain length. +, anhydrous 1,2-diacyl-DL-phosphatidylethanolamines; •, 1,2-diacylphosphatidylethanolamines in water; △, anhydrous 1,2-diacyl-L-phosphatidylcholine monohydrates; ○, 1,2-diacyl-L-phosphatidylcholine monohydrates; ×, 1,2-diacyl-L-phosphatidylcholines in water. For comparison the melting points of some normal paraffins, □, are shown. [From Chapman, D., et al., *Chem. Phys. Lipids* 1:445 (1967)]

1,2-dilauryl-DL-phosphatidylethanolamine drops rapidly to less than 0.1 gauss, indicating a sudden very sharp rise in motion.

The above work was carried out on polycrystalline samples unsuitable for complete X-ray crystallographic studies, but diffraction patterns are observed from such samples, although the information obtained is limited. Anhydrous lipids show diffuse diffraction bands[14] that can be classified into "short spacing" and "long spacing" bands, with the former arising from the lateral perodicity in the arrangement of the hydrocarbon chains within the bilayer and the latter from the periodic stacking of bilayers. The interpretation of these diffraction patterns has lagged well behind that given for the more interesting hydrated lipids, but at least one important qualitative conclusion can be reached, namely, that in the neighborhood of T_t, the presumed chain melting temperature, the bilayer thickness drops drastically — from ~60 Å to ~40 Å in the

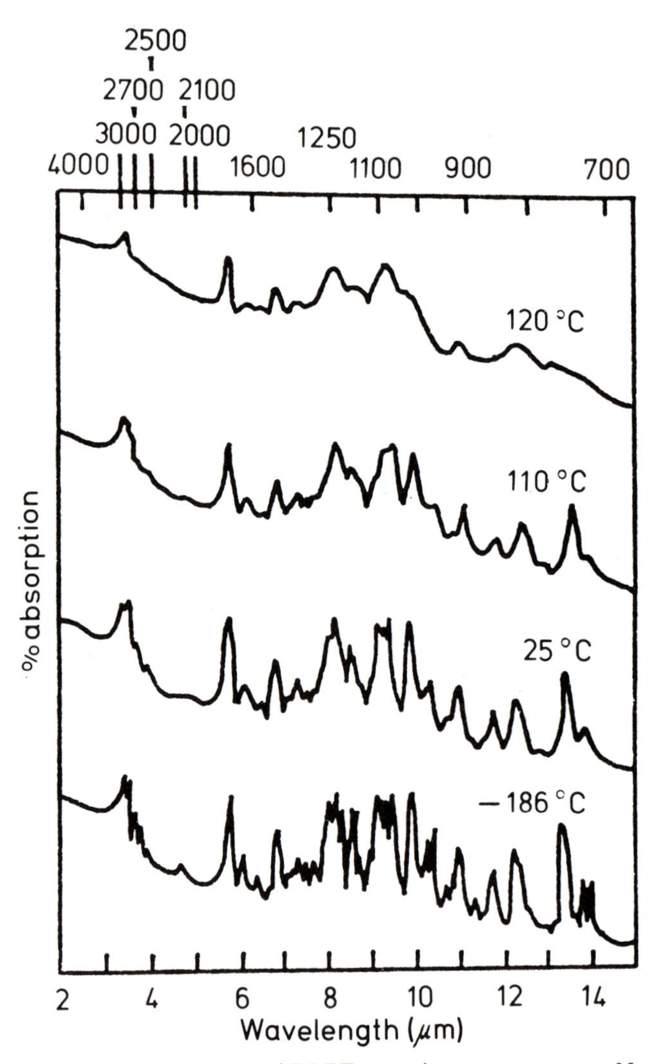

Figure 17. Infrared spectrum of DLPE at various temperatures. Note the broadening of the lines as the temperature rises. Sharp IR lines are usually associated with solids and broad lines with liquids. [From Chapman, D., et al., *Proc. Roy. Soc.* A290:115 (1966).]

case of DPPC. This collapse is to be expected if the rigid, maximum-length all-trans chains crumple into floppy, effectively shortened conformations. Later we document the overwhelming spectroscopic evidence for this picture in the case of hydrated lipids.

Other physical properties show more or less sudden changes at the so-called gel–liquid crystal transition, but we leave them unlisted since our point has been made.

We have ignored the minor peak in the DTA plot (Figure 15) that occurs at a temperature just above the main transition. The area under the peak indicates that the transition enthalpy is small. The nature of the corresponding phase change is not known but may involve a minor change in the ordering of the head groups.

The main message of this chapter is that among the different phases of anhydrous phospholipids we can expect to find at least one that is typified by a liquid-like hydrocarbon core sandwiched between two relatively rigid polar sheets. This seems a reasonable model for part of the fabric of the cell membrane but a disturbing point is that the values of T_t for the anhydrous forms of naturally occurring phospholipids are way above physiological temperatures. This does not fit in well with our intuitive demand for a fluidlike core for a membrane bilayer. The rigidly packed and disciplined all-trans chains of the gel state are not friendly enough to adapt to the awkward shapes of protein and other molecules. In fact, bilayers containing water display much lower chain melting temperatures than the corresponding anhydrous compounds. This brings us to the subject of hydrated lipids, thus moving us significantly closer to the living cell.

3 Hydrated Bilayers

Since life depends on water, we now turn to the properties of hydrated lipids in the hope that they provide a better model for cell membranes than do their anhydrous counterparts. Membranes, fortunately, do not dissolve in water, and we do not expect phospholipids to be water soluble. Addition of water to anhydrous phospholipids results, as we will see, in an initial adsorption of water molecules into and onto the preexisting bilayers. The structure of the resulting systems depends on the temperature, but is also sensitive to other factors such as the presence of metal ions. The various phases of systems containing water and lipids have fascinated scientists for half a century and continue to provide a field in which, despite a huge amount of experimental data and the impressive generalizations of the theoreticians, our ignorance is considerable. Since the main motivation in studying hydrated lipids is the belief that they share many properties of the cell membrane, a great deal of attention has been paid to phases consisting of planar or near-planar bilayers and our treatment will reflect this bias. However, hydrated lipids exist in a wide range of structures and it would be unwise to dismiss the biologic relevance of even seemingly exotic structures.

In the first part of this chapter we deal in some detail with the phase diagram of a single hydrated phospholipid in which all the phases are planar or near-planar and are produced solely by variations in temperature and the percentage of water. Our concern will be with structure, and no attempt will be made to explain the forces, geometric influences, or entropic factors that stabilize phases, neither will we examine the motion of the constituent molecules. These matters come later. The second part of the chapter will be mainly concerned with the orientation of the head groups and the location of water in hydrated bilayers.

PHASE DIAGRAM OF A HYDRATED LIPID

To illustrate the way in which phase diagrams of phospholipids have been constructed, we present a précis of a paper by Janiak, Small, and Shipley[15] in which differential scanning calorimetry and X-ray diffraction were used to investigate the temperature and compositional dependence of the structure of hydrated dimyristoylphosphatidylcholine, (DMPC) which the authors call dimyristoyl lecithin.

The type of information resulting from this study is typical of that obtained on many hydrated lipid systems and for this reason we take some time over our exposition of what is a particularly thorough study. Differential scanning calorimetry (DSC) is a technique similar to DTA. A sample of the compound under study is electrically heated simultaneously with a reference sample. When a phase change occurs, the temperature of the compound tends to drop below that of the reference, but equality of temperature is maintained by supplying extra electric power to the compound. It is the difference between the power supplied to the reference and that to the compound that is plotted against temperature. The areas under the peaks are proportional to the transition enthalpy. The DSC thermograms shown in Figure 18 indicate that for the various hydration states of DMPC there is always one dominant transition, labeled T_3, which occurs at temperatures at which the X-ray diffraction pattern changes to that expected from a liquid crystal (melted-chain) structure. Thus T_3 is associated with chain melting and we see from Figure 19 that above ~25% water T_3 remains constant, independent of composition. Unlike the behavior at higher concentrations of water, that at concentrations of <15% water is not even approximately reversible, as seen from Figure 18. If, however, samples in this range are maintained at −20°C for several hours, they return on their initial behavior showing T_1, T_3, and T_4 transitions as in the upper curve of Figure 18. The complete data for the observed transitions, as plotted in Figure 19, are in fact a partial temperature–composition phase diagram, and form the basis of all but the vertical lines in Figure 20. The vertical lines indicate the percentages of water above which bulk water and hydrated lipid form separate phases. When water is added to anhydrous

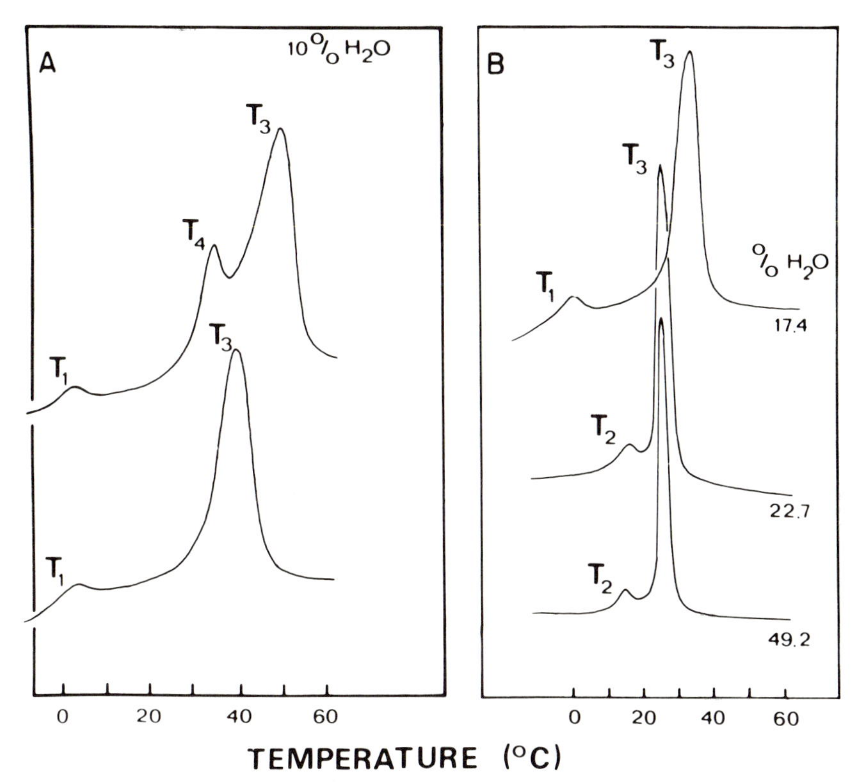

Figure 18. Differential scanning calorimetry thermograms for DMPC. **A**: Curves typical of those found for samples containing 9 to 14% water. The upper curve is that for a first heating and the lower for a subsequent heating of the sample. If, after the first heating, the sample is held at −20°C for several hours and then heated, the upper curve is recovered. **B**: Curves typical of samples containing above 15% water. [From Janiak, M.J., et al., *J. Biol. Chem.* 254:6068 (1979).]

lamellar lipid, it is adsorbed into the head group lattice, resulting in a swelling of the bilayers. Above a certain percentage no further water is adsorbed, but this so-called swelling limit, indicated by the vertical lines in Figure 20, depends on the lipid phase concerned. In Figure 20, which has been slightly idealized from that in the original article, the phases have been given their traditional labels but as yet we have not given the experimental basis for their structures. This comes from X-ray diffraction, which has such an important role in the history of membrane science that we digress to comment on the use of the method in the structure determination of noncrystalline materials.

Diffraction methods depend on the presence of order. The kind of repetitive spatial order represented by perfect crystals is the basis of our

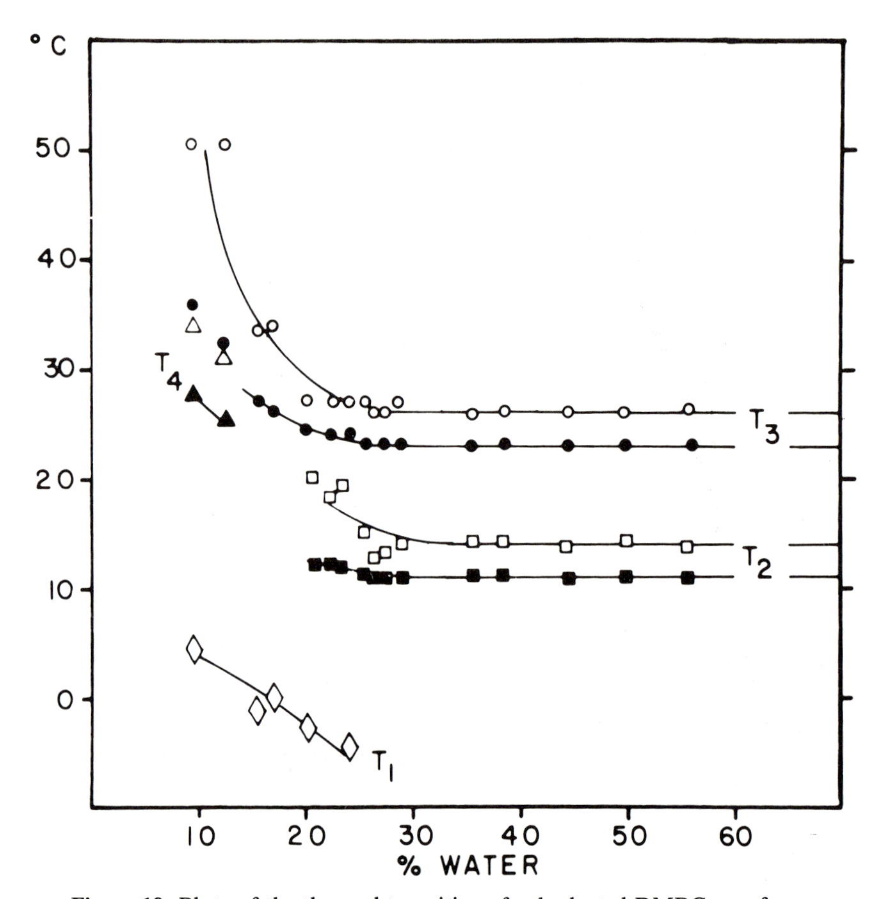

Figure 19. Plots of the thermal transitions for hydrated DMPC as a function of percent water. ◇, T_1; □, T_2; ○, T_3; △, T_4. Compare Figure 18. Solid symbols indicate the temperature of onset of the transition; open symbols the maximum of the transition. This figure, from Janiak et al., is in fact a phase diagram. Compare Figure 24.

ability to pinpoint the atoms in crystals of sodium chloride, cholinesterase, or DMPC. In liquids there is no obvious order, but there is a poor man's alternative — correlation of the positions of molecules over a distance of several molecular diameters. In oversimplified terms, the fact that I am at a given point in space prevents my neighbors from entering a certain volume around me and so our positions are not independent. On the molecular level this correlation is enough to result in destructive and constructive interference in scattered radiation, interference that results, not in the sharp diffraction patterns of crystals, but in broad diffuse bands of the kind shown in Figure 21, from which pair correlation

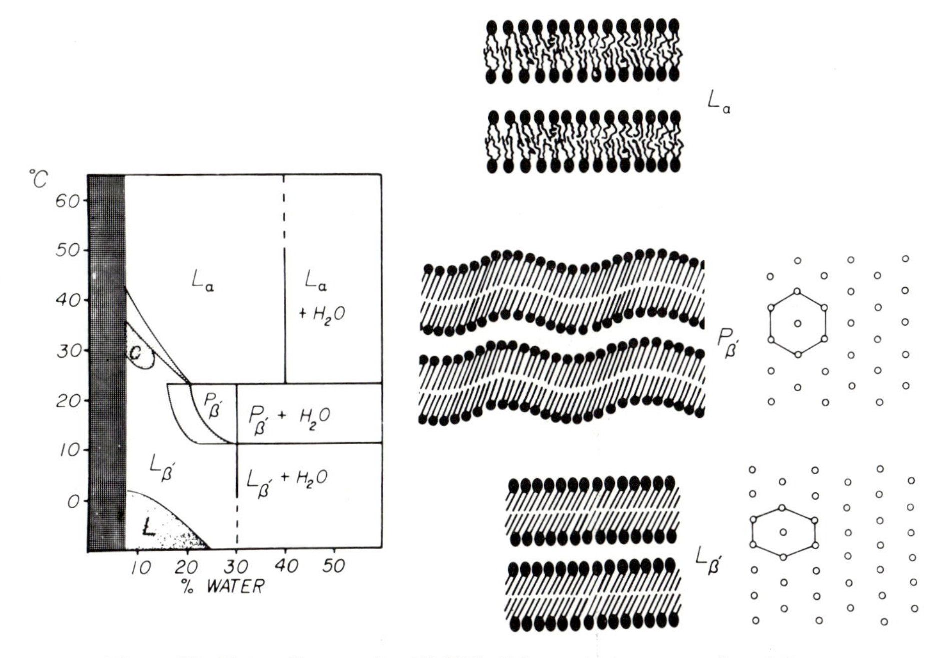

Figure 20. Phase diagram for DMPC. Schematic representations of some of the phases are shown together with the chain-packing pattern for the hydrocarbon chains. [From Janiak, M. J., et al., *J. Biol. Chem.* 254:6068 (1979).]

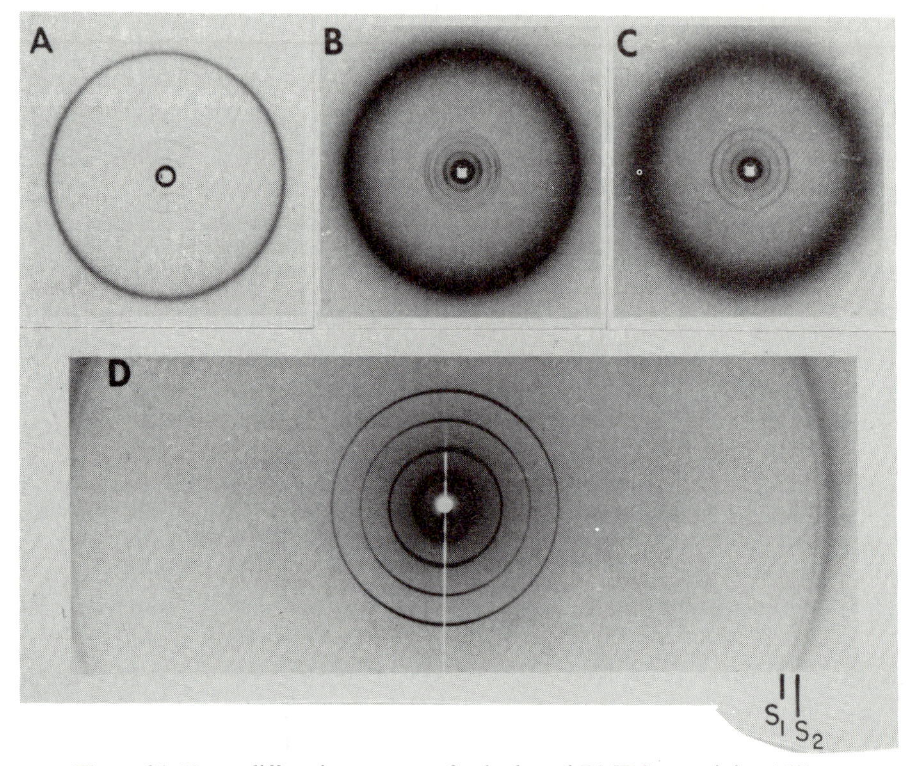

Figure 21. X-ray diffraction patterns for hydrated DMPC containing 16% water. **A.** 10°C. The sharp outer ring is produced by the smallest spacing, 4.2 Å, which is associated with the regular array of hydrocarbon chains. The smaller fainter rings at the center of the pattern are due to the low angle scattering produced by the spacing between bilayers. **B.** 37°C. The outer ring is now more diffuse; the chains are 'waiting' to melt. There are now two sets of lamellar reflections (small rings)—one set is identical with those of the low-temperature sample shown in A, the other is similar to that found in C. **C.** 60°C. The high-angle region, 4.5 Å, is now very diffuse, revealing the disorder in the hydrocarbon chains of the L_α phase. **D.** 20°C. Two short spacings, 4.16 Å and 4.25 Å, are observable corresponding to the distorted hexagonal lattice of the L_{β}' phase; compare Figure 20. [From Janiak, M. J., et al., *J. Biol. Chem.* 254:6068 (1979).]

functions can be derived. The X-ray diffraction patterns of liquid longchain hydrocarbons are typified by a diffuse band at $(4.6\ Å)^{-1}$. This has been interpreted in terms of hexagonal packing of the chains. Although one can argue about this conclusion,[16] it fits in comfortably with the picture of roughly rodlike molecules, continually flickering between conformations but on the average packed like piles of parallel cigarettes.

We now look at diffraction patterns obtained for DMPC above and below the chain-melting transition temperature. The pattern in Figure 21C is from DMPC containing 16% water at 60°C, that is, in the region of the phase diagram labeled L_α. The sharp lines near the center of the pattern arise from ordered stacks of parallel bilayers and allow the lamellar repeat distance to be deduced. The diffuse outer band centered at $(4.5\ \text{Å})^{-1}$ is interpreted, by analogy with the $(4.6\ \text{Å})^{-1}$ band of liquid hydrocarbon, as giving the spacing of liquidlike but hexagonally packed chains. The pattern in Figure 21D, taken at 20°C, again shows long-spacing lines at the center, but also, in addition to a broad band at $(4.25\ \text{Å})^{-1}$, a sharp ring at $(4.16\ \text{Å})^{-1}$. These reflections arise from a deformed hexagonal lattice of tilted chains, with each chain having four neighbors at 4.87 Å and two at 4.77 Å; see phase L_β in Figure 20.

We see that the structural information from these diffraction patterns is much cruder than that from single crystals. We speak in terms of the spacings of molecules whose forms we cannot deduce and of layers whose structure is hidden. However, more than this is revealed by Fourier-transforming the diffraction pattern, a process that produces the electron-density profile in going across the bilayer (Figure 22). The particular profiles shown in the figure are similar in form to those obtained from many other lipid bilayers, including biologic membranes. The electron-density profile is as near as we can get to a structure determination, and although it gives a very smeared picture, it is nevertheless valuable. The local electron density is dependent on the types of atoms and their packing. Figure 23 allows a comparison of the density in a bilayer with that in some common substances. Of the atoms in bilayers, phosphorus has by far the highest atomic number and makes a large contribution to the electron density in its neighborhood. The peaks in Figure 22 are in the head-group regions and d_{p-p}, the distance between the two points of maximum electron density in each bilayer, is termed the bilayer thickness. The deep valleys are at the center of the bilayers where the methyl groups of opposing chains are comparatively loosely packed since they are separated by distances appropriate to van der Waals rather than covalent bonds. The high valleys fall between bilayers. The differences between the two profiles in Figure 22 are informative. The addition of water to lipid in the L'_β phase containing 12% water does not increase d_{p-p}, but does increase the repeat distance d between bilayers from 52.4 to 61.6 Å. Thus the swelling of the multilamellar lipids is here produced by water pushing in between the individual bilayers. Notice that the high valleys deepen in going from 12 to 40% water, the bottom of the valley falling to a density appropriate to that of bulk water. If we assume that

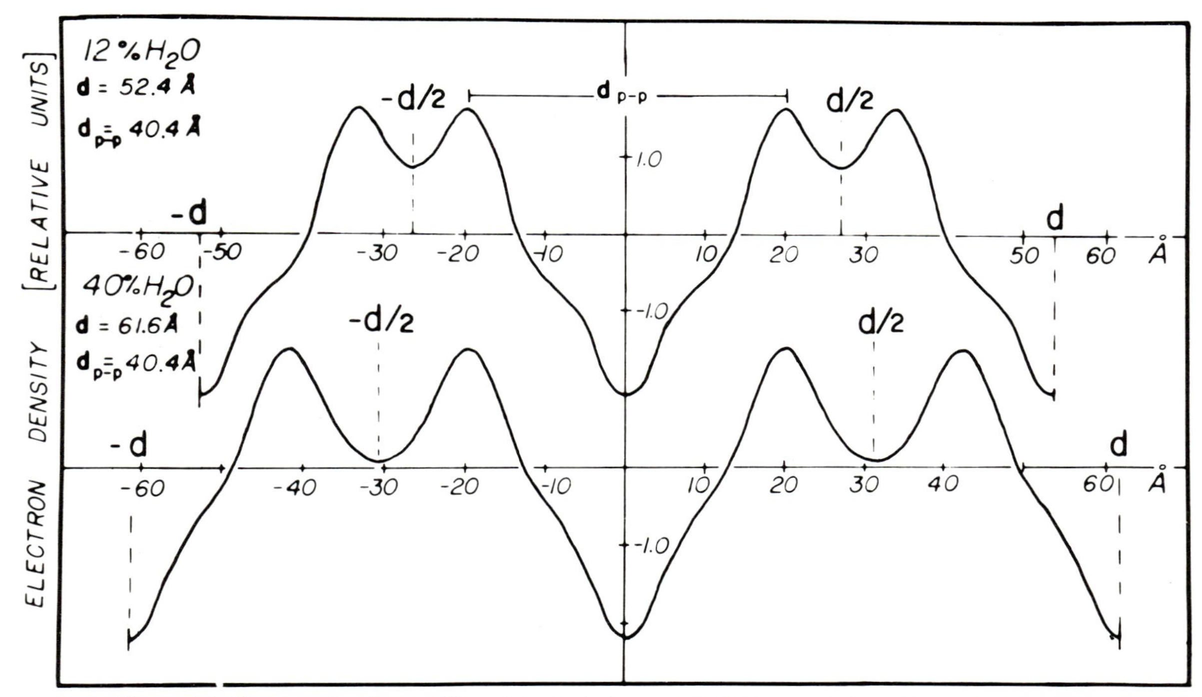

Figure 22. Electron-density profiles of the L_{β}' phase of DMPC containing 12% and 40% water. The distance between $-d/2$ and $d/2$ is the unit cell length here taken between two points situated in the layer of water between neighbouring bilayers. The deep minimum in the center of the profile lies in the middle of a single bilayer, at the center of the hydrocarbon core. The maxima are roughly centered on the lipid head groups so the d_{p-p} is the bilayer thickness. [From Janiak, M. J.,et al., *J. Biol. Chem.* 254:6068 (1979).]

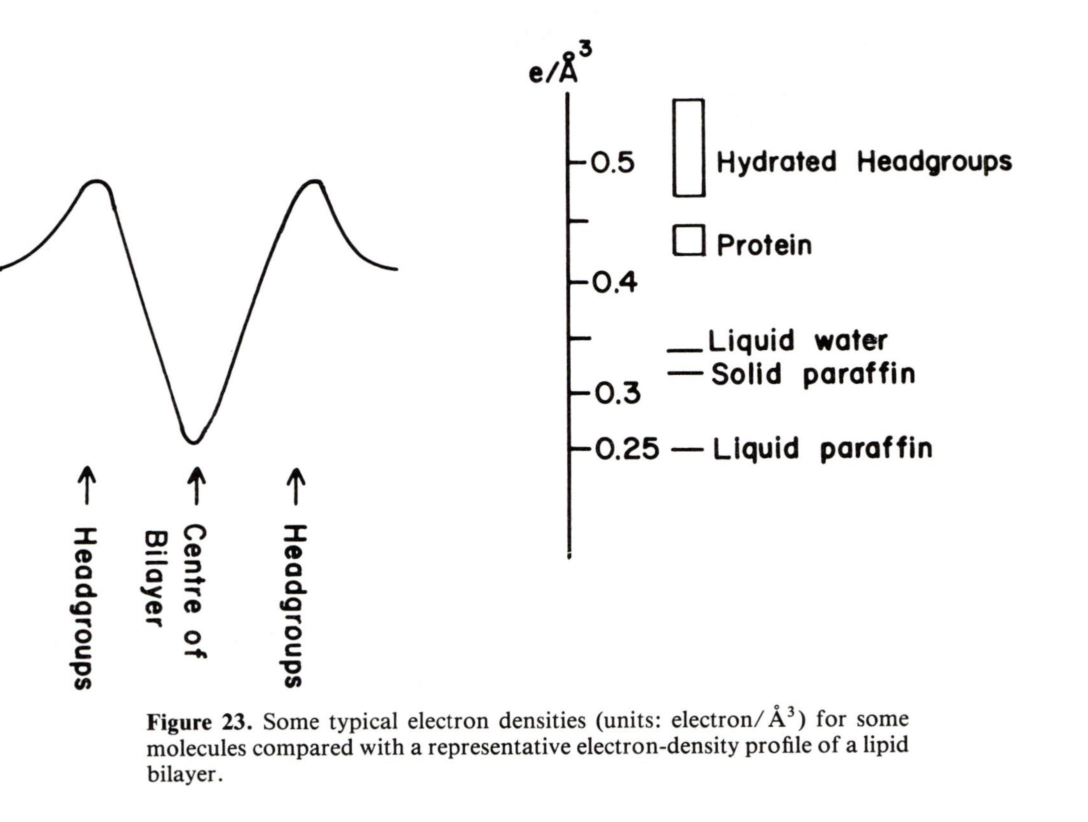

Figure 23. Some typical electron densities (units: electron/$Å^3$) for some molecules compared with a representative electron-density profile of a lipid bilayer.

the hydrocarbon chains are all-trans, rigid, and perpendicular to the surface, the estimated bilayer thickness based on molecular models is 51.6 Å, which is considerably above the value of 40.4 Å for d_{p-p} seen in Figure 22. The difference is explicable in terms of the tilting of the chains in the L_β' phase by about 35 degrees as confirmed by analysis of the diffraction pattern.

Spacings and electron-density profiles supply enough information to construct a good first sketch of phase structures. In the case of DMPC, the phases of biologic relevance are those with high percentages of water — L_β', P_β', and L_α — which are symbolized in Figure 20. Of these the low-temperature phase L_β', with its fully extended trans-chains and planar lamellae, is nearest to the crystalline state. The high-temperature phase L_α, with its melted chains, is nearest to our intuitive ideas of a biologic membrane. Janiak, Small and Shipley, on the basis of other experimental studies and chemical common sense, propose a general scheme for the phase diagram of the even-carbon phospholipids (Figure 24).

It is interesting that the swelling limit of the four lecithins concerned is affected by the length of the hydrocarbon chains even though, as we shall see, hydration occurs only in the head-group region. At the swelling limits the number of water molecules associated with each lipid molecule is 23 for DLPC, 25 for DMPC, 27 for DPPC, and 29 for DSPC.

It is indicative of the dynamic state of our knowledge in this field that the L_β' gel form of DPPC has recently been shown to be metastable.[17] Incubation at low temperatures converts the L_β' phase to a more crystalline form in which the hydrocarbon chains lose their rotational disorder and assume as yet not clearly defined conformations and orientations with respect to their neighbors. This form is stable below 14°C; above this temperature the usual phase changes $L_\beta' \rightarrow P_\beta'$ and $P_\beta' \rightarrow L_\alpha$ occur.

The intermediate phase P_β' (Figure 20) deserves further comment. The ripples proposed to account for the X-ray diffraction patterns have been observed much more directly by the electron microscope (Figure 25), which reveals the presence of two kinds of ripple. The nature of the chain packing and the physical causes for the formation of these curious phases are still open questions. Larsson[18] noted that if the cross-sectional areas of the chains, $n\Sigma$, and the head groups, S, differ, tilting will solve this geometric problem, but in general will not be compatible with the optimal packing of neighboring all-trans chains. Figure 26 illustrates tilting compatible with optimal chain packing, but there is no reason to expect that the tilt angle will satisfy the relation $S \cos \phi = n\Sigma$. Larsson regards the competition between these factors as the source of the rippled phase, and his model is supported by Sackmann et al.,[19] who find it to be not inconsistent with the two rippled phases that they observe in

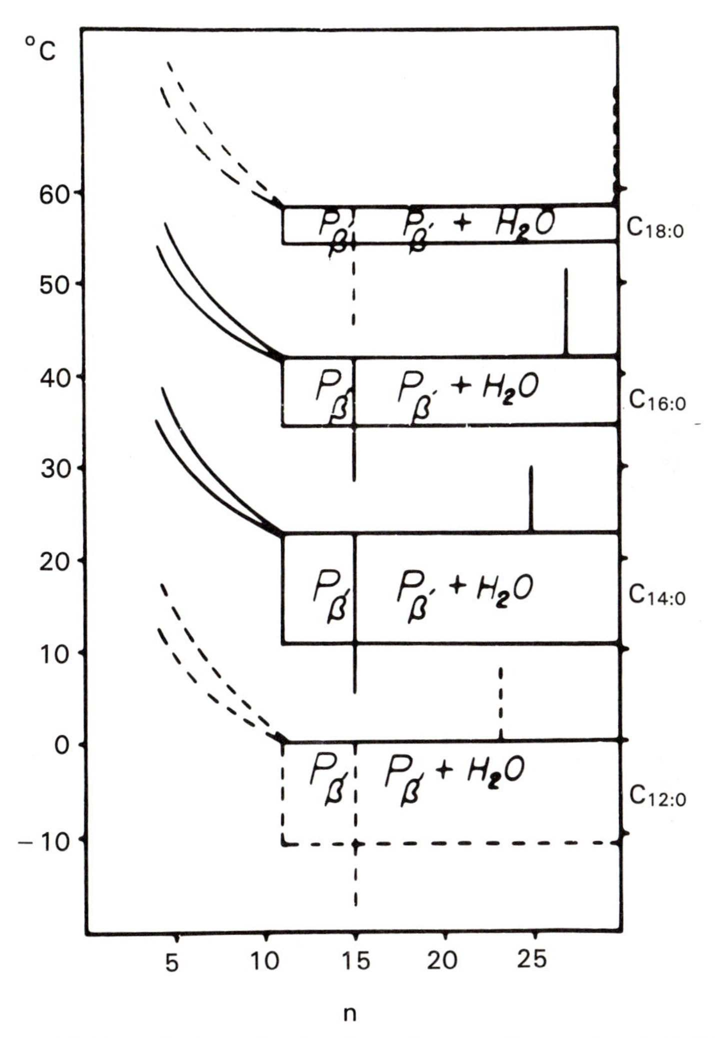

Figure 24. General scheme for the phase diagram of some phospholipids as proposed by Janiak et al. $C_{12:0}$ = DLPC, $C_{14:0}$ = DMPC, $C_{16:0}$ = DPPC, $C_{18:0}$ = DSPC. *n* is the number of water molecules per lecithin molecule. The solid lines are approximate experimental phase boundaries. The dashed lines are hypothetical.

DMPC. It is only fair to say that other theories have been proposed.[20]

The foregoing account of hydrated lipids has put the emphasis on changes induced by heat — *thermotropic* changes. It is instructive to examine *lyotropic changes,* those produced by altering the proportion of water in the system. We follow a horizontal line at 37°C on the temperature–water phase diagram, concentrating particularly on the dimensions of the multilamellar stacks. From Figure 27 we see that:

1. The surface area of each lipid molecule increases from ~50 Å^2 at 10% water by weight to a limiting value of 62.2 Å^2 ~25% water. This expansion of the head group lattice is evidence of the infiltration of water between the phosphatidylcholine groups.

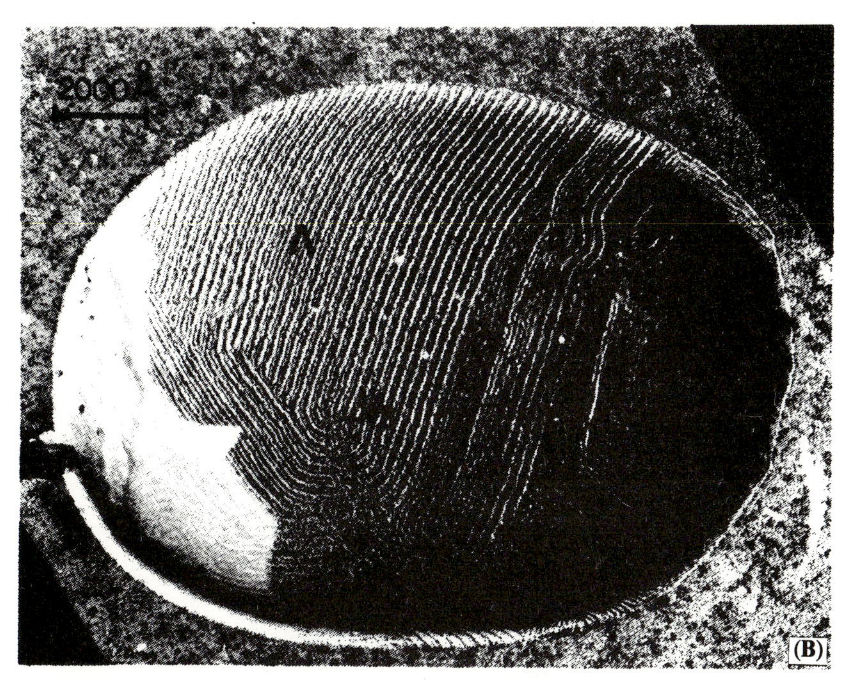

Figure 25. Electron microscope photograph of "rippled" phases in DPPC produced by cooling large vesicles from the $P_{\beta'}$ phase. Two types of textures are apparent, the "Λ phase" with ripple distance Λ and the "Λ/2 phase." A wall defect and two other types of packing faults ($S = \pm \frac{1}{2}$) are indicated. [From Sackmann, E., et al., in *Liquid Crystals of One- and Two-Dimensional Order*, Springer-Verlag, Berlin, 1980, p. 314.]

2. The bilayer thickness decreases to a limiting value of 35.5 Å from 44 Å at 10% water. This apparent shrinkage is offset by the lateral expansion noted above so that the volume occupied by the hydrocarbon chains is roughly constant; they merely become more crumpled. The possibility that the decrease in bilayer thickness is due to chain tilting is ruled out by the X-ray diffraction patterns, but the thinning of the hydrophobic core is consistent with the picture of melted chains suggested in the previous chapter and elaborated upon in the next.

3. Hydration of anhydrous phospholipids results initially in the adsorbtion of all added water into the head groups, a process that continues up to ~21% water in the case of DMPC. Beyond this point free water begins to appear between the bilayers, which start to separate. For phosphatidylcholines the distance between bilayers reaches a maximum, the swelling limit, which depends on the chainlength and the temperature, with further hydration only resulting in the appearance of bulk

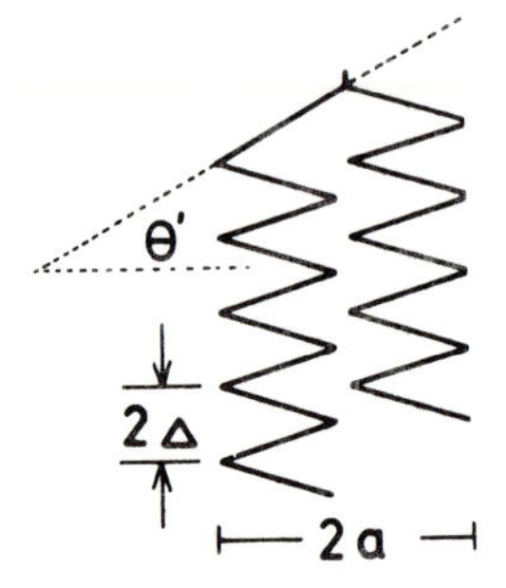

Figure 26. Packing of all-trans chains compatible with chain tilting. [From Sackmann, E., et al., in *Liquid Crystals of One- and Two-Dimensional Order*, Springer-Verlag, Berlin, 1980.]

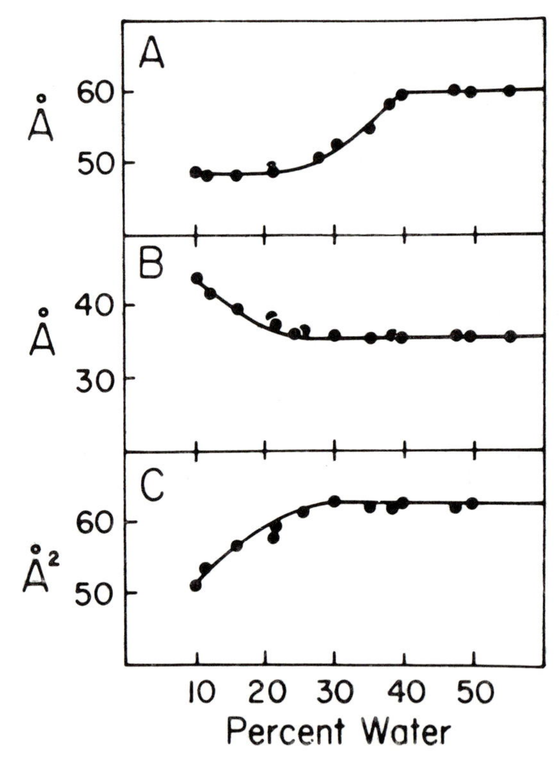

Figure 27. Dependence on % water in DMPC samples of: **A**, the bilayer repeat distance *d* (see Figure 22); **B**, the thickness of the lipid chain layer within the bilayer; **C**, the area of each DMPC molecule at the bilayer surface. The bilayer is in the L_α phase at 37°C. [From Janiak, M. J., et al., *J. Biol. Chem.* 254:6068 (1979).]

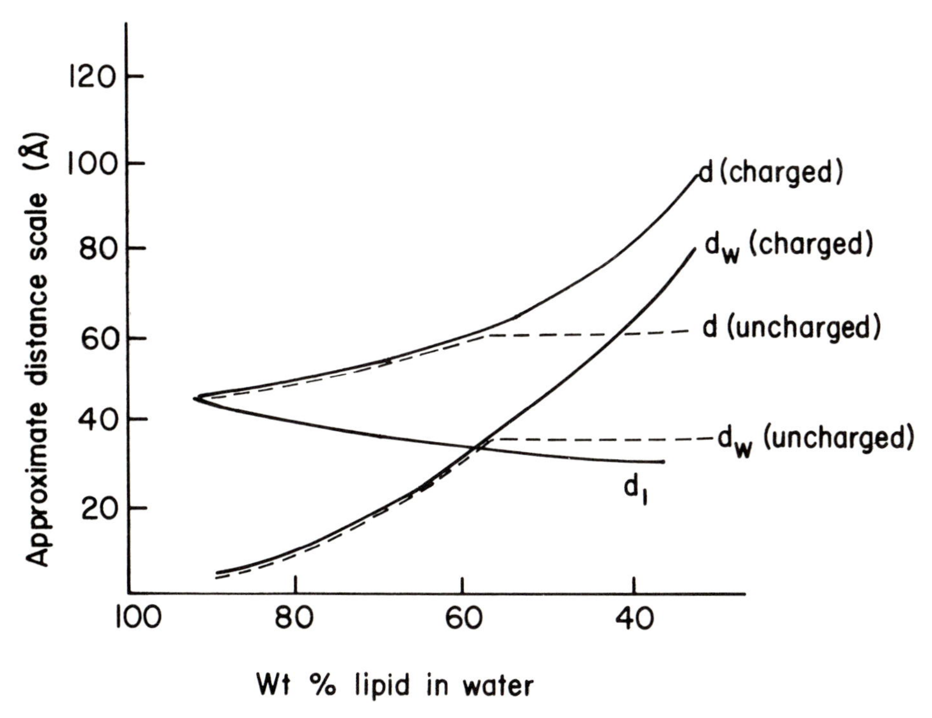

Figure 28. Schematic curves showing the interlayer separation for charged and neutral bilayers as a function of water content.

water situated outside the stacks of bilayers. The fact that there is an equilibrium separation between adjacent layers is striking evidence of the existence of attractive forces between hydrated bilayers, and also of repulsive forces since otherwise the bilayers would not move apart. The bilayer separation can be altered by playing with a variety of parameters, including ionic strength, pH, pressure, and dissolved cations. From such studies much has been learned of the strength and distance dependence of interbilayer forces (Chapter VII). Those lipids that have charged head groups do not have an equilibrium separation in excess water but separate indefinitely with addition of water due to the overwhelming effect of electrostatic repulsion. The behavior of neutral and charged phospholipids on hydration is summarized schematically in Figure 28.

Support for the supposed near-invariance of volume of the lipid chains on crumpling comes from studies on the volume of bilayers undergoing thermotropic melting. Thus in a study of DPPC the volume change on going from the gel state to the liquid crystal state was 3700×10^{-5} ml g^{-1}, with most of this being attributed to changes in the volume of the hydrophobic core rather than in the head-groups.[21] Even if the chains are the sole source of the volume change, a 4% expansion is considerably smaller than the corresponding values found for the melting of most solids. The clamping effect of the head groups prevents the hydrocarbon chains from spreading into the kind of living space that they have in simple liquids.

ORIENTATION OF HEAD GROUPS IN HYDRATED BILAYERS

Electron-density profiles provide a rough means of localizing the head groups of phospholipids in hydrated systems, but as in the case of hydrocarbon chains, the information obtained is a long way from the precise localization of atoms that is possible for crystals. The most clear-cut indicators of head group conformation have come from neutron diffraction and nuclear magnetic resonance. We give here a brief account of the neutron diffraction results, postponing the discussion of NMR to Chapter 5.

Accelerated neutrons of energy around 0.1 eletron-volt such as those used in neutron diffraction, have a wavelength of about 1 Å, within the range of wavelengths used in X-ray diffraction. The difference is that neutrons are diffracted by nuclei and X rays by electrons. Neutrons interact with the magnetic moments of nuclei, but this will not interest us here. A measure of the diffracting power of an atom is the atomic scattering amplitude, which for electron, X-ray, and neutron diffraction have typical values of 10^{-8}, 10^{-11}, and 10^{-12} cm respectively. Thus neutrons are a less sensitive probe than X rays but they have a distinct advantage for our purposes in that the scattering cross section for the deuteron is $+6.7$ fermi units compared with -3.7 for the proton. This fact has been used to determine, among other things, the location of water in bilayers and the orientation of head groups. Neutron diffraction patterns can be transformed to give nuclear scattering length density profiles which, as in the case of X-ray diffraction, show high peaks in the head-group region and a central valley. The exploitation of the high scattering amplitude of deuterium is illustrated in Figure 29 which shows the profiles for the L_{β}' gel phase of DPPC deuterated at the C5 and C15

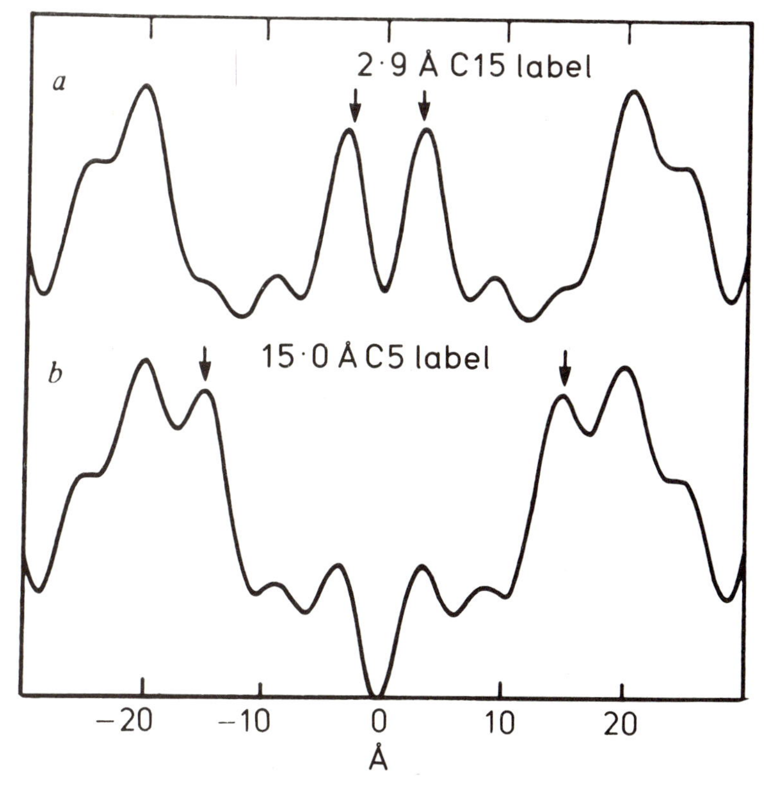

Figure 29. Neutron diffraction scattering profiles for the L_{β}' phase of DPPC deuterated at the C5 and C15 positions in both hydrocarbon chains. [From Büldt. G., et al., *Nature* 271:182 (1978).]

positions of both hydrocarbon chains.[22] One can almost see the deuterium nuclei. Analysis of the profiles in terms of models shows that the two deuterated C15 segments are out of step by approximately one and a half carbon–carbon bonds. This staggering is not unexpected in the light of the inequivalence of the two chains displayed in the crystal state (Chapter 1) and revealed by deuterium labeling[23] of the carbon segments adjacent to the acyl groups in each chain, that is, on carbon atoms 22 and 32 in Fig. 1B. The 22 segment is about 1.9 Å further away from the center of the bilayer than the 32 segment. The most interesting results are those obtained by deuteration on carbon atoms 11, 12, and one of the methyl groups of the choline moiety. The distances of these segments from the bilayer center are almost identical, proving that the head group is folded down parallel to the membrane surface, which is true at water concentrations of 5–6 weight percent (wt%) or 25 wt% at 20°C, in the gel state, or 50°C in the liquid crystal L_α phase (Table 2) In the light of the considerable motional freedom possessed by hydrocarbon chains in the liquid crystal state of phospholipids, the reader has the right to expect a degree of flexibility in the head groups. In particular, the aqueous environment of hydrated bilayers should provide little resistance to the swinging of a head group through positions roughly normal to the bilayer surface. The fact that neutron diffraction and X-ray results show a folded-down conformation can be taken to indicate either that the head groups are fairly rigid or that they move in such a way as to be largely restricted to folded-down conformations. As we shall see later when discussing NMR studies in Chapter 5, the second interpretation is more likely. The conformations adopted by the head groups are probably mainly determined by internal steric considerations and hydrogen-bond or electrostatic interactions between head groups.

The consistency of X-ray and neutron diffraction results is demonstrated by the fact that in the L_{β}' phase at 25 wt% water the deuterium-labeled head groups are closer to the bilayer center than for 5–6 wt % water; compare columns 1 and 2 of Table 2. The figures fit in with X-ray data that show that in going from 5 to 25 wt% water the tilt angle of the hydrocarbon chains increases from 16 to 30 degrees, thus decreasing the bilayer width. Incidentally this change in tilt angle is presumably due to an expansion of the head-group lattice, which has geometric consequences derivable from the equation $S \cos \phi = n\Sigma$. The folded-down conformation of the head groups has been confirmed by ^{31}P NMR, as we shall see later.

In the same way that carbon segments can be pinpointed, it is possible to localize water molecules. If the profile across a hydrated bilayer as deduced from the neutron diffraction pattern is subtracted from the profile of a similar bilayer hydrated with D_2O, the difference gives the

Table 2
Distance, x_o, of Deuterium Labels from Center of DPPC Bilayer

	$L_{\beta'}$ phase, 6% (w/w) H_2O, 20°C			$L_{\beta'}$ Phase. 25% (w/w)H_2O, 28°C	L_α Phase. 25% (w/w) H_2O, 50°C
Carbon Atom*	x_o(Å) Expt.	(x_o from Space-Filling model) Head Group Parallel to Surface	Head Group Perpendicular to Surface	x_o(Å) Expt.	x_o (Å) Expt.
Choline methyl	25.1 ± 0.6	24.4	29.6	24.4 ± 0.6	21.8 ± 0.6
groups					
32	24.8 ± 0.7	25.3	28.0	24.1 ± 1.0	21.2 ± 1.0
31	24.5 ± 0.7	24.0	27.2	23.6 ± 1.0	21.0 ± 1.0
3	23.1 ± 1.0	23.5	23.5	21.6 ± 1.5	17.4 ± 1.5

*The nomenclature is that used in Figure 1B. From Büldt, G., et al., *Nature*, 271:182 (1978).

location of the water deuterons[22,23] (the protons of the hydrocarbon chains are not exchangeable under the experimental conditions). Figure 30 shows that water is localized in the polar head group area, as it is in the crystal of hydrated DMPC.

We now discuss the localization of cholesterol in bilayers since it is an important component of biologic membranes and has a major influence on the freedom of motion of hydrocarbon chains (Chapter 5). Electron-density profiles of bilayers with and without cholesterol were determined for phosphatidylcholines with 12-, 16-, and 18-carbon atom chains.[24] Accepting the limitations set by resolution, it was found that in all cases one end of the cholesterol molecule is located near the carbonyl groups of the lipid and the rest is buried in the hydrocarbon core. Since the only hydrophilic region in cholesterol is the hydroxyl group, it is probable that this group is situated near the polar head groups of the bilayer. For 12- and 16-carbon atom chains cholesterol increases the width of the bilayer in both the gel and liquid crystal state. This can be understood in terms of the effect on neighboring chains of the inflexible ring system of cholesterol. The elecron-density profiles indicate that the bilayer head group lattice is effectively undisturbed by the intrusion of cholesterol. The rigid molecule nevertheless can be accommodated by a decrease in the tilt of the all-trans chains of the gel state and a consequent thickening of the bilayer. In the liquid crystal state the melted chains are already, on the average, extended perpendicularly to bilayer surface, but are, as we have seen, shortened by rotational disorder. The presence of the stiff pseudo-planar ring system of cholesterol acts as a surface on which the neighboring melted chains are pulled into configurations containing a heightened proportion of trans

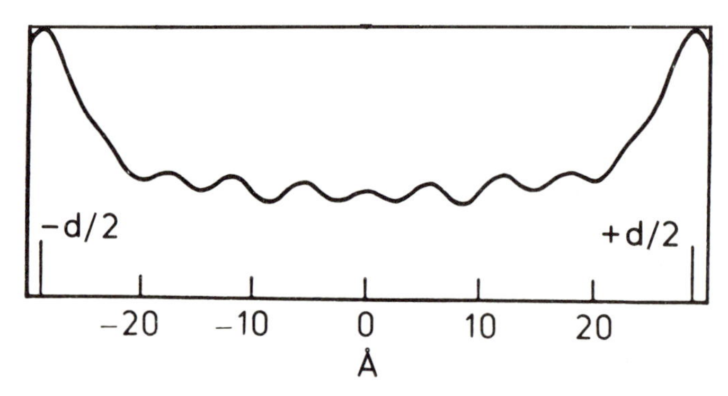

Figure 30. Difference neutron diffraction scattering profile of bilayers hydrated with H_2O and D_2O. [From Büldt, G., et al., *Nature* 271:182 (1978).]

sections, and therefore a longer average length. Cholesterol reduces the width of the 18-carbon bilayer, possibly due to the curving of the tails of the long chains to caress the ends of the shorter cholesterol molecules. A rather different way of looking at the packing of cholesterol is presented in the next chapter.

The scattering density profiles obtained for a given system by X-ray and neutron diffraction are not similar in form since X-ray scattering amplitudes are proportional to the atomic number, and therefore to electron density, but no such simple relationship holds for neutron scattering. In membranes containing lipids, proteins, and water, neither method reveals unambiguously the distribution of these components. However, the scattering densities associated with different membrane components are known, both for X rays and neutrons, and using this knowledge, together with both the X-ray and neutron-density profiles, it is possible to set up and solve a series of simultaneous equations that allow the determination of the distribution of lipid, protein, and water across the membrane. This method, originated by McCaughan and Krimm,[25] would seem to hold considerable potential.

Our discussion of phospholipids has been limited to saturated hydrocarbon chains but it is known that unsaturated chains occur in biologic membranes, particularly in mitochondria. Figure 31 shows the gel-to-liquid-crystal transition temperatures of a series of hydrated monounsaturated phosphatidylcholines, all having 18-carbon chains.[26] Two points are obvious: The values of T_t are all well below that of the corresponding saturated lipid (54.9°C) and T_t goes through a minimum for lipids having the single cis bond near the center of the chain. It was also found that the transition enthalpy for the unsaturated compounds was lower than that for the saturated lipid. The double bond must have an effect on the packing of the chains in the gel state, and although crystal-structure studies for unsaturated lipids are not available, it seems likely that the packing is looser for these compounds than in the case of all-trans chains. The comparatively low transition temperatures of unsaturated lipids appear to have direct biologic implications. In particular, the percentage of these lipids in the membranes of certain bacteria depend on the temperature at which the cultures are maintained. At low temperatures the fraction of unsaturated lipids synthesized by the cell is raised.[27] This seems to be a regulatory mechanism preventing the membranes from freezing into the gel state, a condition not conductive to biologic activity. In general it appears that the lipid composition of the membrane adjusts to the environment in such a way as to maintain the "fluidity" of the membrane, a concept which we later expand.

In this chapter we have limited ourselves to structure, but it should be realized that much of the information derived from physical

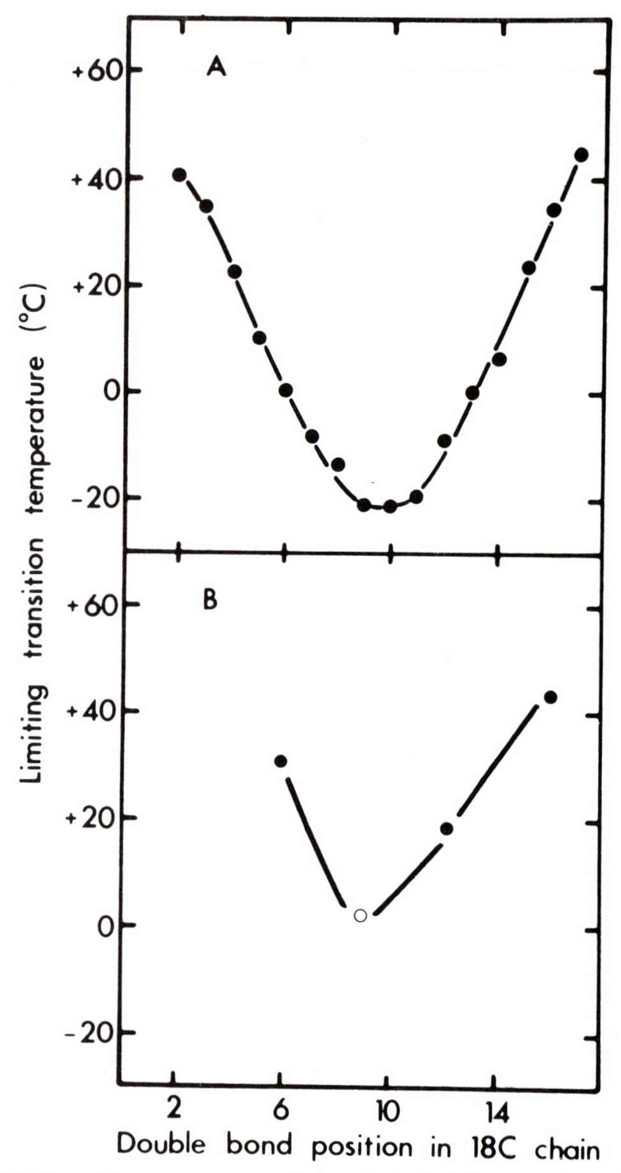

Figure 31. Transition temperatures for synthetic lecithins as a function of the position of a double bond in the hydrocarbon chain. **A.** Both chains contain a double bond in an 18-carbon chain. **B.** One chain is saturated; the other contains a single double bond. [Barton, P. G., and Gunstone, F. D., *J. Biol. Chem.* 250:4470 (1975).]

measurements on liquid crystalline phases, and even on gel phases, is time-averaged. As we will later see in more detail, all spectroscopic methods are characterized by a time scale, which separates us from ultimate reality. Thus it is clear that if an X-ray diffraction pattern takes a few seconds to record, the derived atomic positions will be averages over billions of vibrations. (However, see reference 28.) Subtler limitations hold for optical and magnetic resonance spectroscopy. The atoms in the head groups and chains in hydrated phospholipids meander over distances well outside those typical of atomic vibrations, but nevertheless, as we will see in Chapter 5, we can quantify this motion to a certain extent, especially in the case of planar hydrated bilayers.

4 Micelles, Vesicles, and Bilayers – Steric Factors

Apart from the phases discussed in the previous chapter, lipids aggregate in aqueous solution to give a number of forms, the best characterized of which are now described. In Figure 32, diagrammatic representations of some phases are presented.

CURVED BILAYERS – VESICLES

Many phospholipids in water form closed spherical shells consisting of one bilayer.[29] These so-called vesicles can be produced by a number of methods; the most common is the sonication of an aqueous suspension of phospholipid to give a fairly homogeneous population, in terms of size, with an average vesicular diameter of 200 to 300 Å, the figure depending on the particular lipid (Figure 33). Large vesicles of diameters in the region of 1000 Å have been produced by other methods. Closed multilayered structures probably exist under certain conditions.

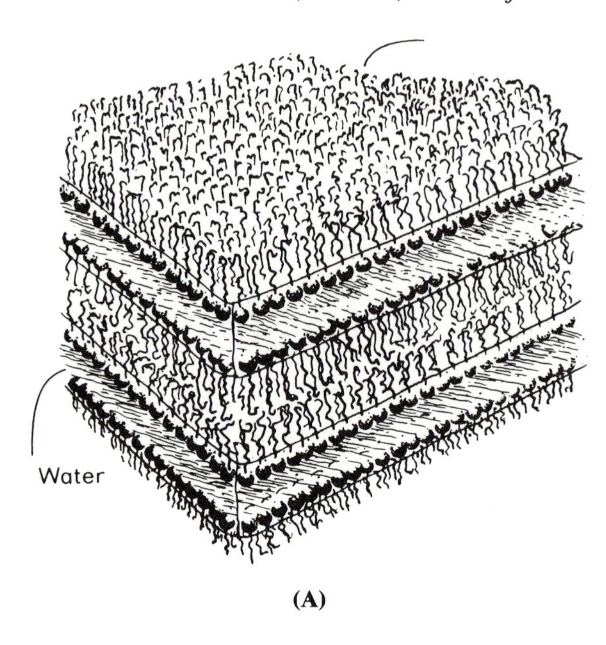

(A)

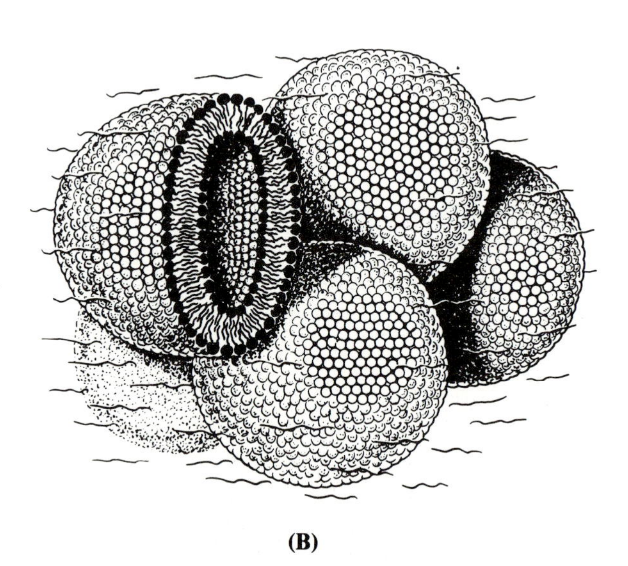

(B)

Figure 32. Diagrammatic representation of some common lipid phases. The spheres indicate the head groups and the wiggly lines the lipid chains of phospholipids. **A.** Planar bilayers. **B.** Vesicles. **C.** The hexagonal H_{II} phase. **D.** Micelles. (From Khetrapal, C. L., et al., *Lyotropic Liquid Crystals*, Springer-Verlag, Berlin, 1975).]

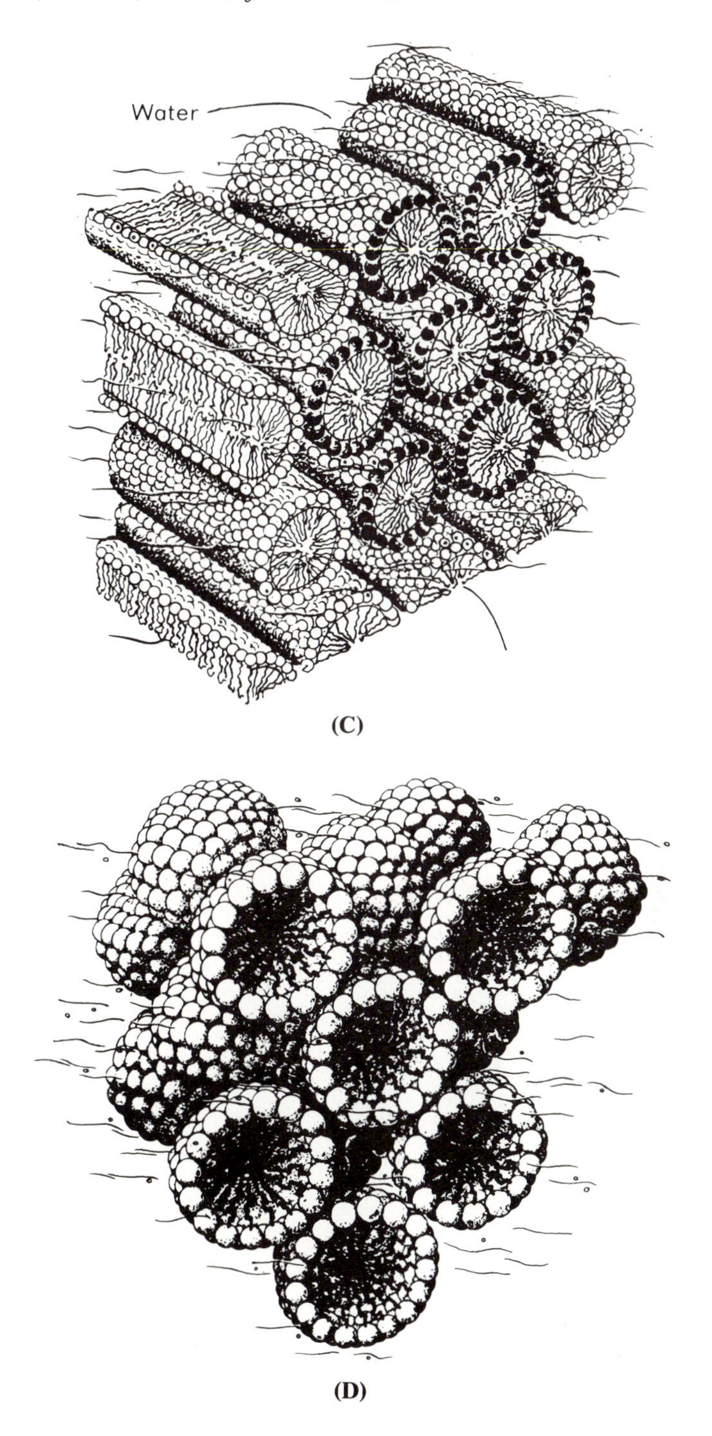

(C)

(D)

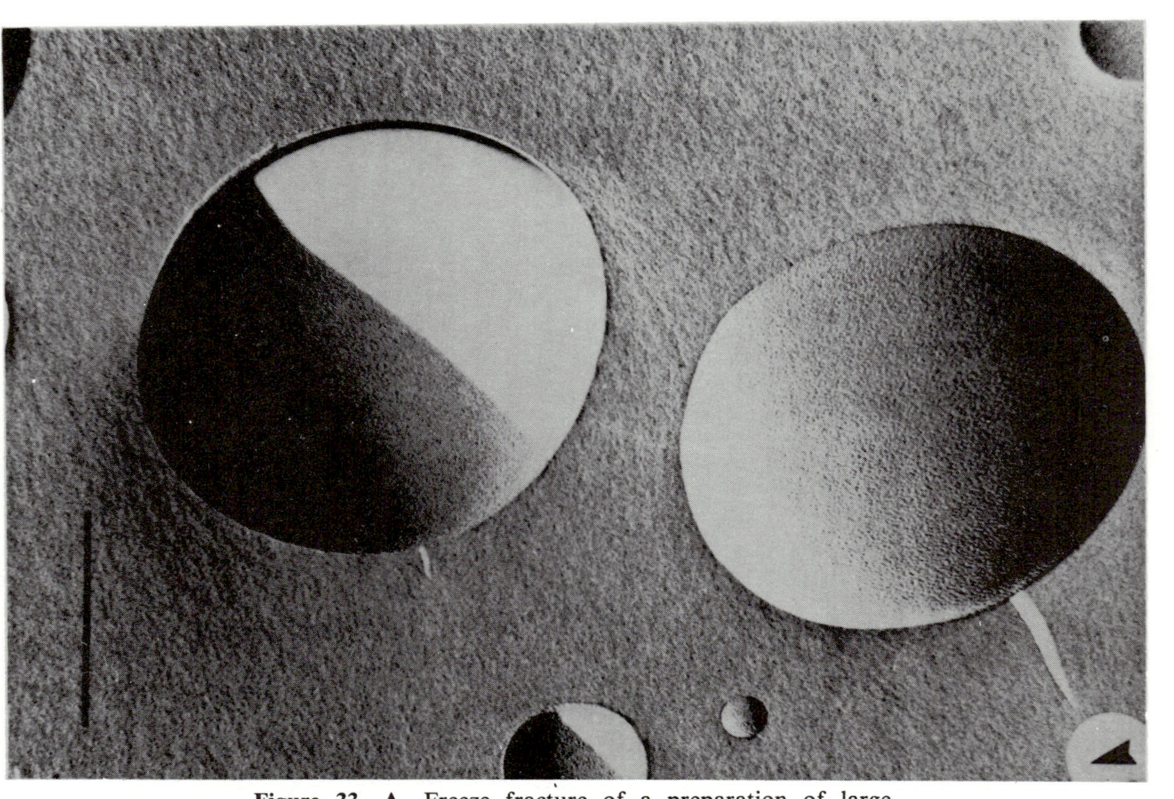

Figure 33. A. Freeze fracture of a preparation of large unilamellar vesicles. Bar, 100 nm.

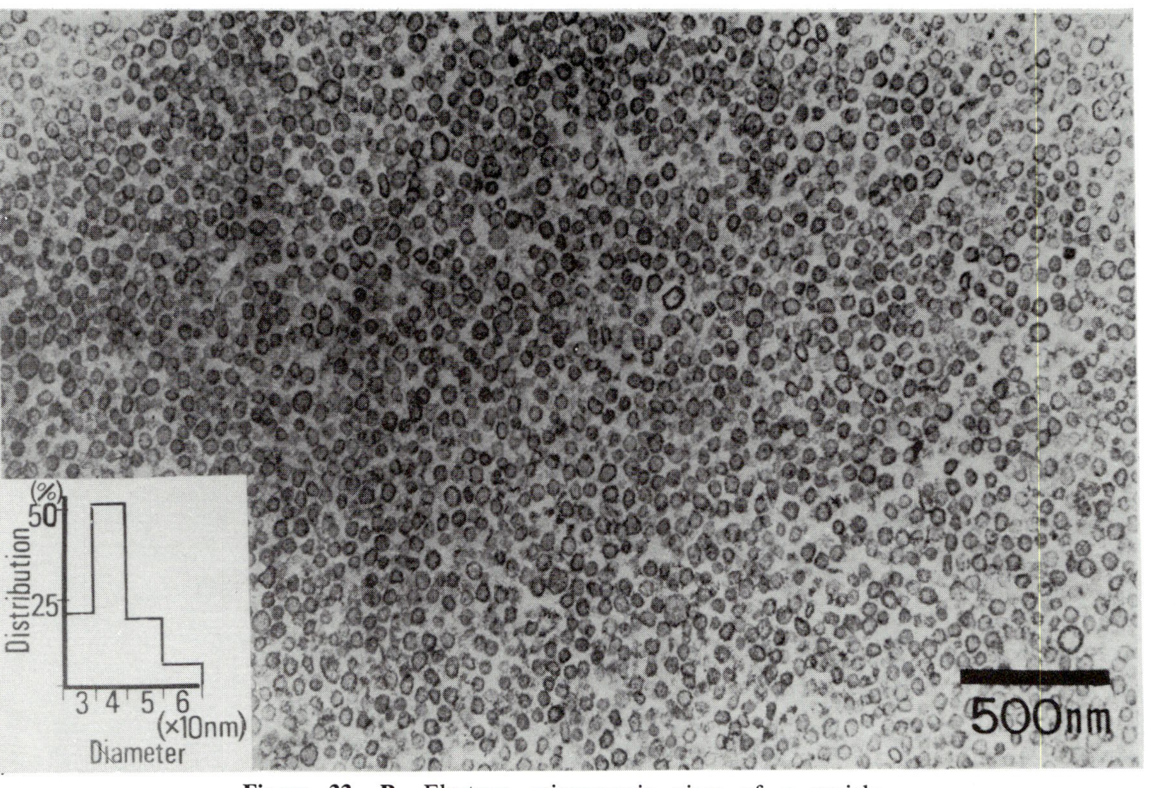

Figure 33. B. Electron microscopic view of a vesicle preparation together with the size distribution. It is probable that single-lipid vesicles in solution are spherical. [From Ohsawa, K., et al., *Biochim. Biophys. Acta* 648:2061 (1981).]

It is not clear whether vesicles represent a thermodynamically stable form; we return to this point in Chapter 6. Vesicles have been used medically as microcapsules, enclosing molecules to which they are impermeable and which they release in the body due either to rupture of the vesicle or to changes in its permeability. Hopes have been entertained that by a suitable choice of lipids and other constituents of the vesicle it will be possible to target vesicles to specific sites in the organism.[29]

MICELLES

Detergent solutions are the most familiar examples of micelles. In general, it is single-chain lipids that form micelles, their hydrophobic chains driven together into the interior of the micelle by factors that seem clear in their crude nature but, as we will see, are far from understood in detail. The lifetime of molecules in micelles is short, 10^{-3} to 10^{-5} seconds,[30] compared with measured lifetimes of the order of hours for many double-chained lipids in bilayers or vesicles. The average number of molecules in a micelle, its mean aggregation number, is strongly dependent on the nature and number of ions in solution. Spherical, globular, and cylindrical micelles have been characterized.

INVERTED MICELLES

Some lipids in aqueous solution cluster with their polar head groups in close consultation and their fatty tails waving outward. The trouble with inverted micelles is that their exterior is hydrophobic, a problem that they overcome by aggregating, usually in the hexagonally packed H_{II} phase (Figure 32) in which the lipid molecules form cylindrical structures containing an aqueous core. The hydrophobic exterior of each cylinder contacts its own kind in the six neighboring cylinders. Lipids that normally form other types of aggregate sometimes can be induced to change to the inverted micelle form by the addition of divalent cations such as Ca^{++}.

Systems consisting of mixed lipids show all of the above phases in the right conditions, but also display phenomena, such as lateral phase separation in bilayers and assymetric distribution of lipids between the inner and outer layers of vesicles, that may have considerable biologic importance.

STERIC CONSIDERATIONS

The reasons for which a given lipid prefers to form a certain structure in water must include, of course, the nature of the intermolecular forces involved, which is discouraging in that we have a rather incomplete understanding of the interactions between large molecules in solution. Fortunately, as we now show, we can make sweeping generalizations concerning the phase preferences of lipids solely on the basis of steric considerations.[31]

If we attempt to gift-wrap lipid molecules, we find that we need boxes of three shapes — cylindrical, conical, and truncated conical — depending on the lipid (Figure 34). Lipids with single chains and large heads need conical boxes; those with single chains and smallish heads fit into truncated conical boxes, as do lipids with double chains and small head groups. Many double-chain lipids with large heads and fluid chains slip best into cylindrical containers.

We now consider the problem of packing large numbers of such boxes, taking into account the two restrictions that the head groups are adjacent and the hydrocarbon chains have maximum mutual contact. Both conditions make chemical common sense in that in aqueous solution, as in the crystal state, we expect the hydrophobic and hydrophilic portions of amphiphiles to obey a molecular apartheid policy. The consequences are summarized in Figure 35.

Conical molecules pack best as spherical micelles, as observed for

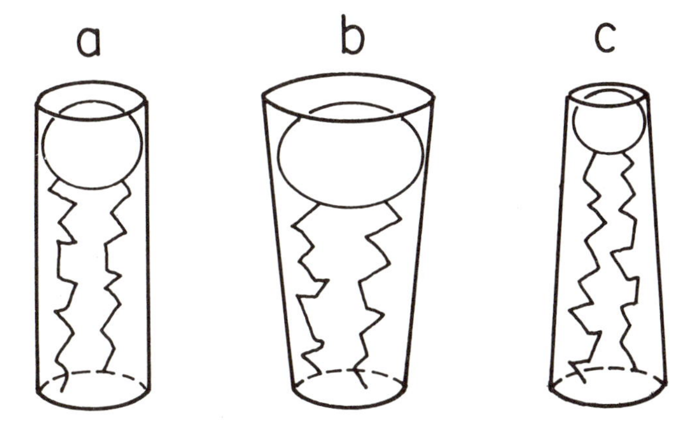

Figure 34. Gift-wrapped lipid molecules showing the crude shape of the molecule in the liquid crystal state. The relative size of the head groups and the lipid chains are, as explained in the text, partially determined by the charge state of the head group and the degree of disorder of the chains.

Shape	Critical Packing Parameter $v/a_o l_c$	Lipids	Examples	Structures Formed
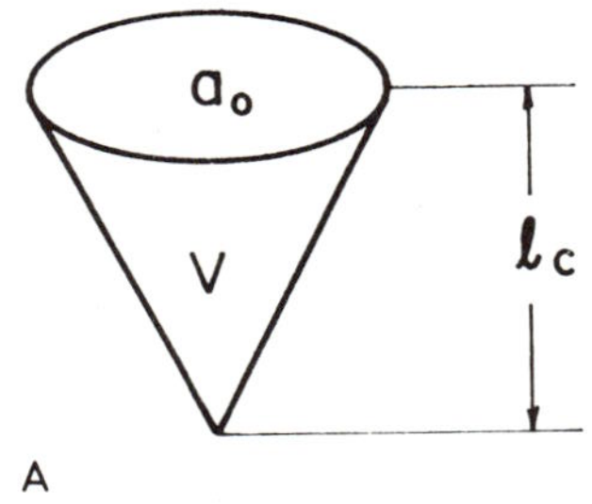 A a_o, V, l_c	$< 1/3$	Single-chain lipids with large head group areas	Detergents especially at low ionic strengths	Spherical micelles
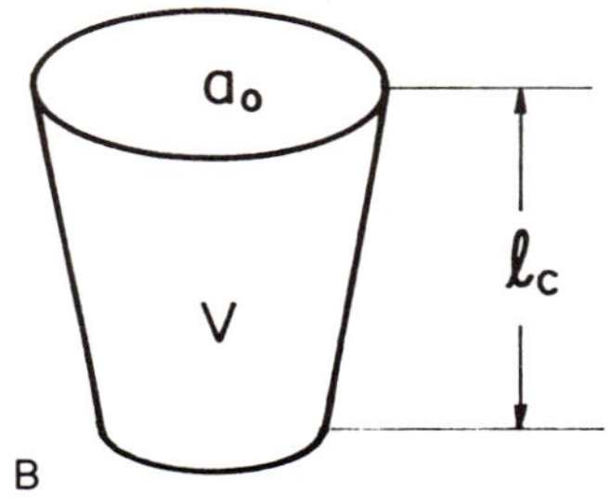 B a_o, V, l_c	$1/2 - 1$	Double-chain lipids with large head group areas Single-chain lipids with very small head groups	DPPC, DPPS, phosphatidic acid, Sphingomyelin	Vesicles

Shape	Critical Packing Parameter $v/a_o l_c$	Lipids	Examples	Structures Formed
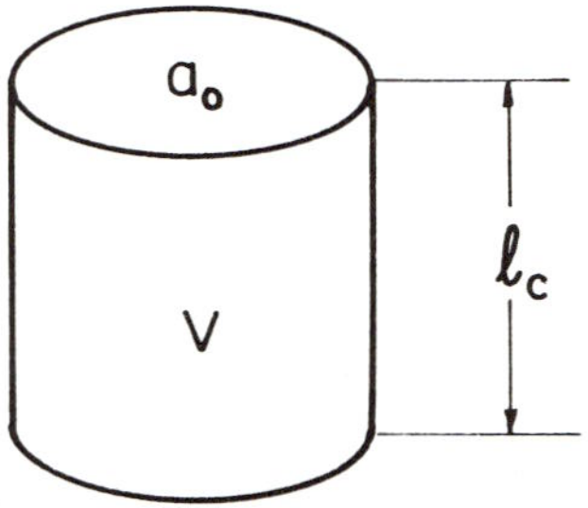 C (a_o, V, l_c)	1	Double-chain lipids with small head group areas	DPPE, DPPS + Ca^{2+}	Planar Bilayers
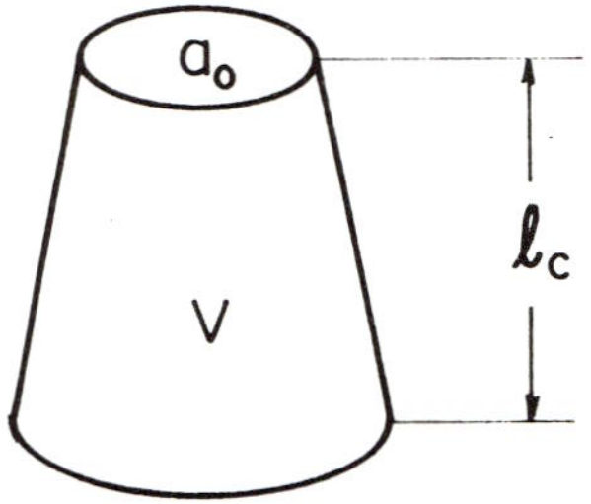 D (a_o, V, l_c)	>1	Double-chain lipids with small head group areas	Unsaturated DPPE, Cardiolipin + Ca^{2+} Cholesterol	Inverted micelles

Figure 35. Molecular "shapes," packing parameters, and preferred aggregated forms for amphiphiles. [Adapted from Israelachvili, J. N., et al., *Q. Rev. Biophys.* 13:121(1980).]

many detergent molecules and for some lysophospholipids. Most micelle-forming molecules have charged head groups, giving a large effective head group area. Added salts reduce electrostatic repulsion by shielding the charges and simple geometry shows that larger micelles can be formed, but since the interior of a micelle is hydrophobic and essentially free of water, large spheres will be unfavored energetically and so nonspherical micelles are often formed on salt addition. It also follows that the average number of molecules in a micelle is strongly dependent on the nature and concentration of salts in the solution.

Truncated-cone molecules for which the cross section of the head group is small compared with that of their chains tend to form *inverted* micelles that commonly form the hexagonal H_{11} phase referred to above. Truncated-cone molecules for which the head group is at the base of the cone are similar to conical molecules in high salt solution in that they could, in principle, pack into large spherical micelles but prefer to form other (e.g., cylindrical) shapes so as to exclude water from their interiors.

For near-cylindrical or cylindrical shapes, bilayers can be formed. These are flexible, and in some cases form closed structures — vesicles — that can have large diameters since the head groups lining the interior of the bilayer are hydrophilic and have no objection to the occlusion of water within the vesicle.

Primitive though the above considerations appear, they provide a fairly trustworthy guide to phase preferences. If we wish a slightly more quantitative basis, we can use simple solid geometry and show that the so-called packing parameter $v/a_o l_c$ will determine the preferred structure in solution.[32] Here v is the volume taken up by the hydrocarbon chains of one molecule, a_o is the effective head group area, and l_c is the so-called critical chain length. This last parameter is the maximum length that the chains assume at a given temperature, in the sense that chain lengths above l_c cause a sudden increase in the free energy of the system. This definition is a little vague; it attempts to limit us to liquid chains, avoiding the fully extended all-trans conformation. Perhaps l_c is best taken from the literature, in which case it turns out to be a bit less than the fully extended length. For a conical molecule $v/a_o l_c = 1/3$; for a cylindrical molecule $v/a_o l_c = 1$. Figure 35 shows the shapes, critical packing parameters, and preferred phases of amphiphiles. In Chapter VI we delve deeper into energetic and entropic factors.

THE STRUCTURE OF VESICLES

Vesicles are important as membrane models and as research tools in biology and medicine and we therefore devote more attention to them than to micellar structures. Vesicles produced by sonication of lipids above the phase-transition temperature are usually not more than 2 to 300 Å in diameter. Large vesicles are unfavored by entropy while a lower limit on the diameter is roughly set by the energetic requirements associated with an increase in head group cross-sectional area. The "vesicular weight" is in the region of 10^6 to 10^7 daltons. The average size in a given preparation depends on the lipid or lipids involved and on any other variable one cares to name: temperature, ionic strength, pH. Nevertheless, single-lipid vesicles produced by sonication have sizes that vary within narrow limits with values that can be fairly well predicted from the "shape" of the molecule and simple geometric considerations. The most studied lipid, DPPC, gives spherical vesicles, with inner and outer radii of around 60 to 110 Å respectively.

Much of our knowledge of the physical parameters characterizing vesicles has come from the studies of Huang and co-workers,[33] who have used the ultracentrifuge and viscometer to good effect. Some of their results for DPPC vesicles are shown in Table 3. The internal volume was determined by sonicating the vesicles in a solution of say $K_3Fe(CN)_6$ and passing the resulting suspension through a sepharose column that removed the bulk $K_3Fe(CN)_6$, leaving only that occluded by the vesicles that are impermeable to the triply charged ferricyanide ion. The addition of propanol disrupted the vesicles and the released $K_3Fe(CN)_6$ was determined spectrophotometrically. Knowledge of the initial concentrations of $K_3Fe(CN)_6$ and phospholipid allows the trapped volume to be determined. A consequence of the spherical form of vesicles is that the volume of the inner monolayer in the bilayer is smaller than that of the outer monolayer. For a spherical shell with internal and external radii of 60 Å and 110 Å and two monolayers of equal thickness, the volumes of the inner and outer layers are ~ 1.67 and $\sim 3.00 \times 10^6$ Å^3 respectively, that

Table 3
Some Physical Characteristics of Egg-Yolk Phosphatidylcholine Vesicles

Molecular weight	1.9×10^6 dalton
Number of molecules per vesicle	2400
External diameter	210 Å
Internal diameter	100 Å
Partial specific volume	0.9885 ml g^{-1}
Diffusion coefficient, $D°_{20}$	1.87 cm^{2}sec$^{-1} \times 10^7$
Sedimentation coefficient, $S°_{20,w}$	2.10 S

[From Huang, C., and Mason, J.T., *Proc. Nat. Acad. Sci.*, 75: 308 (1978).]

is, the ratio of inner to outer volume is ~1 : 1.8. It is possible to estimate this ratio experimentally. One way relies on the fact that in NMR spectra the chemical shift of a given nucleus is influenced by the environment of the molecule, as evidenced, for example, by solvent shifts. Figure 36 shows the ^{31}P NMR spectra of egg lecithin vesicles in D_2O. The two peaks, of differing intensity, can be tentatively assigned to the head groups of the inner and outer layers of the vesicle. This interpretation is validated by adding the paramagnetic Co^{2+} ion, which results initially in a broadening and slight shift of the larger peak. At high concentrations of Co^{2+} the peak is so broadened as to be effectively undetectable, show-

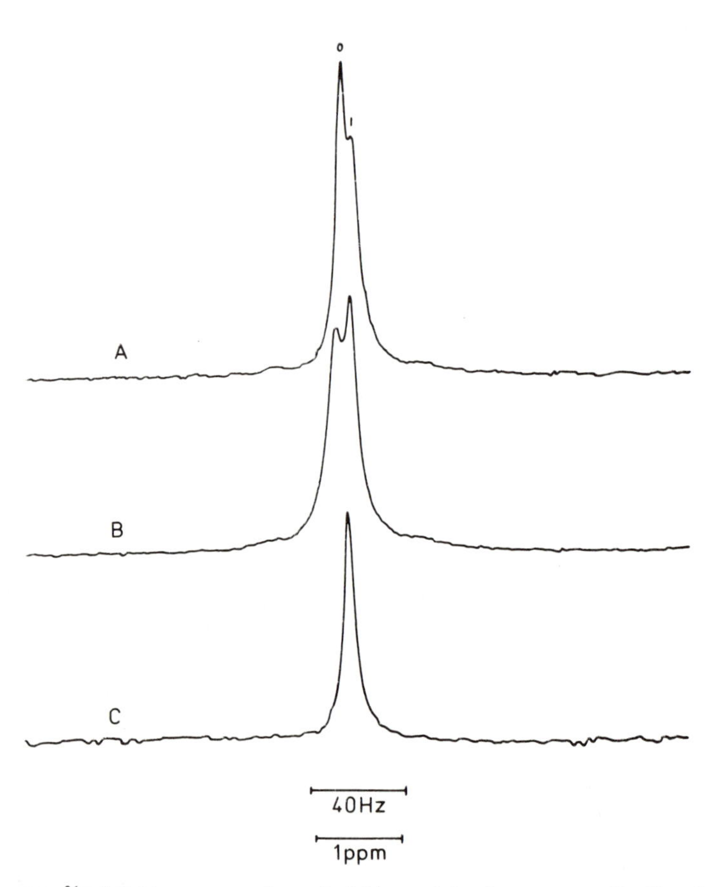

Figure 36. ^{31}P NMR spectra of egg lecithin vesicles in aqueous (D_2O) solution at pH 7.6 at 27°C. In descending order the solutions contain: no additions, 1 mM $CoCl_2$, and 4 mM $CoCl_2$. The inner and outer head groups give separate resonances. The outer resonance is broadened, finally beyond recognition, by the addition of the paramagnetic Co^{++} ion.
[From Berden, J. A., et al., *Biochim, Biophys. Acta* 375:186 (1975).]

ing conclusively that this peak arises from head groups on the external surface of the vesicle.The observed ratio of 1.9 agrees well with the calculated figure. However, the assumption, made above, that the inner and outer monolayers are of equal thickness is suspect. Mason and Huang, making the more credible assumption that the volumes occupied by individual molecules in the two layers are identical, found values of 16 Å and 21 Å for the widths of the inner and outer layers of egg phosphatidylcholine vesicles having an external radius of 99 Å. The data they used were the ratio X of the number of molecules in the two layers as determined by NMR, and the radius R_C of the vesicle and the total number of molecules per vesicle N_T, both determined by viscosity and ultracentrifuge measurements. Defining N_O and N_I to be the number of molecules in the outer and inner layers, and V_O and V_I to be the volumes of individual molecules in the two layers, the following equations can be set up:

$$N_O/N_I = X$$

$$N_O + N_I = N_T$$

$$V_O = \frac{4}{3\pi}\,(R_C^3 - R_B^3)/N_O$$

$$V_I = \frac{4}{3\pi}\,(R_B^3 - R_A^3)/N_I$$

where R_A and R_B are as shown in Figure 37. To proceed, the authors relied partially on the experimental finding that the density of a bilayer increases by only about 3.5% for the gel-to-liquid-crystal phase change. Since it can safely be assumed that the two monolayers, both in liquid crystal form, are unlikely to differ in density more than do two different phases, Mason and Huang put $V_O = V_I$, and using experimental data derived the substantially different thicknesses of the two monolayers quoted above. They estimated the head group areas for molecules in the inner and outer layers at 60 Å^2 and 72 Å^2, respectively, which can be compared with the value of 69 Å^2 for the head groups of DPPC in the L_α phase at room temperature as estimated by Tardieu.[34] The compression of the head groups in the inner layer presumably has sufficient effect on conformations to account for the different ^{31}P chemical shifts of the inner and outer layers and the separate ^{1}H NMR signals (intensity ratio 1.9) arising from the inner and outer choline methyl groups.[35] The differing environments in the two monolayers also affect the chemical shifts of chain atoms, an effect that was detected in a ^{19}F NMR study of vesicles in which the lipids were labeled with gem difluoro groups at selected positions.[36] Spectra showed two peaks of differing intensities.

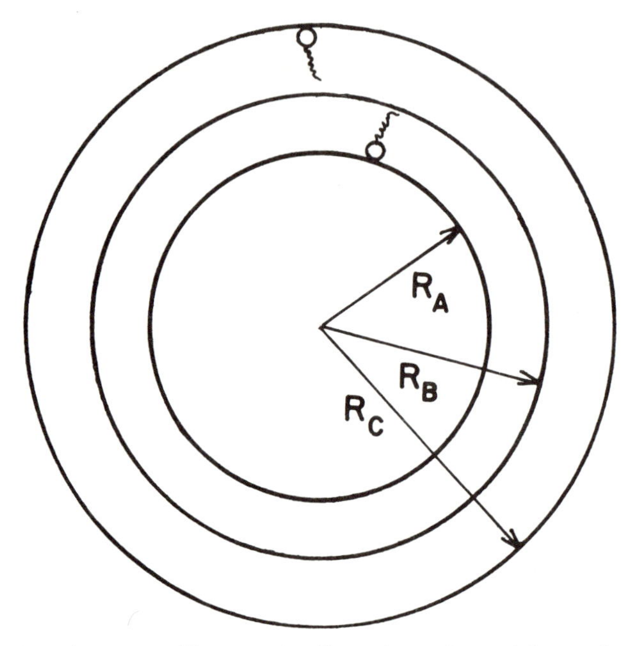

Figure 37. Diagram to illustrate the dimensions of a vesicle as referred to in the text.

Addition of Mn^{++} to the suspensions broadened the up-field resonance while Mn^{++} occluded in the vesicle broadened the down-field line, thus unambiguously assigning the peaks.

Vesicles containing lipids with different head groups often display separate ^{31}P resonances from the two components, allowing a determination of the chemical composition of each monolayer. The results are of theoretic interest and biologic relevance. For example, in vesicles formed from a mixture of DPPE and DPPC in the molar ratio 1.08 : 1, the following ratios were estimated:[35] total outside lipid/total inside lipid, 1.41; inside DPPE/inside DPPC, 1.38; outside DPPE/outside DPPC, 0.92; outside DPPE/inside DPPE, 1.17; outside DPPC/inside DPPC, 1.76. It is clear that the distribution of lipids is assymmetric in that DPPC shows a preference for the outer layer in vesicles containing DPPE. It is tempting to generalize and conclude that lipids with smaller head groups than DPPC will prefer the inner layer for simple packing reasons. Indeed sphingomyelins, which have bigger head groups than DPPC, induce a preference of DPPC for the inner layer of their mixed vesicles. That charge also has a role is shown by the pH-dependant distribution observed in vesicles composed of DPPC and DPPS.[35] At low pHs the normally negative charge of the serine head group at physiologic pH is reduced, as is the electrostatic repulsion between head groups. The effective lowering of the head group area results in an increased preference by DPPS for the inner layer of vesicles as compared with that found at neutral pH. Asymmetric distributions of lipids have been found in a number of cell membranes.[37]

The effects of shape and charge are, of course, not limited to vesicles, as evidenced by the following studies on the effects of temperature, Ca^{++}, cholesterol, pH, and detergents on bilayers.

Phosphatidylethanolamines with unsaturated chains undergo a phase change from the bilayer to the hexagonal H_{II} form as the temperature is increased. The transition, which has a very low enthalpy, usually occurs within $\sim 10^{\circ}C$ of the gel-to-liquid-crystal transition temperature T_t. These lipids have smallish heads and at temperatures around T_t can be regarded as approximately cylindrical with a slight tapering toward the head group. As the temperature of the bilayers is raised, the chains become more and more disordered and occupy an increasing effective volume. The tapering becomes more accentuated until the inverted truncated cones find life easier in the hexagonal form. The greater the number of unsaturated bonds in the chains, the greater is the disorder compared with saturated chains at the same temperature, the more tapered the cones, and the lower the bilayer–hexagonal transition temperature.[38]

Cardiolipin, although having a smallish head compared with its four chains, has two negative charges, which in the pure lipid, due to electrostatic repulsion, result in a large enough effective head group cross section to cause the formation of bilayers and vesicles rather than inverted micelles. Addition of Ca^{++} ions or other divalent cations that complex at the head groups results in a partial neutralization of the negative charge and a consequent reduction in inter-head group repulsion. The head-group cross section drops and cardiolipin becomes a sufficiently tapered inverted truncated cone to form the hexagonal H_{II} phase (Figure 32) as detected by ^{31}P NMR, X-ray diffraction, and freeze-fracture studies.[39] Unsaturated phosphatidic acids behave in the same way in the presence of Ca^{++}.

Measurements of monolayers of cholesterol give a molecular cross-sectional area of about 38 $Å^2$, yet in films or bilayers of DPPC containing up to about one-half mole fraction of cholesterol, the cholesterol has an area at the surface of less than 20 $Å^2$. In the same systems DPPC has a surface area of ~ 71 $Å^2$ per molecule, essentially the same value obtained for the outer surface of small DPPC vesicles.[40] The apparent "condensing effect" of cholesterol on DPPC and similarly shaped lipids can be rationalized in terms of the fitting together of the truncated cone

DPPC with an inverted truncated cone cholesterol (Figure 38). As the mole fraction of cholesterol increases, the apparent surface areas of the two molecules remain unchanged but the average DPPC molecule is pushed into an increasingly tapered form so that the chains become more ordered (see Chapter 5 for confirmation) and the bilayer thicker. At a certain critical mole fraction of cholesterol, if the system is forced to remain planar, as in bilayers on a solid surface or monolayers on water, the cross-sectional area of cholesterol should rise and begin to approach that of pure cholesterol monolayers. This occurs experimentally at a mole-fraction of cholesterol of about 0.75, which can be compared with the theoretical value of 0.77 obtained from calculations based only on geometric considerations.[31b]

Variation of pH can, as we have seen in the case of DPPS, cause changes in the charge of ionizable groups and, through the effect on electrostatic repulsion, changes in effective head group area. In fact, at

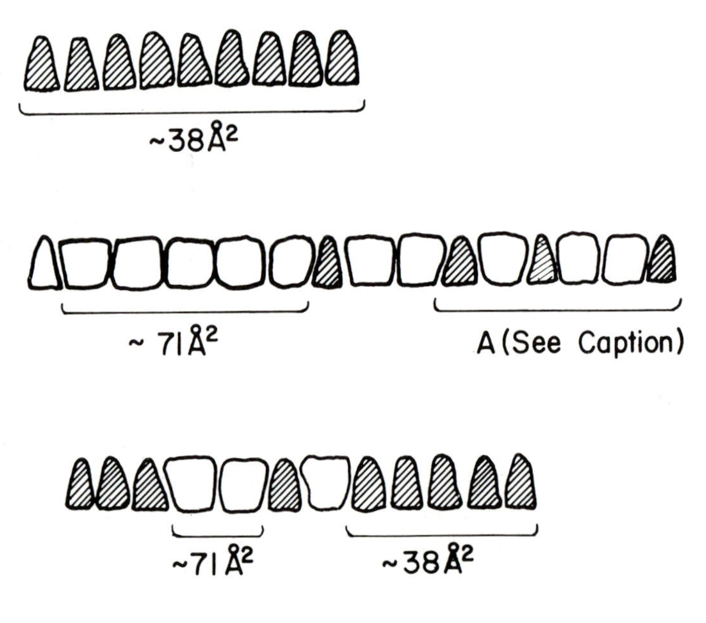

Figure 38. Schematic illustration of the "condensing effect" of cholesterol on lipid bilayers. Cholesterol molecules (shaded) project less surface area when they fit into the crevices of a DPPC layer (A) than when they are part of a cholesterol monolayer or a cholesterol-rich area in a mixed bilayer. The figures indicate the apparent surface area of lipids in different regions of a bilayer containing cholesterol.

pH 2.5 DPPS adopts the hexagonal H_{II} phase preferred by many small-headed lipids, while at pH ~ 5 the planar bilayer phase is stable. Another type of pH-induced change is found in the normally negatively charged lipid phosphatidylglycerol, which occurs in both plant and animal cells.[41] At pH 8 the phosphate group is fully ionized, and at pH 1.5 it is effectively completely protonated. The gel–liquid-crystal transition temperature T_t is pH dependent, but we concentrate on another experimental finding, namely, the pH dependence of chain tilt. At a water/lipid ratio of 0.24 at pH 8, the lamellar repeat distance is 62.4 Å, compared with 67.3 Å at a water/lipid ratio of 0.22 at pH 1.5. The area per lipid molecule in the plane of the bilayer, S, is 48 $Å^2$ at pH 8.0 and 37 $Å^2$ at pH 1.5. The cross-sectional areas of the molecules in the plane perpendicular to the chain axes, that is, $n\Sigma$, are almost identical, 40.5 $Å^2$ and 39.2 $Å^2$ at pH 8.0 and pH 1.5 respectively. Using $S \cos \phi = n\Sigma$, we find that at pH 1.5 the tilt angle ϕ is apparently zero while at pH 8 the chains are tilted by about 32°. These experiments were carried out at 20°C in the gel phase, the chains being essentially all-trans. For untilted chains an increase in the electrostatic repulsion between head groups cannot be compensated by a simple increase in the chain–chain separation since the forces between chains are too sharply dependent on distance. However, chain tilting can give an expanded head-group lattice while leaving interchain interactions relatively unaffected. The small difference in chain–chain overlap in the tilted chains nevertheless should reduce the force between chains slightly, and this well may be a major contribution to the decrease of 15 degrees in the chain melting temperature in going from pH 1.5 to pH 8.

Detergents do not form bilayers or vesicles and are often used to disrupt biologic membranes by solubilizing the lipids in micelles. Nevertheless, in certain cases they *stabilize* bilayers, a phenomenon that fits in well with our steric approach. For example DPPE is an inverted cone lipid that, although forming bilayers when fully hydrated, undergoes a change to the hexagonal H_{II} phase at about 30°C. DPPE and detergents have complementary shapes in the sense that truncated cones and inverted truncated cones can be mixed to give planar structures. Thus it was predicted that detergents should stabilize the bilayer structure of DPPE with respect to the hexagonal phase, and indeed the temperature T_{BH} of the bilayer–hexagonal transition is substantially raised by the addition of detergent.[42] For example, T_{BH} for fully hydrated egg phosphatidylethanolamine is 28°C, but rises to 40°C when sodium deoxycholate is present in a ratio of 0.05 for detergent to phospholipid. It should be noted that the same stabilizing effect is produced by naturally occuring single-chain phospholipids such as egg lysophosphatidylcholine.

It has not been our objective in this chapter to review the many properties of vesicles and bilayers, but rather to show that crude considerations of molecular shape provide a satisfying and unifying explanation of a wide range of disparate phenomena occurring in these two common structures. Nevertheless, our steric criteria are too simple to handle the details of molecular conformation and dynamics made available by spectroscopy, which has, in particular, laid bare the nature of what we have been calling "melted" or "crumpled" chains. In the following chapter we take a look at the private lives of melted chains and hydrated head groups.

5 Order and Disorder

Molecules never rest. The formally rigid all-trans chains in crystalline phospholipids have a measure of vibrational freedom that increases toward the methyl group tail of the chain (Figure 9). In liquid crystal phases the chains are disordered, any given segment rapidly changing its orientation. We expect the motion of a given segment to be more limited near the tightly packed head groups than at the tail ends of the chains but since the orientation of a given segment is time-dependent and unpredictable, we can only make statistical statements about motion. In this chapter we show that a variety of spectroscopic techniques allow us to make such statements since they provide quantitative measures of *order*, a term we shortly shall define. The theoretic framework that we use was invented to deal with the physical properties of liquid crystals.

Certain molecules can exist in phases that are liquid in the sense that the material flows but crystalline in the sense that the sample has anisotropic physical properties.[43,44] Conventional liquids are isotropic. Of the many liquid crystal phases, we focus on the *smectic* phase. In the smectic phase the centers of the molecules are arranged in parallel planes, but not necessarily in a definite pattern within the plane. The orientations of the molecules are, at any given moment, similar, but not often identical (Figure 39). Since science is obsessed with numbers, we now attempt to give a quantitative flavor to the qualitative statement that "a degree of orientational order exists" in smectic phases, and indeed in all liquid crystals. First we must realise that Figure 39 represents one moment in the life of a smectic crystal. The molecules have two-dimensional translational freedom. This freedom exists for bilayers in the liquid crystal state even though they display a time-averaged hexagonal ordering (Chapter 3). Molecules in the smectic phase also have a degree of *rotational* freedom. For rod-shaped molecules rotations about the long axis are practically unhindered and have frequencies falling in the range of 10^7 to 10^{10} Hz so that the molecule has effective axial symmetry, again on the time scale of most experiments. Complete rotations perpendicular to the long axis are strongly discouraged since the molecules get in each other's way. Nevertheless, the long axes of rigid rodlike molecules in liquid crystals fluctuate over a range of angles determined largely by the available space. These fluctuations affect macroscopic and microscopic physical properties of the liquid crystals, the measurement of which, mainly by spectroscopy, has allowed quantitative estimates to be made of the degree of angular disorder of the molecules. The results are usually expressed in terms of *order parameters*, which we now define.

We first fix a cartesian coordinate system in the molecule, convention requiring that the z-axis be aligned with the long axis of the molecule. We now define a unique axis in our smectic sample chosen so that the movement of molecules has cylindrical symmetry about this axis, the so-called *director*. In a vibrating box of uncooked spaghetti the z-axis would be along the individual spaghetti "molecule" and the director would be a line parallel to the long edges of the box. There is no need to specify

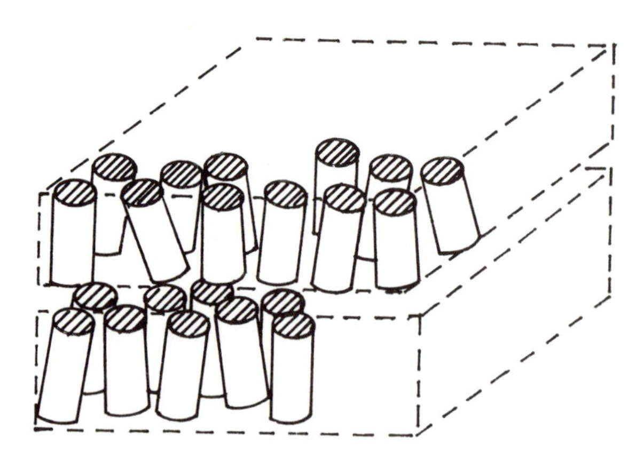

Figure 39. Orientation of molecules in smectic liquid crystals. The molecules are arranged in sheets but their positions are not ordered two-dimensionally within the sheet. In our example—a smectic A crystal—the directions of the molecules fluctuate about the normal to the sheet.

where in the box the director lies since only its orientation has significance. We will use the symbol z' to label the unit vector along the director. In flexible molecules such as lipids it is usual to focus on the orientation of individual bonds or chain segments, but for the moment we continue to deal with rigid molecules and we define a set of order parameters[45] by the equations:

$$S_{ii} = \tfrac{1}{2}(3\,\overline{\cos^2 \theta_i} - 1) \qquad i = 1, 2, 3 \quad x, y, z \tag{5.1}$$

where θ_i is the instantaneous angle between the molecular Cartesian coordinate i and the director z'. This angle varies with time due to fluctuations and $\overline{\cos^2 \theta_i}$ is a time average for one molecule. [The theoretically inclined may note that S_{33} is the Legendre polynomial $P_0^2(\cos\theta)$ and is a component of a second-rank irreducible tensor.[46]] It is the determination of the values of S_{ii} and their interpretation that will occupy much of this chapter. We now examine some general properties of S_{ii}.

1. If the molecule is permanently aligned along the director, then θ_3 is zero and $S_{33} = 1$, its maximum possible value. Also $S_{11} = S_{22} = -\tfrac{1}{2}$.
2. Because, by the orthogonality relation between cosines, $\Sigma \cos^2\theta_i = 1$, it follows that $\Sigma\, S_{ii} = 0$.
3. In performing the average in (5.1) θ is allowed to vary from 0 to Π. This domain is sufficient to cover all possible orientations of the z-axis because we have assumed that the movement of the molecule is cylindrically symmetric about the director.
4. It is commonly assumed that $S_{11} = S_{22}$ that is, $S_{xx} = S_{yy}$. This implies either that the molecule has cylindrical symmetry about its z-axis or that it rotates about that axis in a time short compared with the time scale of the experiment. Both experimental and theoretical work suggest that for lipid chains in bilayers the x- and y-axes are not equivalent as far as motion is concerned (Chapter 8).
5. For a molecule undergoing isotropic tumbling, $S_{33} = 0$. This result can be obtained by noting that θ_3 ranges over all angles between 0 and π but the probability of θ_3 falling in the interval θ_3 to $\theta_3 + d\theta_3$ is given by $\sin\theta_3 d\theta_3$ (see Figure 40). Then

$$\begin{aligned} S_{33} &= \tfrac{1}{2}(3\,\overline{\cos^2 \theta_3} - 1) \\ &= \frac{1}{2}\left(3\int_0^\pi \cos^2\theta \sin\theta d\theta - 1\right) \\ &= \frac{1}{2}\,\frac{[-\cos^3\theta]_0^\pi}{[-\cos\theta]_0^\pi} - 1 = 0 \end{aligned}$$

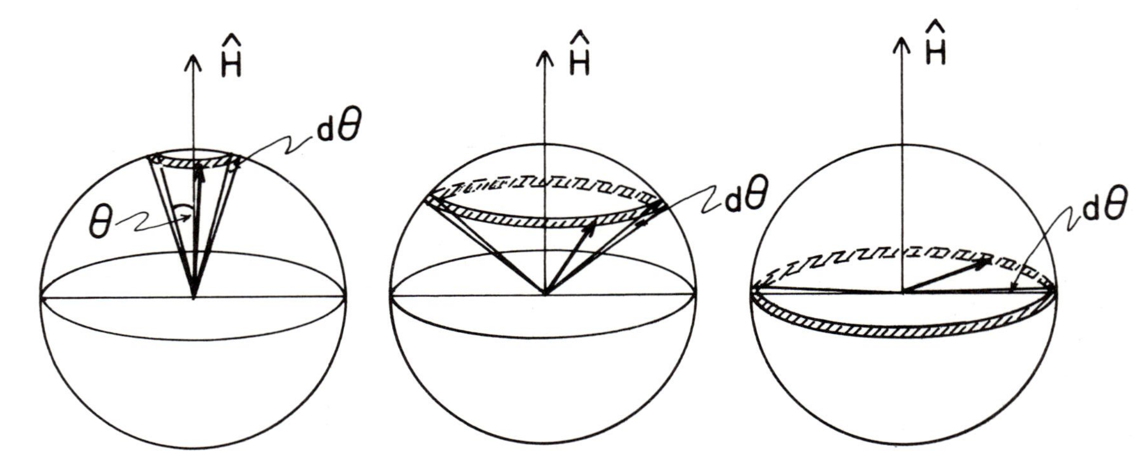

Figure 40. To illustrate that for isotropically distributed molecules the probability of finding a given axis in a molecule at an angle between θ and $d\theta$ to a vector, for example a magnetic field is proportional to the corresponding band on the surface of a sphere, that is, to $\sin\theta\, d\theta$. The probability of finding the axis near the plane normal to the field is far greater than finding it near the field vector.

Since $\Sigma S_{ii} = 0$ and we are assuming that $S_{11} = S_{22}$, it follows that S_{11} and S_{22} are also zero.

6. The definition of order parameters is operational. Someone has to do the experiment. Two techniques may give different answers if they have different time scales. The time scale of NMR is such that if we are observing one bond in a hydrocarbon chain of a bilayer in the liquid crystal state, then the average over θ_3 will encompass a huge number of conformational changes in the chain. On the time scale of vibrational spectroscopy, on the other hand, the chain may appear to be effectively static and the range of θ_3 correspondingly limited. Problems of this kind are rare but not unknown, as a reading of Chapter 10 will confirm.

A word of caution is needed concerning the director. A macroscopic sample of a liquid crystal may consist of a number of microdomains, each having its own director. The collection of directors is not necessarily colinear and this sometimes complicates the interpretation of experimental data.

We now consider the practical problem of measuring order parameters. In the context of this book it is the hydrocarbon chains of phospholipids that will be of overriding interest. In the crystalline phase these chains, as we saw in Chapter 1, are rigidly ordered, apart from thermal vibrations. In liquid crystal phases the chains have limited freedom of motion, and it has been possible to concentrate selectively on individual segments in the chains and determine their order parameters. The most popular and powerful experimental methods have been magnetic resonance techniques, in particular, deuterium nuclear magnetic resonance (^{2}H NMR), and electron spin resonance (ESR). We now sketch the principles involved in interpreting the spectra of membranes and follow up with a review of the main results. Subsequently we will have a look at the way in which spectroscopy has provided insights into the orientation of head groups.

The measurement of order depends on the anisotropy of the spectra of molecular fragments. We concentrate on the use of ^{2}H NMR and consider the spectrum of of deuterium in a C—D bond. Deuterium has a nuclear spin quantum number $I = 1$, which implies that the nucleus possesses a quadrupole moment. Without ploughing through the theory,[47] we go straight to the practical consequences, which are that the ^{2}H NMR spectrum of deuterium in a C—D bond (or any other bond) depends on the relative orientation of the bond to the magnetic field $\hat{H}$ of the NMR spectrometer (Figure 41). The spectrum consists of a doublet, with the distance between the lines, the so-called quadrupole splitting $\Delta\nu_Q$, being at a maximum when the bond is colinear with $\hat{H}$. At other orientations the splitting, to a very good approximation, depends

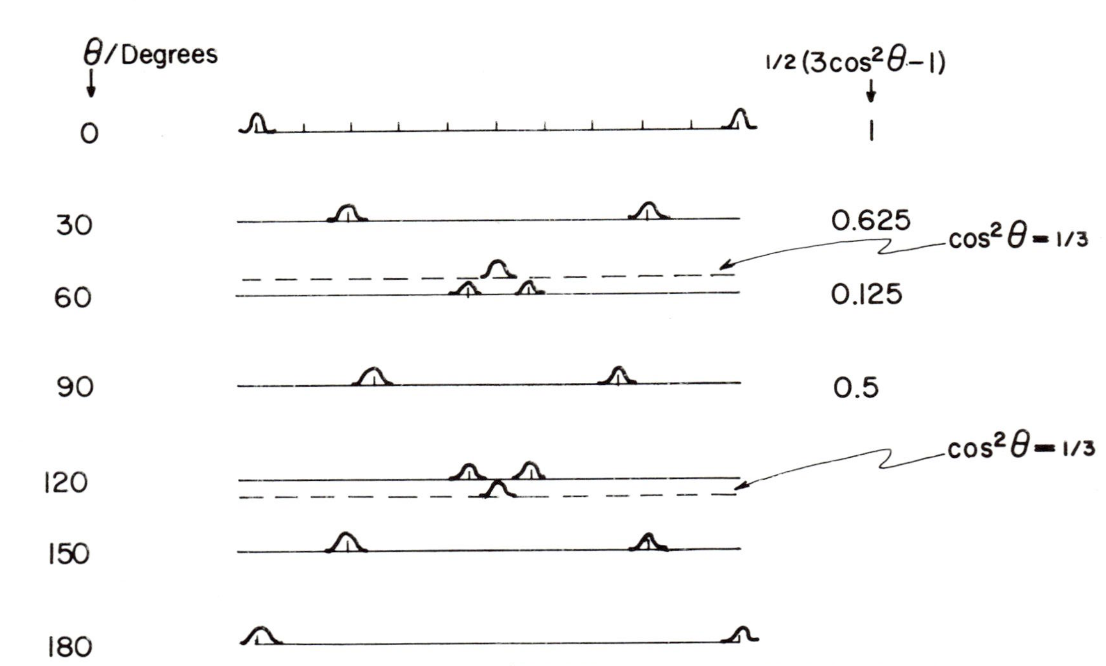

Figure 41. Dependence of the ^{2}H NMR spectrum of a C—D bond on the angle between the bond and the external magnetic field. For $\cos^2\theta = 1/3$, the "magic angle," the quadrupole-induced doublet collapses.

only on the angle θ between the C—D bond and $\hat{H}$. (For connoisseurs, the quadrupole coupling tensor is nearly axially symmetric.) The dependence of the splitting on θ is given by

$$\Delta\nu_Q(\theta) = \frac{3}{2}\left(\frac{e^2qQ}{h}\right)\left(\frac{3\cos^2\theta - 1}{2}\right) \tag{5.2}$$

where e^2qQ/h is the quadrupole coupling constant and is a combined property of the deuterium nucleus and the gradient of the electric field that it experiences due to its electronic environment.

For the C—D bond e^2qQ/h is not very sensitive to the nature of the molecule, a low value being that for acetonitrile, 167 kHz, and a high value that for benzene, 193 kHz. For C—D bonds in paraffins 170 ± 2 kHz is the accepted figure. From (5.2) it is seen that when $\theta = 0$, the C—D bond parallel to $\hat{H}$, $\Delta\nu_Q = (3/2)\ e^2qQ/h = 255$ kHz. For $\theta = \cos^{-1}\ 1/3 = 54.74$ degrees, $\Delta\nu_Q = 0$, and for $\theta = 90$ degrees, $\Delta\nu_Q = (-3/4)\ e^2qQ/h = -127.5$ kHz. (The negative sign has no experimental significance in everyday experiments.) The angular dependence means that the spectrum can reveal information on the orientation of a C—D bond, as we see from Figure 41. Furthermore, the splitting, through (5.2), gives us a value for $1/2\ (3\cos^2\theta - 1)$ that is the function appearing in the order parameter S_{33} (5.1). Life is not that easy; for liquid crystals and biologic material, we can hardly expect the C—D bonds of all the molecules in the sample to be so regimented that all simultaneously make the same angle $\hat{H}$. By focusing on one specific C—D bond in one specific lipid chain in a bilayer we can expect to observe, in the liquid crystal phase, a perpetual shifting of position and angle, which, although apparently random, has been shown experimentally in hydrated lipid bilayers to display symmetry about the normal to the bilayer. In our previous terminology, the normal is the director z'. The ^{2}H NMR spectrum shows a certain splitting when $\hat{H} \parallel z'$ and at other angles $\Delta\nu_Q$ depends only on β, the angle between $\hat{H}$ and z'. At any orientation of the bilayer to $\hat{H}$ the spectrum is time-averaged over the motion of the C—D bond, since the molecular motion is very fast on the time scale of the NMR experiment. If $\hat{H}$ is parallel to z', we can write, for an axially symmetric quadrupole tensor,

$$\Delta\nu_Q(\theta) = \frac{3}{2}\left(\frac{e^2qQ}{h}\right)\left(\frac{\overline{3\cos^2\theta - 1}}{2}\right) \tag{5.3}$$

where θ is the angle between the C—D bond and $\hat{H}$ (i.e., z' in this case) and the bar indicates a time average. (If the C—D bond is comparatively static and effectively aligned along the director, then $\Delta\nu_Q = 3/2\ e^2\ qQ/h$

as we found before.) Equation (5.3) can be rewritten as

$$\Delta\nu_Q(\theta) = \frac{3}{2}\left(\frac{e^2qQ}{h}\right)S_{33} \qquad (\hat{H} \parallel z') \tag{5.4}$$

where S_{33} refers to the ordering of the direction of maximum splitting, that is, the direction of the C—D bond. The order parameter thus can be obtained directly from a single measurement. The splitting at other angles β between $\hat{H}$ and z' is given by

$$\Delta\nu_Q(\theta\,;\,\beta) = \frac{3}{2}\left(\frac{e^2qQ}{h}\right)S_{33}\left(\frac{3\cos^2\beta - 1}{2}\right) \tag{5.5}$$

An example is shown in Figure 42. In this study the splitting for $\beta = 0$ was found to be 56.6 ± 0.6 kHz. Using (5.4) and the value of 170 kHz for e^2qQ/h, we obtain $|S_{33}| = 0.22$. The structural interpretation of this result is based on the acceptable assumption that the long axis of the hydrocarbon chain is, on the average, close to the normal z', and therefore the C—D bond is compelled to lie roughly perpendicular to z'. For C—D at 90° to z', $S_{33} = -0.5$. It therefore can be supposed that the observed value of S_{33} is negative, that is $S_{33} = -0.22$, which, since it falls below the rigidly perpendicular result, can be taken to indicate a degree of freedom for the C—D bond with respect to movement out of the plane perpendicular to z'. S_{33} can here be renamed S_{CD}, the order parameter for the C—D bond, because the maximum value of the quadrupole coupling coincides with the direction of the C—D bond. (The C—D bond vector is a principal axis of the coupling tensor.) Note that S_{CD} tells us nothing about the azimuthal angle of the bond since motion around the director is fast on the time scale of NMR.

Much experimental effort has gone into selectively labeling hydrocarbon chains with deuterium at various positions and recording the ^{2}H NMR spectra. From such studies it has been possible to construct order-parameter profiles for lipid chains in bilayers and membranes. Often the results are given not in terms of S_{CD} but of S_{mol}, the order parameter of the normal to the plane containing the two C—H bonds of a methylene group (Figure 43). The parameters are related by

$$S_{mol} = -2S_{CD} \tag{5.6}$$

so that for an all-trans chain normal to the bilayer $S_{CD} = -0.5$ and $S_{mol} = 1$ (neglecting vibrations). The reader should check that for a C—D bond in the terminal methyl group of such a chain $S_{CD} = -1/3$. Figure 44 shows order parameters for some bilayers and membranes. The general picture is of a plateau reaching to carbon atom 8 after which S_{mol} falls steadily toward the end of the chain. Order parameters decrease with rising temperature. Parameters for different lipids fall close to each other

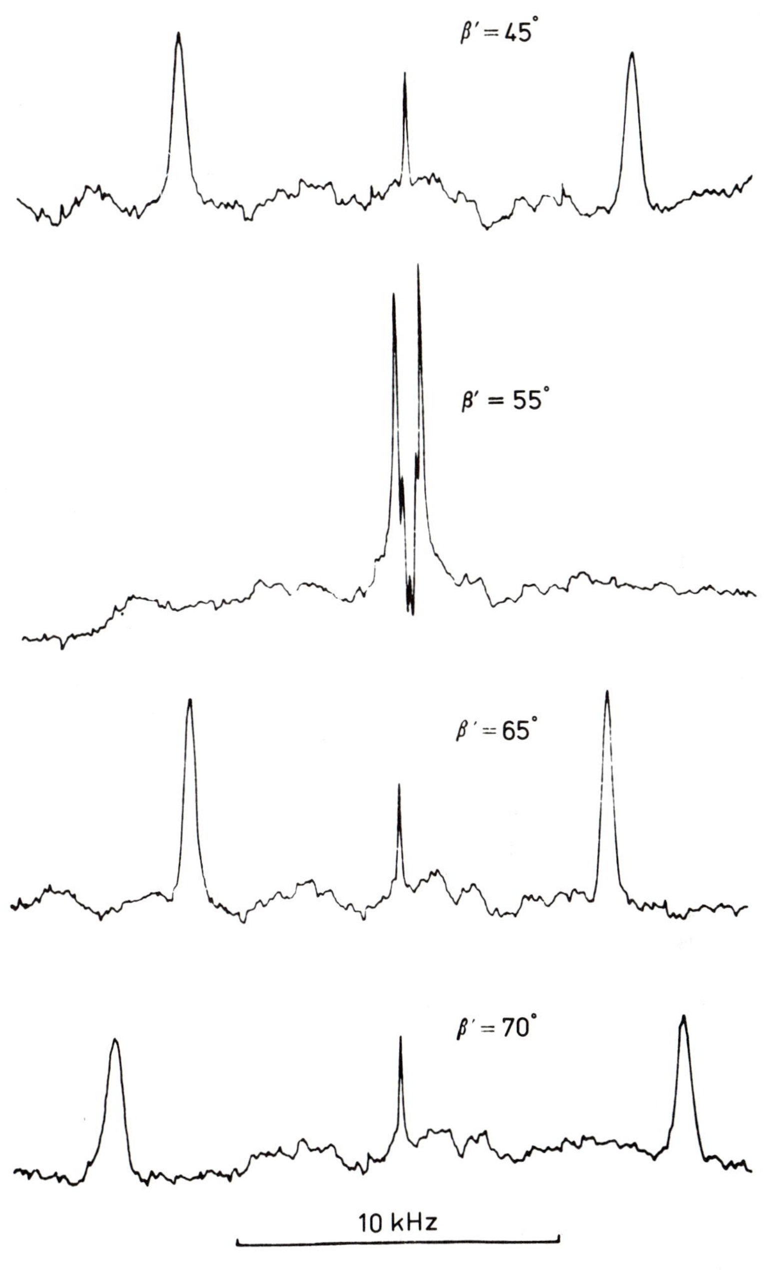

Figure 42. The ^2H NMR spectrum of bilayers deuterated in one methylene group and oriented on stacked parallel glass plates. The quadrupole splitting, which is dependent on the angle between the magnetic field and the bilayer normal, approaches zero near the magic angle of 54°44′. [From Seelig, J., and Neiderberger, W., *J. Am. Chem. Soc.*, 96:2069 (1974).]

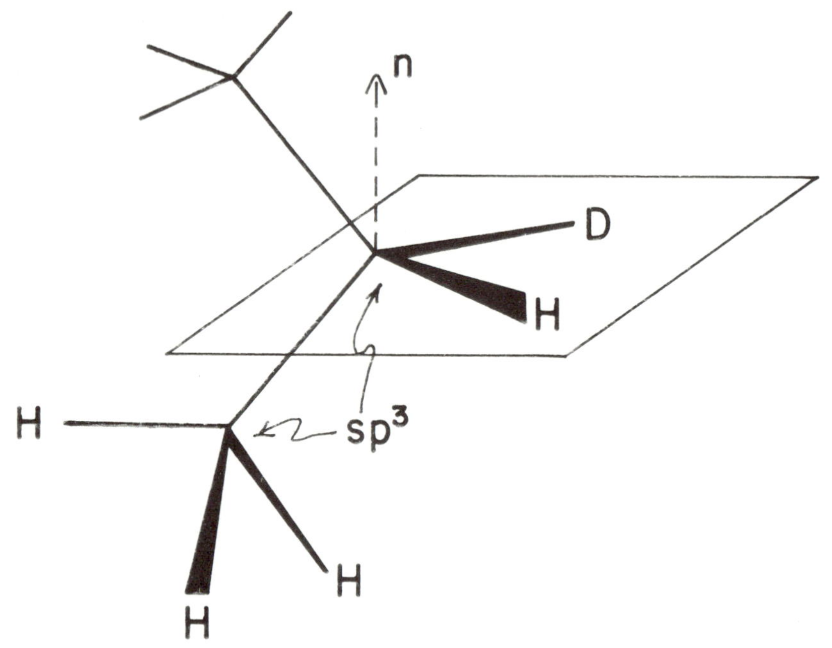

Figure 43. Illustrating the orientation of a methylene group with respect to the chain axis in an all-trans chain. The vector n will coincide with the bilayer normal if the chain is perpendicular to the bilayer surface. In the liquid crystal state, both the C—D bond and n will have time-dependent orientations with respect to the director. The average orientation of a C—D bond determines S_{C-D} and the average orientation of n determines S_{mol}.

when plotted as a function of a reduced temperature $\theta = (T - T_t)/T$, where T_t is the gel-to-liquid-crystal transition temperature. The use of a reduced temperature, an idea borrowed from the theory of liquids, is an attempt, apparently successful in this case, to define a similar physical state for lipids with widely different transition temperatures (Figure 45).

When deuterium-labelled lipids are incorporated into biological membranes, the observed order parameters are similar to those found in lipid bilayers, which is encouraging for those who believe in bilayers as good models for the lipid areas of biological membranes (Figure 46). [In Figure 46 the order parameters for carbon atoms 8 to 12 in the oleic acyl chain are explicable in terms of the orientation of the cis double bond of the unsaturated chain (Figure 47).] We return to natural membranes in Chapter 9 in the general context of protein–lipid interactions.

Another experimental approach to the ordering of hydrocarbon

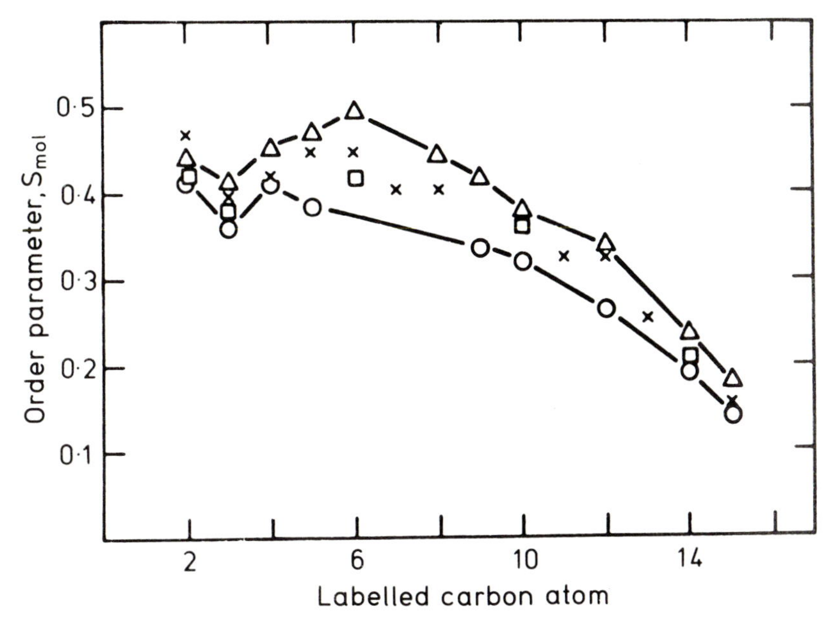

Figure 44. Order parameter profiles for different bilayers. The results are for a common reduced temperature $\theta = (T - T_t)/T_t = 0.061$.
○, 1,2-dipalmitoyl-glycero-3-phosphocholine (DPPC),
△, 1-palmitoyl-2-oleoyl-glycero-3-phosphocholine (POPC),
□, 1,2-dipalmitoyl-glycero-3-phosphoserine (DPPS),
×, Acholeplasma laidlawii.
[From Seelig, J., and Browning, J. L., *FEBS Lett.* 92:41 (1978).]

chains is via the ESR spectra of paramagnetic probe molecules in bilayers and membranes. The overwhelming majority of studies have made use of probes containing the N—O moiety, which, when neighbored by bulky, organic groups, is comparatively chemically stable (Figure 48). The ESR spectra of nitroxyl radicals consist of three lines with a hyperfine splitting dependent on the relative orientation of the radical and the magnetic field of the spectrometer. The center of gravity of the spectrum is also orientation dependent. Without burdening our readers with formal developments, we ask them to believe that the spectral anisotropy of the nitroxyl group can be used in much the same way as that of deuterium to provide details of ordering. (The similarity of many of the equations to those for ^2H NMR is a consequence of the fact that the quadrupole tensor of ^2H and the g− and hyperfine tensors of free radicals are all second-rank tensors.[46])

The hydrocarbon chains of lipids have been *spin-labeled* in different

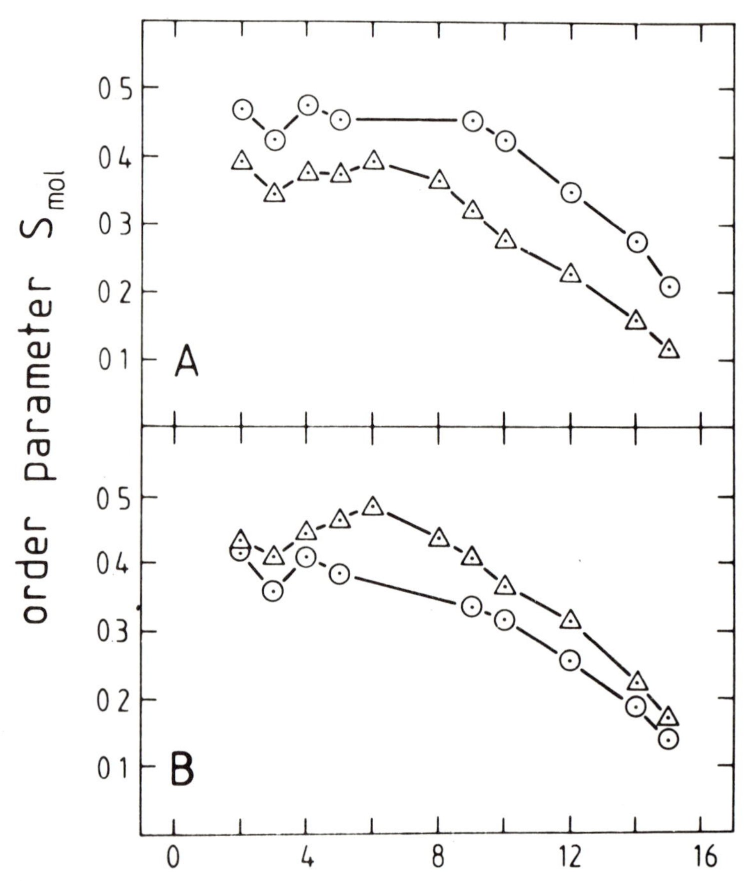

Figure 45. Order parameter S_{mol} for: △, 1-palmitoyl-2-oleoylphosphatidylcholine (POPC) and ○, 1,2-dipalmitoylphosphatidylcholine. **A.** At the same temperature. **B.** At equal temperatures relative to their phase transitions, that is, POPC at 14°C (T_t + 19°C), and DPPC at 60°C (T_t + 19°C). [From Seelig, A., and Seelig, J., *Biochem.* 16:45 (1977).]

positions by the covalent attachment of nitroxyl spin labels, and the labeled lipids have been incorporated into bilayers and membranes. Additionally, spin-labeled fatty acids and cholesterol derivatives have frequently been used as probes since they dissolve readily in the hydrocarbon core of bilayers. Much information obtained by the spin-label method has been inaccessible by other techniques, but as far as order parameters are concerned, it must be admitted that although the results roughly confirm those derived from ^{2}H NMR, there are differences and doubts. There is every reason to believe that spin-labeled molecules, because of the bulky label, behave differently from their

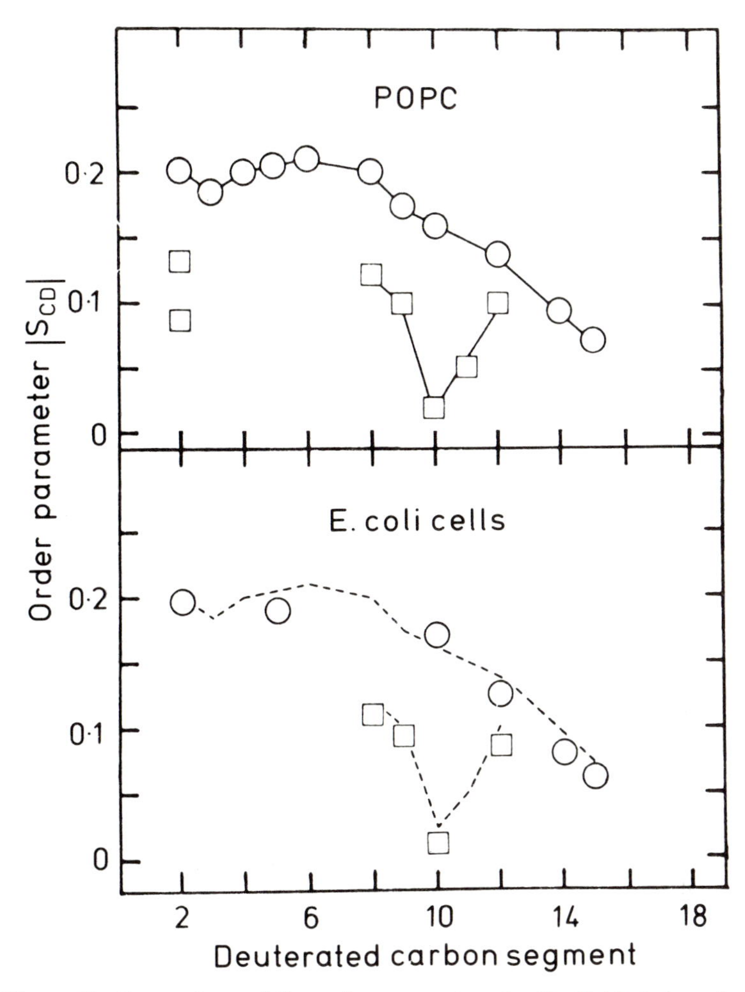

Figure 46. Comparison of the order parameters for the lipid chains of a biologic membrane and a synthetic unsaturated lipid, POPC. The samples contain 50% by weight of lipid. The membrane is that of *E. coli* cells grown on selectively deuterated media. ○, deuterium label in the palmitic chain, □, label in the oleic chain. Both natural and synthetic bilayers show a discontinuity in the order-parameter curves at the position of the double bond. [From Seelig, J. and Seelig, A., *Q. Rev. Biophys.* 13:1(1980).]

unlabeled counterparts, and also sterically perturb their environment. As someone put it,[48] the probe may be making the news, not reporting it. This should be borne in mind in contemplating Figure 49. Nevertheless, spin-label studies of order do reveal trends, even if the results have limited quantitative significance. By using oriented bilayers, usually

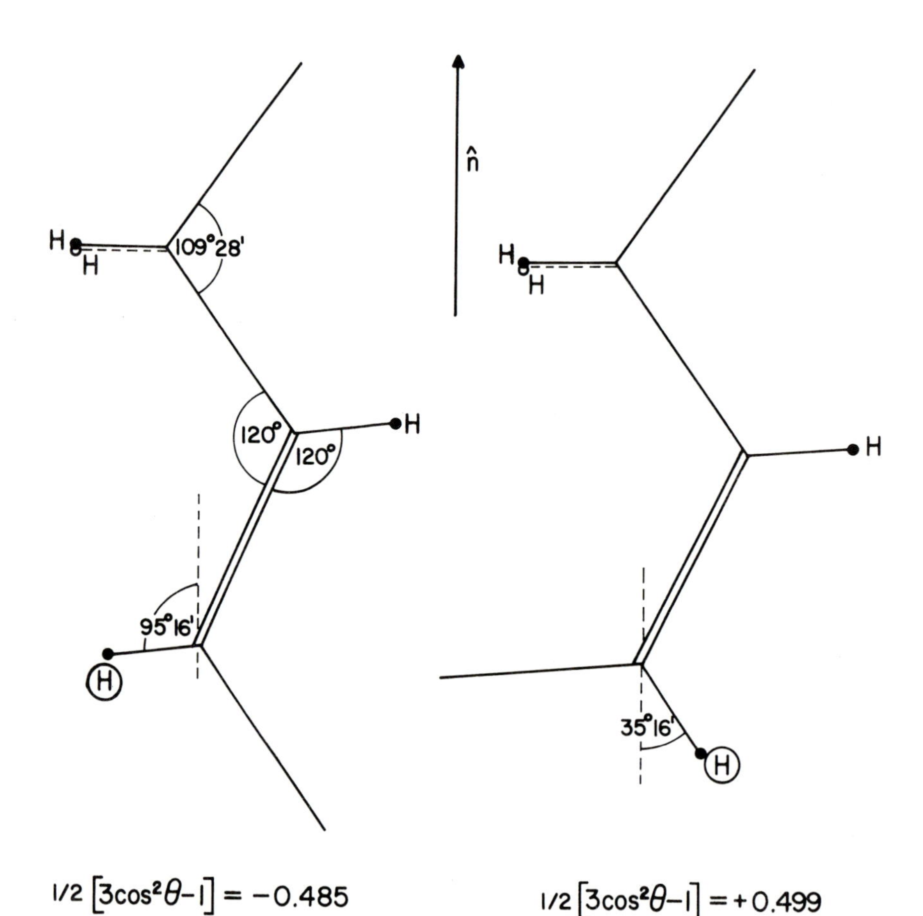

Fig 47. Two allowed orientations for a C—H bond attached to a double-bonded carbon in an otherwise saturated and all-trans chain. If the marked hydrogen atom spends equal time in the two states and movement is rapid, then the observed order parameter will be close to zero. Compare Figure 46.

multilayers spread on glass, it has been possible to determine the effects on the *probe* of a great variety of variables, including temperature, pH, and ionic strength.[48] The data are far too abundant to allow even partial summary here, but we note in passing one result that is relevant to our earlier mention of cholesterol. The addition of cholesterol to egg lecithin bilayers in the liquid crystal state and containing 3-doxylandrostane spin probe resulted in an increased ordering of the probe, which can be taken to indicate a straightening of the lipid chains and a reduction in their freedom of motion. On the other hand, the same experiment performed on DPPC in the gel state results in a decrease in the order parameters.[49] These results confirm those obtained by other methods showing that cholesterol increases the "fluidity" of the gel state and decreases the "fluidity" of the liquid crystal state. We will soon have more to say about fluidity.

We have been careful to refer to the order-parameter profile of the

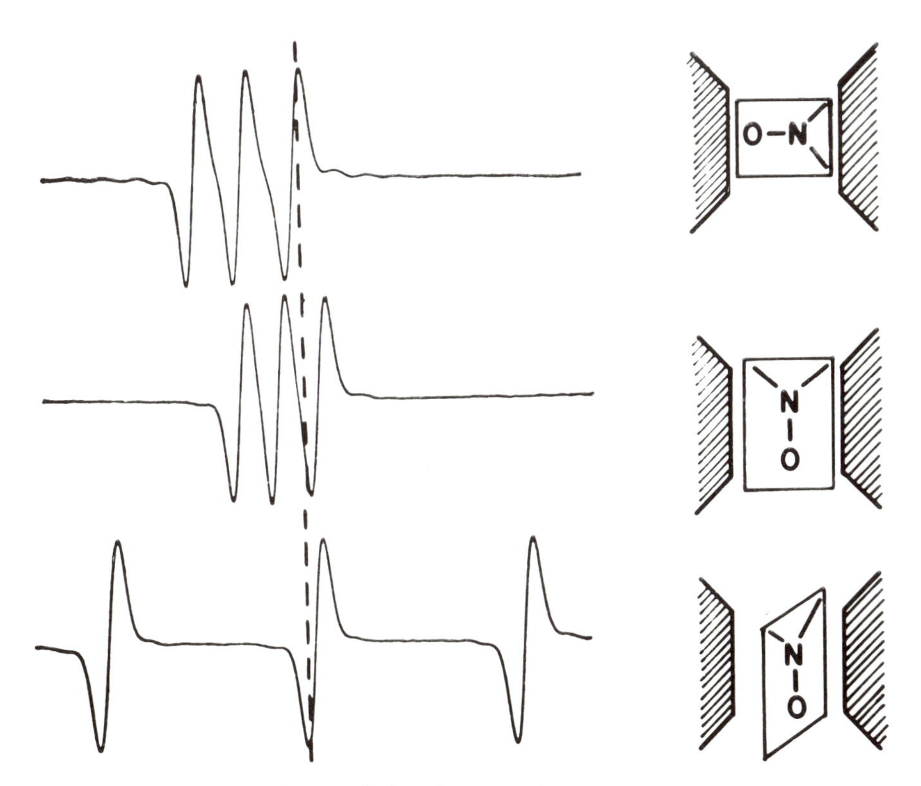

Figure 48. Dependence of the electron spin resonance spectrum of a nitroxide spin label on the orientation of the N—O bond to the external magnetic field.

chains, not of the membrane. There is no reason to suppose that in the liquid crystal state the distance of the carbon atoms from the head groups at a given moment depends linearly on the ordinal number of the carbon atom in the chain. The increasing disorder toward the end of the chains is certainly the reflection of an increasing range of motion that includes extensive excursions of the chain ends into regions removed from the center of the bilayer. Thus at any given point in the bilayer a variety of differently numbered carbon atoms can appear and we can define an average order parameter at that point, weighted over the different bonds that contribute. A theoretic estimate of this membrane order profile is shown in Figure 50. Order decreases toward the center of the membrane and it is commonly said that the center is more fluid.

The concept of fluidity and the definition of order parameters both presuppose the existence of chain *motion* but do little or nothing to reveal the nature of that motion. Fluidity is not even precisely defined. In going from lower to higher temperatures, the hydrocarbon chains go from the highly ordered gel state to the loosely organized liquid crystal state, and this transformation is sometimes said to involve an increase in fluidity. Since order also decreases during this process, there is a temptation to associate fluidity with order, an increase in one implying a decrease in the other. Very broadly speaking, this is permissible but there

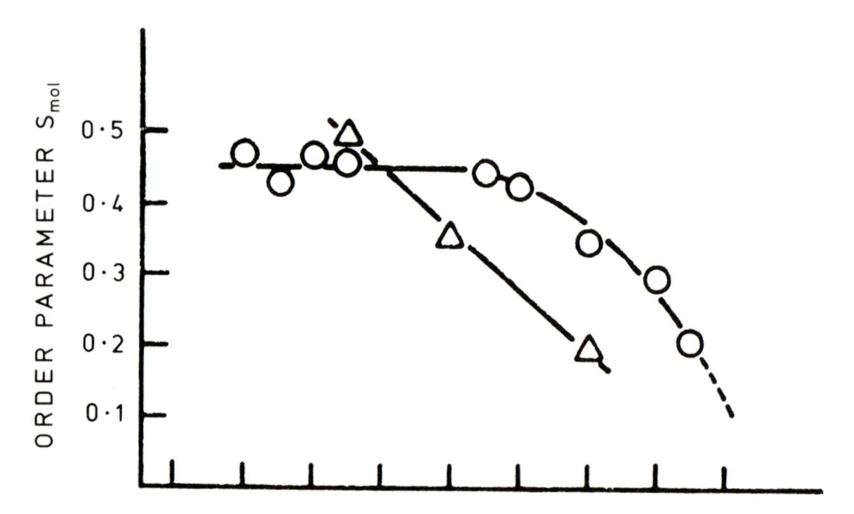

Figure 49. Illustrating the difference between values obtained by ESR and NMR for the order parameters S_{mol} for DPPC at 41°C. [From Seelig, J., *Q. Rev. Biophys.* 10:353 (1977).]

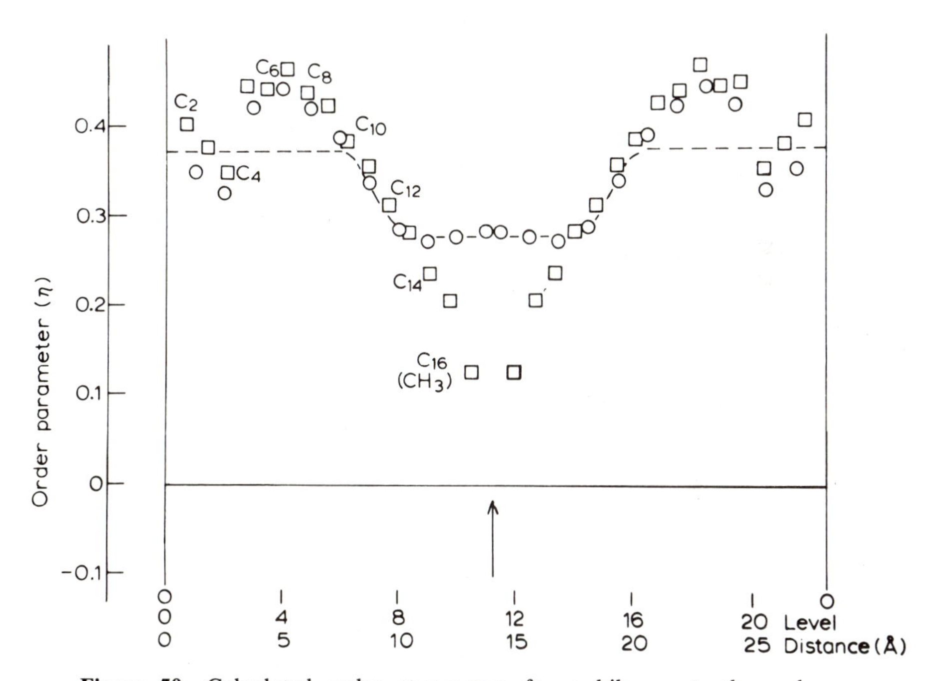

Figure 50. Calculated order parameters for a bilayer. ○, the order parameter profile across the bilayer; □, the order parameter along the *chain* drawn so that the carbon atoms are equally spaced along the bilayer normal. (The dotted line represents the assumed order-parameter profile across the bilayer initially fed into the computer.)

are dangers, as can be appreciated by pondering over a bowl of sleepy earthworms — order is nonexistent but a high degree of fluidity is not apparent either. In any case, since fluidity is not quantitatively defined, it cannot be quantitatively correlated with anything. As for order parameters, they give a feeling for the range of angle over which a given segment moves, but no more than this. The average in (5.1), if it is actually to be carried out, requires a knowledge of the distribution function for the angle θ_3; that is, we have to know $f(\theta_3)$ where $f(\theta_3)d\theta_3$ gives the probability of θ_3 falling in the range θ_3 to $\theta_3 + d\theta_3$. This information is not available and it should be realized that an infinite number of different forms of $f(\theta_3)$ can give the same value of S_{33}. Often $f(\theta_3)$ is taken as a Gaussian distribution but there is no theoretical justification for this choice. In dealing with the order parameters of spin-label or fluorescent probes, use occasionally has been made of a model in which the chosen axis of the probe wobbles uniformly within an imaginary cone so as to give a square-well distribution. Attempts to differentiate between this and the Gaussian model have not been convincing.[50] Of course, even a knowledge of $f(\theta_3)$ tells us nothing about the rate of motion, except that it is fast enough to average the position of the probe on the time scale of the experiment. Two main experimental approaches have been used in trying to learn more than order parameters or "fluidity" can tell us about the dynamic properties of chains. One strategy is to examine the motion of small probes situated in the hydrophobic core. Translational or rotational diffusion constants are used to define a *microviscosity* by using relationships of the Stokes type:

$$\eta = kT/8\pi r^3 D \tag{5.7}$$

where η is the viscosity, r is an effective radius for the probe molecule, and D is the observed diffusion coefficient. The trouble is that different probes tend to give different answers because the shape and chemical nature of the probe affect its rotational reorientation in the membrane core. The concept of microviscosity, although an advance on the hand-waving idea of fluidity, is itself unsatisfactory (compare Chaper 10). A rather different approach is to attempt to determine what might be termed a "motion profile" for the chains by measuring nuclear spin relaxation times (see Chapter 17) and then transforming them to orientational correlation times using models of molecular motion. An early example[51] is the measurement of ^{13}C spin-lattice relaxation times, T_1, for different carbon atoms in the lipid chains of DMPC at 52°C. The results were interpreted in terms of rotational diffusion constants that were converted to viscosities using (5.7). The analysis gives an approximately 80-fold increase in microviscosity in going from the carbonyl to the methyl carbon. This is in the expected direction but the use of (5.7), which applies strictly to spheres in a continuum, seems highly dubious. Indeed in discussing the motion of the chains there seems little justification for using viscosity, except in the case of translational motion (Chapter 10). Even the initial step, the interpretation of relaxation times in terms of motion, rests heavily on the model employed, especially if only one type of relaxation time is recorded. Advances in the technique of magnetic resonance have made possible the determination of a wide range of relaxation times for an increasing number of nuclei. For lipid chains it is now feasible to examine proton, deuteron, and carbon nuclei, and studies have been made of the frequency dependance of T_1 and $T_{1\varrho}$. Although, as we have stressed, the movement of spin-labeled molecules many not faithfully follow the motion of their unlabeled analogs, electron spin relaxation data are still useful additions to our armament and the measurement of electron spin–lattice relaxation times is being carried out in several laboratories. The details of instrumental technique and of relaxation theory are outside our terms of reference but the spectroscopic layperson can appreciate that the more data that have to be accounted for, the less freedom we have in choosing a description of chain motion. An example of the current degree of sophistication in this field is the paper by Pace and Chan.[52] Any acceptable model must, of course, be able to account for both order parameters and relaxation times. In Chapter 8 we will see that theory has succeeded in giving a reasonably convincing picture of the hydrophobic core and of reproducing the order-parameter profile. The problem of chain motion is, however, more intractable and we will see that simple models, although intellectually pleasing, have little chance of representing the truth.

While order parameters are clearly related to motion, they also should be able to tell us something about the average structure of the bilayer. Thus, if all the chains are highly ordered, the bilayer is expected to be thicker than if they are highly disordered. A quantitative analysis follows.

The order parameters emerging from deuterium NMR refer to specific methylene groups and give no immediate picture of the types of overall conformations occurring during the motion of lipid chains. However, the *average chain length* can be estimated as shown by Seelig and Seelig.[53] We follow their argument after first noting a few basic facts about chain conformations.

Research on chain conformations in phospholipids was preceded by much laboratory and paperwork on alkanes. Long-chain alkanes are permitted an enormous number of conformations in the liquid state. Each C—C bond can have a trans or two gauche conformations, symbolized by t, g^+, g^-. Since gauche bonds have higher energies than trans, the all-trans configuration has the lowest energy. Certain conformations are energetically highly improbable due to steric hindrance, for example, g^-g^+ (where the symbols refer to adjacent bonds). Placing one gauche bond in an otherwise all-trans chain results in a bent chain, but the single sequence g^+tg^- (or g^-tg^+), commonly called a "kink" connects two colinear stretches of trans chain. (The reader should play with molecular models.) The sequence g^+tttg^- (known as a "jog"), when introduced into an all-trans chain, also connects two parallel stretches of trans bonds. The jig and the kink have been singled out for mention because, although in the liquid alkane they are not especially noteworthy, they are often considered to be favored "defects" in the liquid crystal state of phospholipids and have been supposed to pack better than, for example, chains containing only a single gauche bond near the head group. This conformation is bent and is expected to encounter considerable steric opposition to its formation. All conformations that include a gauche bond shorten the projection of the chain on the axis defined by the all-trans chain (Figure 51). The shortening produced by a single gauche bond depends on its position in the chain. A kink, wherever it is, gives

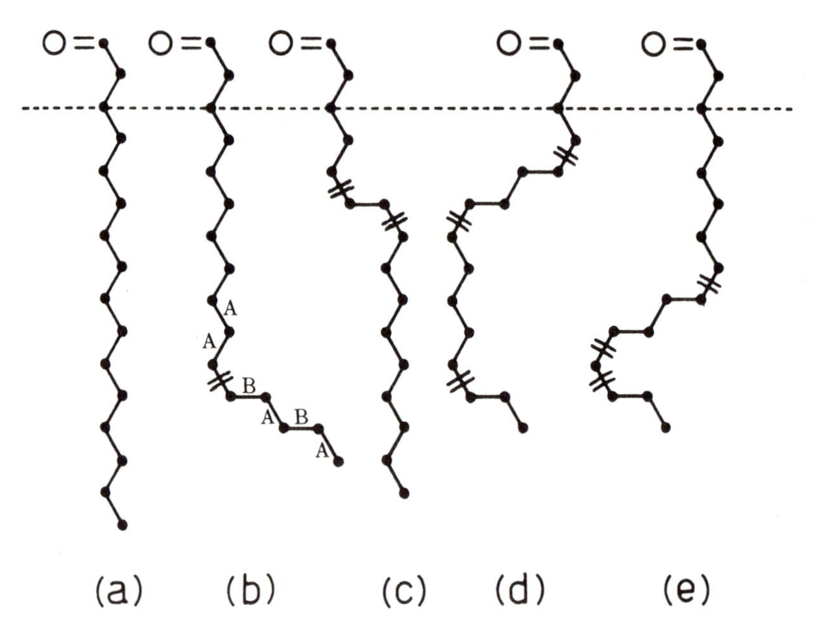

Figure 51. Illustrating the shortening of an all-trans chain due to introduction of gauche bonds. In **B** the symbols A and B label the two bond orientations possible for a model in which only trans and gauche bonds are allowed. **C** demonstrates that a single "kink" gives a shortening of one unit. [Adapted from Pink, D.A., et al, *Biochem.* 19:349 (1980).]

a shortening of one unit, where the unit is the length of the projection of one trans bond on the normal in an all-trans chain, ~1.25 Å. A jog shortens a chain by two units, a fact recorded in that system of nomenclature that denotes a jog by 2g2, where 2g indicates two gauche bonds and the final 2 gives the shortening of the chain in the conventional units. A knowledge of the probabilities of occurrence of every type of defect allows the calculation of average chain length. Seelig and Seelig approach the problem differently, using experimentally determined order parameters. They start by noting that if there is no average chain tilt, then C—C bonds can only have two orientations with respect to the bilayer normal. (In Figure 51 some bonds have been labeled *A* and *B* to indicate these orientations.) The probability of a given bond being in the state *A* is written p_A and elementary probability theory gives

$$p_A + p_B = 1 \tag{5.8}$$

Now from the definition of order parameters we have for a given $-CH_2$ segment:

$$S_{\text{mol}} = p_A \cdot \tfrac{1}{2} \cdot (3\cos^2 0^\circ - 1) + p_B \cdot \tfrac{1}{2} \cdot (3\cos^2 60^\circ - 1) \tag{5.9}$$

This expression follows from the fact that for a bond labeled $A(B)$ the normal to the plane containing the two C—H bonds is inclined at 0 degrees (60 degrees) to the director. Check that for the all-trans configuration with $p_A = 1$ and $p_B = 0$, $S_{\text{mol}} = 1$, as it should. From (5.8) and (5.9),

$$p_B = (1 - S_{\text{mol}})/1.125 \tag{5.10}$$

The effective length along the bilayer normal of an *A* state bond is 1.25 Å ≡ l, and of a B-state bond $l \cos 60^\circ$. The average effective length of a given bond i is then

$$\langle l_i \rangle = p_{iA} l + p_{iB} l \cos 60^\circ = l(1 - 0.5 p_{iB}) \tag{5.11}$$

The total average length of a chain is

$$\langle L \rangle = \sum_i \langle l_i \rangle \tag{5.12}$$

For the case of DPPC this reads:

$$\langle L \rangle = \sum_{i=1}^{15} \langle l_i \rangle = l\left(15 - 0.5 \sum_{i=1}^{15} p_{iB}\right)$$

The p_{iB} can be obtained from the experimental order parameters and equation (5.10). From the NMR data at 50°C a value of $10.52\ l = 13.15$ Å was obtained for $\langle L \rangle$. The all-trans state has a length of 18.75 Å, so that the shortening in going from the presumed all-trans gel to the liquid crystal state is given by 5.6 Å, or 11.2 Å for a bilayer.

The experimental value obtained by X-ray studies on a sample of hydrated DPPC is 11.6 Å. The values of p_A and p_B do not, in general, give the relative fractions of trans and gauche bonds in a chain because, for example, trans bonds containing *B*-type bonds can be found in disordered chains (Figure 51).

A step nearer to a complete understanding of chain behavior is the estimation of the average number of gauche bonds in a chain as a function of temperature. Magnetic resonance is not too helpful here but other forms of spectroscopy have stepped in. The vibrations of molecules are often strongly dependant on molecular conformation and the Raman and infrared spectra of phospholipids have been analyzed to give conformation information complementary to that derived from NMR.

The interpretation of vibrational spectra is not usually straightforward; the frequencies and intensities of vibrational bands in large molecules are usually related to molecular conformations by a mixture of theoretic normal mode analysis and experimental observations on simple systems. Thus studies on n-alkanes have been helpful in understanding the spectra of phospholipids and theoretic work is beginning to provide a basis for quantitative estimates of gauche bonds. The number of studies on phospholipids has been considerable; we stick to the theme of this chapter and concentrate on the use of Raman spectroscopy in elucidating the conformations of lipid chains in bilayers.

The Raman spectra of lipids show three main regions of interest. At ~100–200 cm^{-1} there is a band attributed to a longitudinal acoustic mode (LAM). This band is of importance in that in n-alkanes the LAM band near 110 cm^{-1} appears to occur only in the all-trans conformer and in the conformer having a single gauche bond at the end of the molecule, in which case the frequency is shifted upward slightly. We return to these facts shortly.

In the range 1060–1130 cm^{-1} a skeletal optical mode is observed, and it is this band that will occupy us since it has been widely employed to estimate the number of gauche bonds in lipid chains. A broad band at ~1090 cm^{-1} is ascribed to chain and head-group vibrations. Around 890 cm^{-1} there is a band ascribed to the rocking vibrations of methyl groups at the chain ends when the end is in a trans configuration.

We concentrate on the ~1130-cm^{-1} band, which theory and experiment show to originate in the mode illustrated in Figure 52. This is an in-phase C—C stretching (transverse optical) mode. In the gel state, at temperatures not too far below the pretransition temperature of the chains, the intensity of this band increases linearly with chain length[54] for chains varying from 14 to 20 carbon atoms. On raising the temperature there is an abrupt loss of intensity at T_t and the frequency and intensity of the 1130-cm^{-1} band become effectively independent of chain length. Extrapolation of the experimental intensities in the gel state suggests that the intensity in the fluid state is that to be expected from an 8-carbon all-trans segment. Furthermore, the ratio $I_e = I_{890}/I_{1130}$ decreases with rising temperature, implying that the chain ends become disordered more quickly than the "higher" part of the chain. These results fit in qualitatively (but nowhere near quantitatively) with the ^{2}H NMR results shown in Figure 44, particularly with respect to the initial stretch of comparatively ordered methylene groups. As the temperature of the gel state is lowered, another change in intensity is observed at 20 to 30°C below T_t. This might be correlated with the known change in chain packing from near hexagonal to orthorhombic accompanied by an increase in tilt angle. All this sounds fairly watertight but simplistic. If

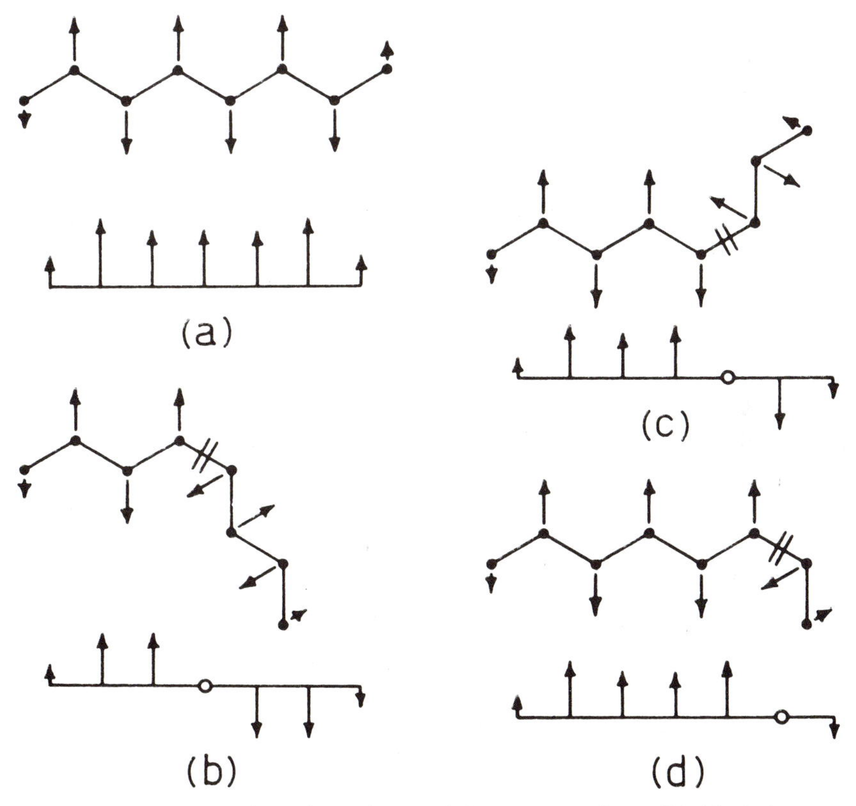

Figure 52. Assumed motions for methylene groups in a C—16 chain exhibiting the transverse optical mode. Chains are allowed to have up to three gauche bonds in the model adopted. The normal modes are shown under each type of chain motion. [From Pink, D. A., et al., *Biochem.* 19:349 (1980).]

the intensity of the 1130-cm^{-1} band depends on the number of trans bonds, then the number and length of *all* the trans segments in the chain should contribute to the total intensity, not just an initial trans stretch attached to the head group. Furthermore, the vibrations of different trans segments cannot be completely independent. On turning to the theoreticians for help, we find two main theoretic approaches to the problem of correlating intensity and conformation:

1. In the theory of Gaber and Peticolas[55] each segment of trans chain is supposed to contribute independently to the 1130-cm^{-1} intensity with a weight proportional to the number of C—C bonds.
2. The theory of Pink, Green, and Chapman[56] takes into account the coupling of the vibrations of trans segments separated by a single gauche bond. The important consequence is that the intensity of the 1130-cm^{-1} band depends on the location of gauche bonds as well as on their number; the intensity is not a simple function of the number of trans bonds.

Which theory is nearer the truth? One test is contained in the study of the 1130-cm^{-1} band in crystalline n-$C_{21}H_{44}$ in the orthorhombic and hexagonal phase.[57] In the low-temperature orthorhombic phase this alkane has all-trans chains and displays a strong 1130-cm^{-1} band; its intensity is constant up to ~30°C, above which there is a sudden loss of intensity associated with the change to the hexagonal form. As the temperature is raised, the intensity of the 1130-cm^{-1} band in the hexagonal phase decreases continuously. Now the concentration of gauche bonds in the hexagonal phase can be estimated from the intensity of the LAM band referred to above. From measurements of this band it can be fairly confidently stated that in the transition from orthorhombic to hexagonal packing, less than one gauche bond per molecule is introduced into the all-trans chains. In returning to the 1130-cm^{-1} band, it is found that the decrease of ~40% in intensity observed at the phase transition corresponds to at least six gauche bonds per chain in the Gaber-Peticolas (GP) model but roughly one gauche bond in the Pink-Green-Chapman (PGC) model. The PGC theory is certainly based on more realistic assumptions than the GP theory and is probably the best means available at present for using intensities to estimate the average number of gauche bonds per lipid molecule $\langle n_g \rangle$. The values of $\langle n_g \rangle$ estimated by Pink *et al.* for some bilayers are as follows: DMPC, 1.06 (0°C) and 7.14 (30°C); DPPC, 1.78 (25°C) and 10.35 (45°C); DSPC, 2.22 (40°C) and 13.98 (60°C). The first temperature in each pair is below the pretransition point and the second is above T_t. [It should be realized that two independent theoretical steps are involved in the PGC theory: (1) the probabilities of different conformations are estimated on the basis of a certain *model* for chains in bilayers, and (2) the relative Raman intensities for the different conformations are calculated, again using approximations.]

HEAD GROUP ORIENTATION

To specify the orientation of head groups we need to know several angles. In crystals these angles are obtainable from X-ray studies, as we saw in Chapter 1. In hydrated phases of phospholipids where both chains and head groups have motional freedom, X-ray, neutron, and electron diffraction are unable to give more than limited — but useful — information. Again it is magnetic resonance that has been the most revealing method, in particular ^{2}H and ^{31}P NMR. The results will be reviewed in this section, but first the principles behind the use of ^{31}P NMR will be detailed since they may be unfamiliar.

The ^{31}P nucleus has an isotopic abundance of 100%. The nuclear spin quantum number of one-half precludes a nuclear quadrupole moment; the ^{31}P NMR of a phosphorus atom in a molecule consists, in the absence of significant spin–spin interactions with its neighbors, of a single line, so the bad news is that we have no doublet to help us find our bearings as we had for deuterium. The good news is that the magnetic field at which the ^{31}P line falls is strongly dependent on the orientation of the phosphorus-containing moiety to the external field. We say that the *chemical shift tensor* σ is anisotropic and it will be this anisotropy that will be our tool. The chemical shift tensor has been determined for two crystalline compounds, barium diethylphosphate,[58,59] which is a diester, and phosphorylethanolamine,[60] a monoester. Single-crystal studies allow the determination of the principal values of σ and the directions of the associated principal axes. Figure 53 shows the orientation of the axes with respect to the phosphate group in the diethyl phosphate anion. Since no similar single-crystal studies have been made on phospholipids, we cannot know the orientation of the principal axes in these compounds. Nevertheless, the principal values of σ can, as we shall see, be found from the spectra of randomly oriented samples and they vary little between different phospholipids (except for phosphatidic acid). It is safely assumed that the orientation of the principal axes are not significantly different from those found in the single-crystal studies of phosphate esters. For the rest of this discussion we take the following values for the components of σ:

$$\sigma_{11} = -80 \text{ ppm}, \quad \sigma_{22} = -20 \text{ ppm}, \quad \sigma_{33} = +110 \text{ ppm}$$

where the shifts are measured with respect to an external sample of 85% H_3PO_4.

In principle, a knowledge of the chemical shift tensor can be used to determine the order parameters of phosphate groups in aligned bilayers in the same way as the quadrupole tensor of ^2H was used to determine the orientation of C—D bonds. In practice, ^{31}P NMR has been used mainly in the study of randomly oriented samples. (Actually the majority of ^2H NMR studies have also been on randomly oriented material, a fact we earlier slyly avoided.) We now take the opportunity to explain the principles behind the analysis of "random spectra," as we will term them, and illustrate by interpreting the ^{31}P NMR spectra of several

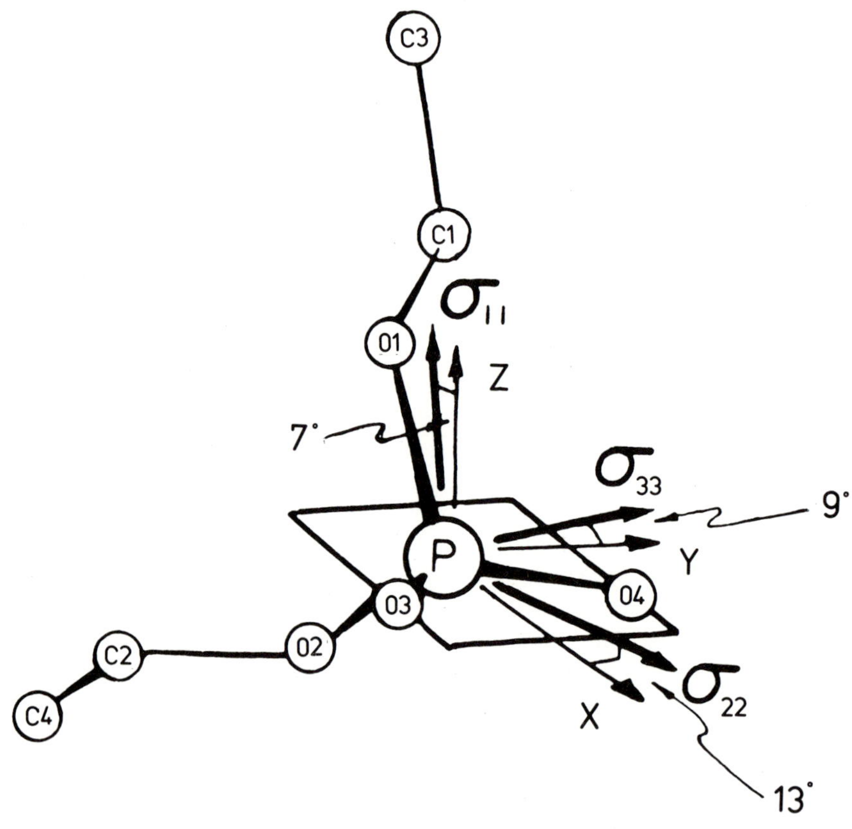

Figure 53. Orientations of the principal values of σ, the ^{31}P chemical shift tensor, for a phosphate group in a dialkylphosphate. The x-axis bisects the O(3)–P–O(4) angle, z is perpendicular to the O(3)–P–O(4) plane and y is orthogonal to x and z. [From Herzfeld, S. J., et al., *Biochem.* 17:2711 (1978).]

phases of phospholipids. To help us visualize the problem, we replace the phosphate group by an imaginary ellipsoid of which the three principal axes have lengths proportional to the three principal values of the chemical shift (Figure 54). Suppose we now take an aqueous dispersion of a phospholipid containing randomly oriented bilayer fragments. The ^{31}P NMR spectrum will be a superposition of all the differently oriented fragments. Although the dispersion is a liquid sample and the individual bits of bilayer are translating and reorienting continually, their movement is so slow on the NMR time scale that they are static as far as the

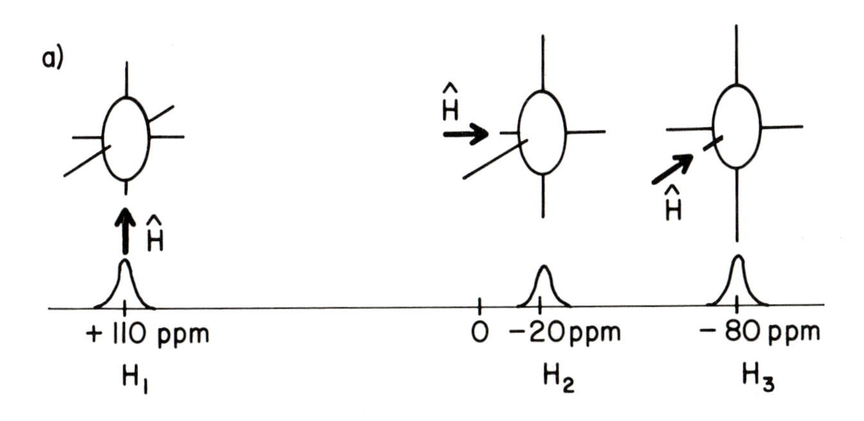

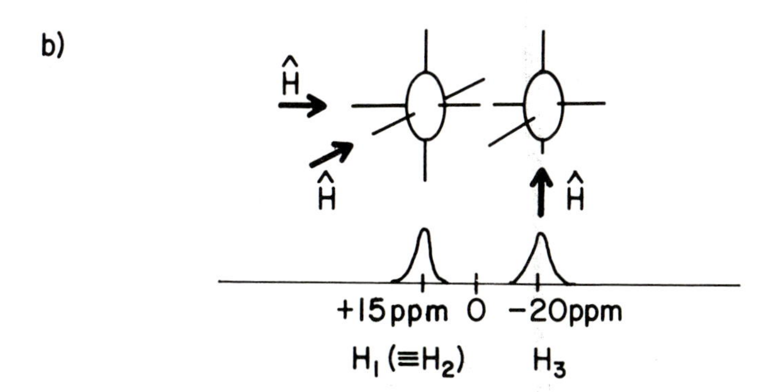

Figure 54. Illustrating the replacement of a -PO_4 group by the ellipsoid representing the chemical shift tensor, σ. If the drawing were to scale, the length of the vector from the center of the ellipsoid to a point on its surface would be proportional to the value of σ observed when the external magnetic field is along that vector in the -PO_4 group.

NMR experiment is concerned. Each orientation of the phosphate group — the ellipsoid — is associated with a resonance line at a field determined solely by the relative orientation of the ellipsoid and $\hat{H}$. The total spectrum is the sum of all these resonance lines, each weighted by the probability of the associated orientation. To simplify the argument initially, we suppose the ellipsoid to be cylindrically symmetric about its long axis and we concentrate on the distribution of this axis with respect to $\hat{H}$. A look at Figure 40 should convince the reader that in a random sample there is far more chance of the long axis lying near the *xy* plane (perpendicular to $\hat{H}$) than of being aligned close to $\hat{H}$. A little thought shows that the probability of an axis falling within the volume defined by the cones θ and $\theta + d\theta$ is proportional to area of the shaded zone in Figure 40. This area is proportional to the radius *r* in the figure and hence proportional to sin θ. Thus the ellipsoids have a large chance of having their long axes on or near the *xy* plane ($\sin \theta \simeq 1$) and a small chance of being aligned along $\hat{H}$ ($\sin \theta \simeq 0$). The spectra will be heavily weighted by contributions from lines near H_1 (Figure 54) and have progressively less intensity as we go in the direction of H_3. A theoretical reconstruction is shown in Figure 55 adjacent to the ^{31}P NMR spectra of an aqueous dispersion of egg yolk phosphatidylcholine. (For the NMR experts the spectra were taken under conditions of broad-band proton decoupling.) By comparison we show, in Figure 56A, the theoretic random spectrum for the noncylindrically symmetric ellipsoid based on the three principal values of the ^{31}P chemical shift tensor. Here there are three characteristic features — a cutoff in the spectra at the two extreme values of the chemical shift and a maximum at the intermediate value. Figure 56B shows the experimental spectra of a randomly oriented sample of anhydrous DPPC.

The conclusions up to now are that a phosphorus atom displaying a completely asymmetric chemical shift tensor (ellipsoid) gives, in randomly oriented samples, a ^{31}P NMR spectrum like Figure 56A whereas for a cylindrically symmetrical tensor/ellipsoid in which, say, $\sigma_{11} = \sigma_{22}$, we expect a spectrum of the type shown in Figure 55. Now we know that σ is completely asymmetric so it is not surprising that anhydrous phospholipids give the spectrum that they do (Figure 56B). The question is why we see other types of random spectra; that is, what is the explanation of Figure 55? In other words, why do the head groups of hydrated DPPC behave as if they had cylindrically symmetric tensors? The answer can be deduced from two observations:

1. If the hydrated lipid is cooled, the spectrum changes to that typical of an asymmetric tensor, that is, Figure 56A.

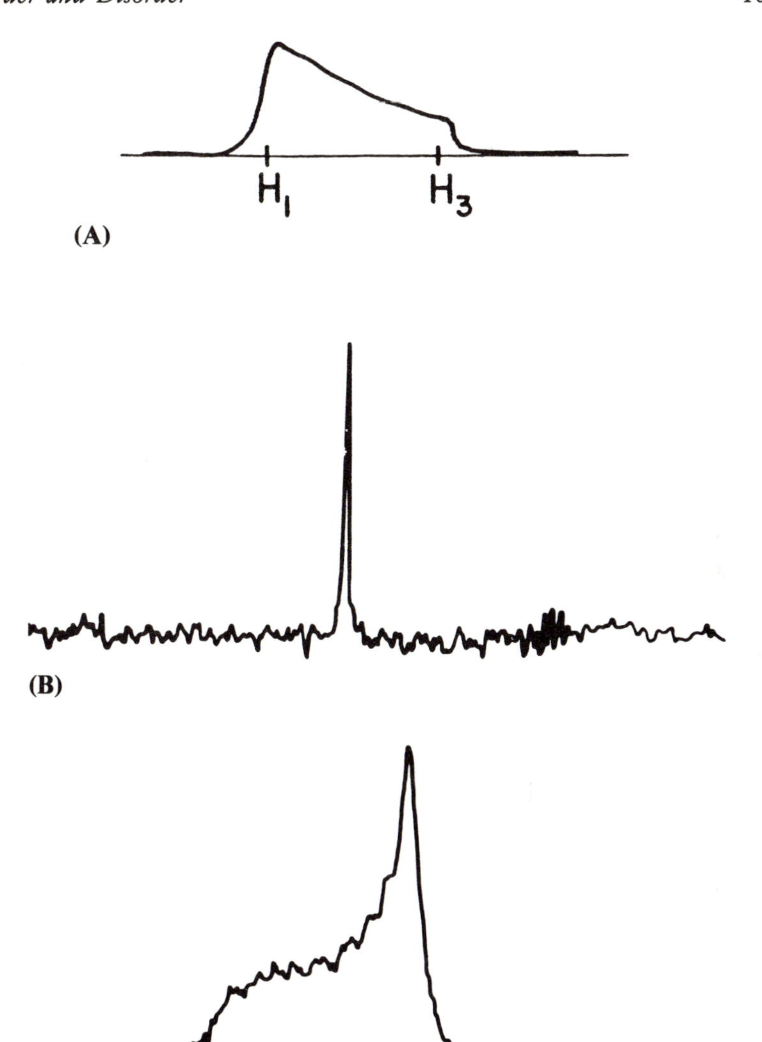

Figure 55. Comparison between (A), the theoretical ^{31}P nmr spectrum of a randomly oriented collection of molecules having axially symmetric ^{31}P chemical shift tensors and (B), an experimental spectrum of egg yolk lecithin. The magnetic field sweeps are oppositely directed in the two spectra. Spectrum (C) is of the same molecule in solution. A single line is observed corresponding to the averaged chemical shift.

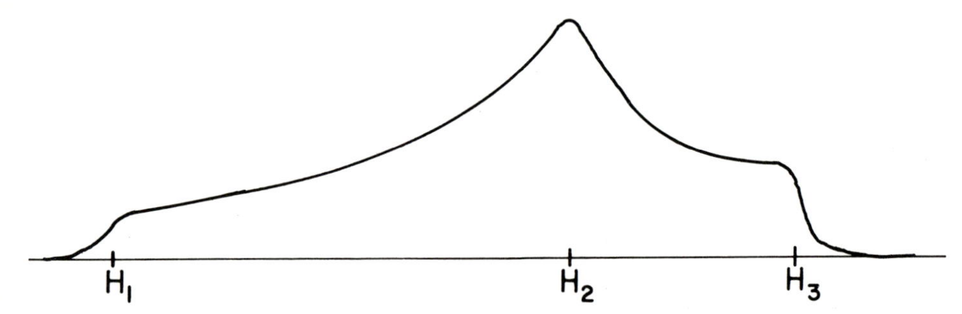

Figure 56. A. Theoretical powder spectrum of a molecule having noncylindrically symmetrical chemical shift tensor.

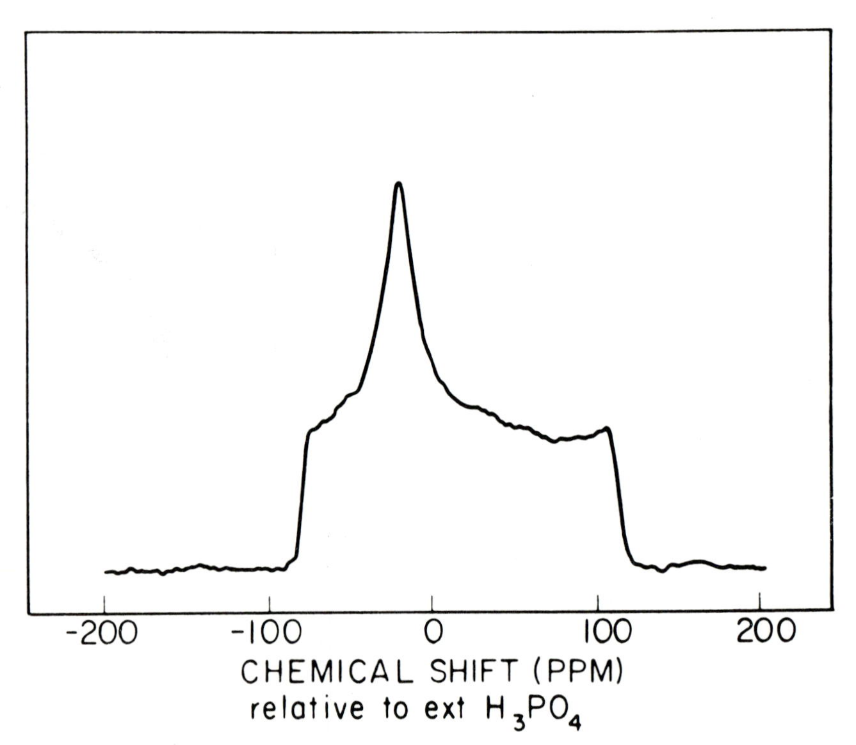

Figure 56. B. Experimental spectrum of randomly oriented multilayers of anhydrous DPPC.

2. If the lipid is dissolved in an organic solvent, the spectrum collapses to a single line whose position is given by the average chemical shift $(\sigma_{11} + \sigma_{22} + \sigma_{33})/3$.

These spectral differences are due to differences in motional freedom. In solution the molecule tumbles rapidly and the instrument records a chemical shift averaged over the motion of the molecule. Although the lipid is not spherical and therefore its motion is not isotropic, the rate of reorientation is so fast that it is effectively isotropic on the time scale of NMR, whose practitioners would say that we have a case of extreme narrowing. On the other hand, the frozen sample of hydrated lipid contains rigid unmoving head groups and all three components of σ are apparent in the spectrum. The spectrum at room temperature is that of ^{31}P displaying only two principal values. This can be explained in terms of a model of the kind illustrated in Figure 57. Suppose that the head groups are oriented so that the σ ellipsoid lies perpendicular to the axis defined by the hydrocarbon chains and that it is rotating rapidly about this axis, the director. To a sleepy-eyed observer the ellipsoid will appear to display the principal values shown in Figure 55 because the two values associated with axes perpendicular to the rotation axis will be averaged: $\frac{1}{2}(+110 - 80) = +15$ parts per million (ppm). The observed, partially averaged, values of σ will be -20 ppm when $\hat{H}$ is along the director and $+15$ ppm for both the other values. The resulting spectrum, calculated using the principles discussed previously for an axially symmetric tensor, will be as shown in Figure 55. The value of $|\Delta\sigma| = |\sigma_{\perp} - \sigma_{\parallel}|$ for our model is 35 ppm, which agrees reasonably well with experimental results for hydrated lipid bilayers, considering the crudeness of our model.

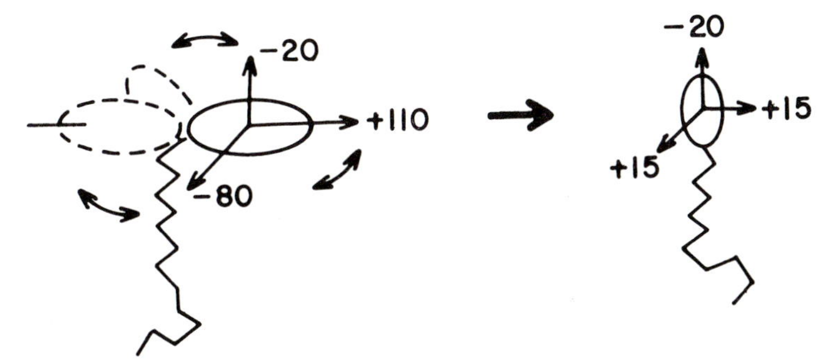

Figure 57. Model explaining the averaging of two of the principal values of the -PO_4 σ tensor by rotation of the σ ellipsoid about the bilayer normal, a type of motion proposed for the head groups of hydrated lipids.

Although the above analysis seems satisfactory, it does not allow us to specify in more than very general terms the motion of head groups. The glib phrase "rotating rapidly" covers up the fact that such rotation is not simple and must involve coordinated changes in a number of bonds. What, for example, is an ethanolamine group doing during this rotation? Partial answers come from ^{2}H NMR studies on deuterium-labeled head groups. Since the principles have been presented previously and the interpretation of the results is still open to discussion, we restrict ourselves to a précis. Deuterium has been introduced into the positions carbon-1, carbon-11, and carbon-12 (where the nomenclature is that of Figure 1B) and into the methyl groups of choline moiety in the DPPC head group. The results, together with those for ^{31}P, have been shown to be roughly consistent with many combinations of torsional angles within the head group but the combinations tend to fall in restricted regions of conformational space.[61] A simpler picture[62] suggests that the head group jumps rapidly between only two enantiomeric conformations (compare Figure 11). All workers agree that there is rapid rotation about the normal to the bilayer and that rotation of bond torsion angles is not free. Perhaps the most precise information is that derived from a study of oriented bilayers in which it was shown that the plane containing the phosphorus atom and the two unesterified oxygens of the phosphate group of DPPC is inclined at about 50 degrees to the bilayer normal.[58] This is consistent with the "folded-down" conformation postulated for the head group on the basis of neutron diffraction (Chapter 3) and other techniques.

Our ignorance of the finer details of head-group rotation does not stand in the way of the use of random spectra as an incisive means for the determination of gross structure in solution. We have seen how the spectra of bilayers can be rationalized in terms of rotation, and in a similar fashion the hexagonal phase of phospholipids give ^{31}P NMR spectra that are explicable if we realize that the effective chemical shift tensor seen in the bilayer, — $+15$, $+15$, and -20 ppm — undergoes an additional averaging due to the lateral diffusion of lipids around the central water-filled channel, which is about 20 Å in diameter. On the time scale of NMR this movement is rapid enough to give an effective tensor of roughly -2.5, -2.5, and $+15$ ppm , with an overall width $\Delta\sigma$ of 17.5 ppm, that is, half the width usually found in bilayers. Figure 58 explains the model behind the principal values of the tensor and Figure 59 shows an experimental spectrum.

In the micellar phase, or in small vesicles, combinations of lateral diffusion, vesicular rotation, and, for micelles, exchange with free lipid in the solvent are so effective in averaging that only a simple narrowed line is observed for the ^{31}P spectrum. The comparative ease with which

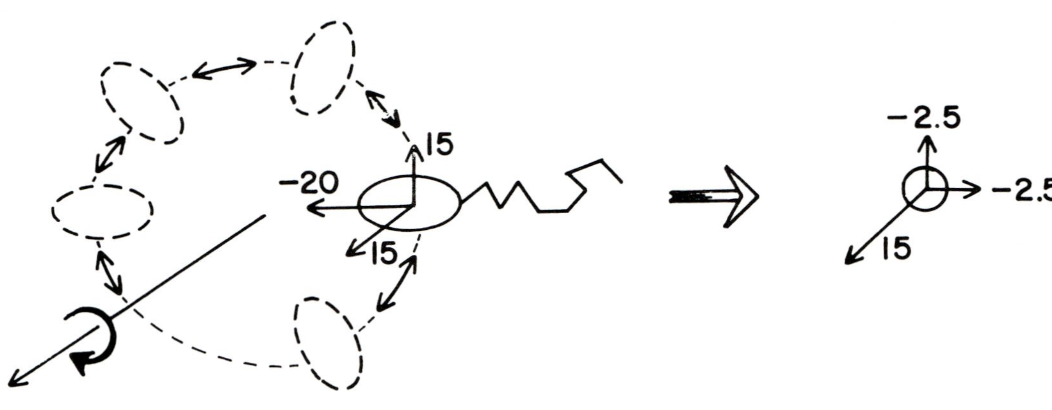

Figure 58. Model explaining the averaging of two of the principal values of the hydrated -PO_4 σ tensor (see Figure 57) by rapid diffusion of the lipid molecules about the axis indicated.

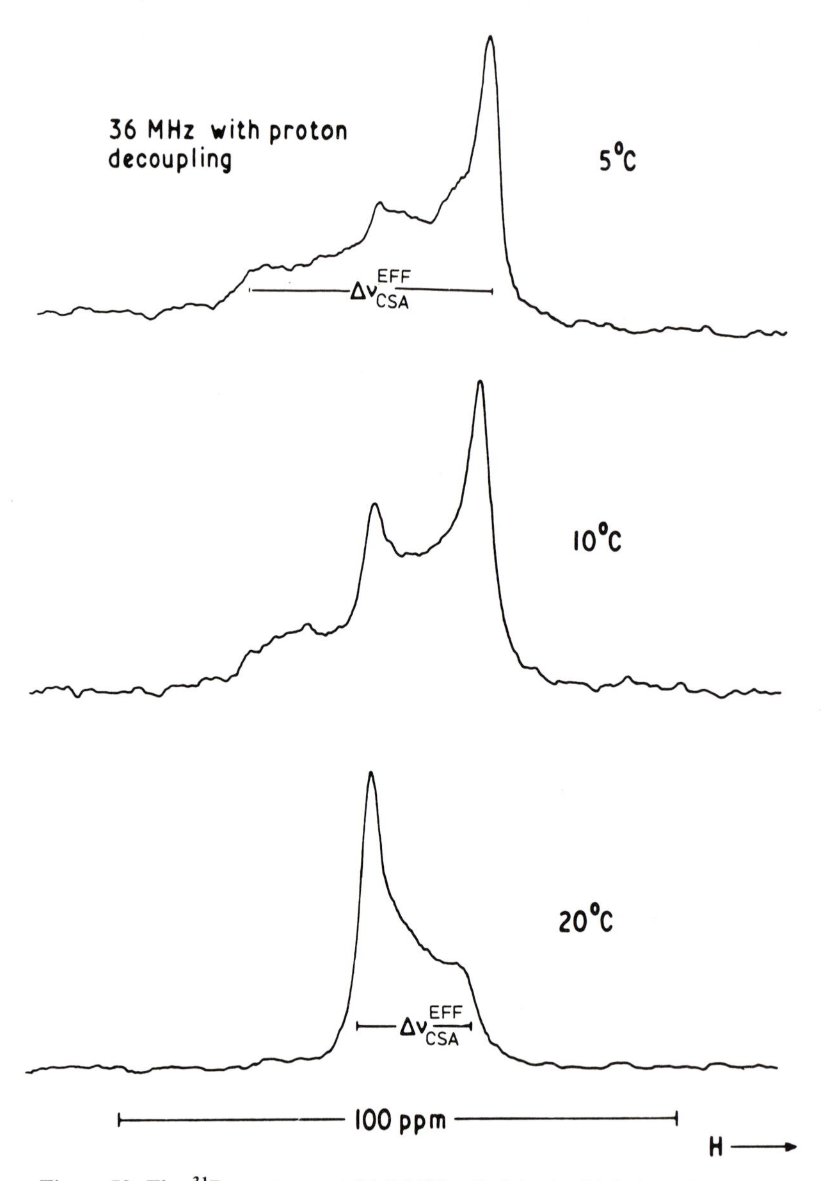

Figure 59. The ^{31}P spectrum at 36.4 MHz of phosphatidylethanolamine in excess water. The spectrum changes from that of a bilayer at 5°C to that of the H_{II} hexagonal form at 20°C. At 10°C the two forms coexist, as confirmed by X-ray diffraction.

^{31}P NMR distinguishes between phases has found extensive use in the monitoring of phase changes induced by temperature, pH, and divalent ions.[63,64]

Our account of phospholipid aggregates has to this point been predominantly descriptive. In the next three chapters theory will take the stand.

6 Thermodynamics

The treatment of phospholipid aggregates in the previous chapters has been mainly descriptive; the one unifying principle employed was the mutual steric compatibility of simplified geometric understudies acting out the roles of real molecules. We now appeal to thermodynamics for aid, limiting ourselves to a hierarchy of three problems that are central to the entire field of lipid aggregates. We first ask why lipid monomers in aqueous systems form aggregates, and then we see what thermodynamics has to say about phase equilibria and phase transitions in single lipids and in binary mixtures of lipids. Finally we examine the roles of energy and entropy in determining the structures adopted by lipid aggregates.

THE HYDROPHOBIC EFFECT

The fact that oil and water do not mix has been known long enough to have become a part of folklore; the reason they do not mix was the subject of learned papers still being published while this book was being written. The contribution of thermodynamics not only illustrates the legendary powers of this discipline, but is an example of the traps laid by common sense, which would surely, but mistakenly, regard the insolubility of, say, butane or argon in water as a reflection of the energetic undesirability of breaking hydrogen bonds in water.

Long-chain phospholipids owe whatever slight solubility they have in water almost entirely to their polar head groups. Since it is clear why the polar moieties are easily hydrated, the efforts of theoreticians and experimentalists have often been directed toward an understanding of the behavior of simple hydrocarbons in aqueous media.

It is almost a century since Traube[65] showed that amphiphiles containing hydrocarbon chains prefer the surface of an aqueous solution as evidenced by their ability to lower the surface tension. The longer the chain, the higher is the surface concentration (in dilute solutions) with respect to the bulk concentration. Of later theoretic significance was Traube's finding that the apparent molal volumes of many organic molecules are about 13 cm^3 lower than for the same molecules in organic solvents.[66] It was subsequently found that the apparent molal heat capacity of homologous series of various amphiphiles in water increases by 105 ± 20 $J\,K^{-1}\,mol^{-1}$ for every methylene group added to the chain. For those cases where the solute exists as a liquid, the corresponding increase per CH_2 group for the pure liquid is in the range of 20–30 $J\,K^{-1}\,mol^{-1}$. The unusual behavior of the molal volumes and heat capacities foreshadowed the stranger findings of Butler,[67] Kauzmann,[68] and many others, who showed that although the free energy of transfer of a non polar molecule from an organic solvent to water was predictably positive, the enthalpy of transfer for many small hydrocarbons was negative. In simple terms, it seems to be energetically advantageous for these non polar molecules to leave their fellows and jump into water. This does not appear to be in accordance with the schoolday dictum that "like goes to like." It is, in fact, a large drop in *entropy* on entering water that saves the day and pushes the free energy of transfer well below zero.[69] We now show how these thermodynamic quantities are related to experimentally available quantities and later summarize telegraphically the current interpretations of their magnitudes and signs.

Consider a liquid hydrocarbon in contact with water. At equilibrium both the concentration of hydrocarbon in the water and that of water in hydrocarbon are very small, and so in dealing with the hydrocarbon phase it is convenient to use a standard state for which γ, the activity coefficient, goes to unity as x, the mole fraction of hydrocarbon, goes to unity. In what follows we use the subscripts H and W to refer to hydrocarbon in organic and aqueous media respectively. Formally we have for hydrocarbon in the hydrocarbon phase:

$$\mu_H = \overset{\circ}{\mu}_H + RT \ln a_H = \overset{\circ}{\mu}_H + RT \ln \gamma_H x_H \tag{6.1}$$

$$(\gamma_H \to 1 \text{ as } x_H \to 1)$$

Neglecting the negligible, we write $x_H \simeq 1$, $\gamma_H = 1$, and

$$\mu_H = \overset{\circ}{\mu}_H + RT \ln x_H = \overset{\circ}{\mu}_H \tag{6.2}$$

By contrast, for the hydrocarbon in the water phase where x_W is very small, it is good strategy to use a different but common convention in which $\gamma_W \rightarrow 1$ as $x_W \rightarrow 0$, giving

$$\mu_W = \overset{\circ}{\mu}_W + RT \ln a_W = \overset{\circ}{\mu}_W + RT \ln \gamma_W x_W \tag{6.3}$$

For $x_W \ll 1$ we put $\gamma_W = 1$ and

$$\mu_W = \overset{\circ}{\mu}_W + RT \ln x_W \tag{6.4}$$

[The above expressions for chemical potentials should include, under isothermal conditions, a pressure–volume term $\bar{v}_i P$ where $\bar{v}_i$ is the partial molal volume of component i and P is the pressure. However, under normal conditions this term can be neglected (but not in the treatment of osmotic pressure). Thus, for example, for water $\bar{v}_i \simeq 18\ \mathrm{cm^3\,mol^{-1}}$ and $P = 1$ atmos $\simeq 10^5$ newton $\mathrm{m^{-2}}$, then $\bar{v}_i P \simeq 1.8\ \mathrm{J\,mol^{-1}}$. For differences in μ° this term will be even less important. Note that we will be dealing with values of $\Delta\mu^\circ$ usually amounting to several kilojoules.]

Expressions (6.2) and (6.4) are both of the form used to express the chemical potential of an ideally behaving component of a solution; moreover, $R \ln x_W$ is that part of the entropy of such a component that depends only on its spatial distribution. This translational entropy is sometimes called the *cratic* contribution to the total entropy and is independent of the nature of solvent or solute. The terms $\overset{\circ}{\mu}_H$ and $\overset{\circ}{\mu}_W$ are termed the *unitary* contributions to the chemical potentials. At equilibrium the chemical potential of hydrocarbon in the two phases must be the same:

$$\mu_H = \mu_W$$
$$\overset{\circ}{\mu}_H = \overset{\circ}{\mu}_W + RT \ln x_w$$

Therefore,

$$\Delta\mu^\circ = \overset{\circ}{\mu}_W - \overset{\circ}{\mu}_H = -RT \ln x_W \tag{6.5}$$

This means that a measurement of the amount of hydrocarbon in the aqueous phase is sufficient to find the unitary free energy difference between hydrocarbon in water and hydrocarbon in the (nearly) pure liquid. The significance of unitary thermodynamic quantities is that differences in such quantities contain only the internal energy and entropy of solute and its interactions with the solvent. In the present case the change in unitary free energy is a molecule's-eye view of what happens to it in going from pure hydrocarbon to water, and this is exactly what interests us rather than the obvious statistical factor (the cratic entropy) that tends to push molecules down concentration gradients.

Another way of looking at a unitary free energy difference is to regard it as the change found when molecules are transferred from fixed positions in one phase to fixed positions in another phase. The unitary free energy can be separated into enthalpic and entropic contributions either by measuring the enthalpy difference $\Delta\bar{H}^\circ$ calorimetrically or by measuring $\Delta\mu^\circ$ as a function of temperature and using the equation:

$$\left[\frac{\partial(\Delta\mu^\circ/T)}{\partial(1/T)}\right]_P = \Delta\bar{H}^\circ \tag{6.6}$$

Extensive quantitative analyses of experimental data on various types of molecules have been presented by several workers; a convenient summary is given by Tanford[70] and a very detailed discussion is contained in a book by Ben-Naim.[71] We list only the major findings and give an illustrative selection of quantitative results in Table 4.

1. The unitary free energy of transfer of hydrocarbons and long-chain amphiphiles from an organic to an aqueous phase is large and positive — oil and water keep faith with popular prejudice. A great many

Table 4
Thermodynamic Data for the Transfer to Water of Some Molecules from the Pure Liquid or from Organic Solvents

	$\Delta\bar{H}^\circ$ (cal mol^{-1})	$\Delta\bar{S}^\circ$ (cal K^{-1}mol^{-1})	$\Delta\bar{C}_P^\circ$ (cal K^{-1}mol^{-1})
From organic solvents			
C_2H_6 (from CCl_4)	−1800	−19	59
C_2H_6 (from C_6H_6)	−2200	−20	59
From pure liquid hydrocarbon			
C_2H_6	−2500	−21	—
C_3H_8	−1700	−22	—
C_4H_{10}	−800	−23	65
C_5H_{12}	−500	−25	96
C_6H_{14}	0	−26	105
C_6H_6	500	−14	54
C_2H_5OH	−2430	−10.7	39
n-C_3H_7OH	−2420	−13.4	56
n-C_4H_9OH	−2250	−15.6	72
n-$C_5H_{11}OH$	−1870	−17.1	84

From ref. 70.
(From Tanford, C., *The Hydrophobic Effect: Formation of Micelles and Biological Membranes*, 2nd ed., Wiley, New York, 1980.)

aliphatic and aromatic hydrocarbons obey a very simple empirical relationship:

$$\Delta\mu^{\circ}_{298} = 6.4 + 1.85\ n_H \qquad (\text{kJ mol}^{-1}) \tag{6.7}$$

where n_H is the number of hydrogens in the molecule. Other empirical relationships that have been proposed for different classes of compound include the following[70]:

$$\begin{aligned}
&\text{For n-alkanes:} && \Delta\mu^{\circ}_{298} = 10.20 + 3.70n_C\\
&\text{For n-alkenes:} && \Delta\mu^{\circ}_{298} = 6.29 + 3.70n_C\\
&\text{For n-dienes:} && \Delta\mu^{\circ}_{298} = 3.78 + 3.60n_C\\
&\text{For aliphatic alcohols:} && \Delta\mu^{\circ}_{298} = -3.49 + 3.44n_C\\
&\text{For aliphatic acids:} && \Delta\mu^{\circ}_{298} = -17.8 + 3.45n_C
\end{aligned} \tag{6.8}$$

where n_C is the number of carbon atoms in the hydrocarbon chain, including the terminal methyl group. The last relationship is for the transfer of acids from n-heptane solution to water and is based on the use of (6.3) for the acid in both phases. The advantage of these expressions over (6.7) is that they show clearly the almost constant increment produced by the addition of each methylene group, and also separate $\Delta\mu^{\circ}$ in alcohols and acids into a positive contribution from the hydrophobic moiety and a negative contribution from the hydrophilic head groups.

2. For small aliphatic hydrocarbons the enthalpy of transfer to water from either the pure liquid or an organic solvent is negative, rising with the addition of methylene groups and becoming approximately zero at C_6H_{14}. For aliphatic alcohols ΔH° is also negative. [$\Delta H^{\circ} = \overline{H}^{\circ}_W - \overline{H}^{\circ}_H$]

3. The unitary entropy of transfer from the pure liquid aliphatic hydrocarbon to water is negative, as it is for aromatic hydrocarbons and aliphatic alcohols.

4. The change in heat capacity on transferring hydrocarbons and aliphatic alcohols to water is positive and rather large. The heat capacity and the enthalpy of transfer are related, of course:

$$\frac{d(\Delta\overline{H}^{\circ})}{dT} = \Delta\overline{C}^{\circ}_P \tag{6.9}$$

5. Although the data are sparse, the apparent molal volume of apolar molecules or long-chain amphiphiles in water is less than in the pure liquid or in organic solvents. (This statement does not invariably apply in concentrated aqueous solutions or at pressures above ~1 kilobar.)

The only unambiguous conclusion is that the negligible solubility of hydrocarbons and the low solubility of long-chain amphiphiles is to be blamed largely on entropic, not energetic, causes. The molecular basis of the large negative unitary entropy of transfer has not been firmly established. The most popular qualitative explanation is in terms of an increase in ordering of those molecules of water that are in the vicinity of the hydrophobic portion of the amphiphile. (Frank and Evans[69] proposed that a "microscopic iceberg" forms around nonpolar molecules and their poetic license has been sometimes taken literally in a way to which they certainly would have objected.) The "frozen" water molecules will have a higher molal entropy than fluid water. Also, since at the hydrocarbon–water surface only weak dispersion forces restrain these molecules, they will, on heating, absorb energy into rotational and translational modes of motion more easily than bulk water — thus accounting for the observed increase in heat capacity. A difficulty accompanying the presumption of a truly icelike structure is the famous increase in volume of water (compared with the room temperature value) on solidifying. This does not fit in with the decrease in apparent molal volume observed when apolar molecules, and incidentally rare gas atoms, enter water. Geometric explanations have been constructed, mostly on the bases of Némethy and Scheraga's suggestion[72] that the nonpolar surfaces of hydrocarbons fit into the open structure of liquid water, thus occupying otherwise empty inter-molecular space. There have been, and undoubtedly will be, variations on this theme, although others have abandoned comparatively simple pictures and sought enlightenment on the peaks of almost a priori calculations.[73] The verdict at present must remain open; we have no convincing molecular explanation of what is commonly called the "hydrophobic effect."

Whatever its causes, the hydrophobic effect literally has vital consequences. At extremely low concentrations of amphiphile translational entropy ensures dispersion, but above very small threshold values phospholipids will form aggregates to lower the total free energy. The unwillingness of such aggregates to break up in water has been attributed in part to a "hydrophobic force." Along with many others, this author feels that unless this expression is employed merely as a convenient label for the packet of energetic and entropic factors that make $\Delta\mu^\circ$ large and positive, its use is a backward step since it suggests the existence of a force additional to those known to physics. (A scathing critique of the concept of hydrophobic force may be found in Joel Hildebrand's paper in the *Journal of Physical Chemistry.*[74])

The hydrophobic effect is normally measured on macroscopic samples of organic and aqueous liquids, but the same factors that discourage the mixing of hydrocarbons and water are certainly relevant to the stability of phospholipid aggregates in water, and in particular to the integrity of bilayers. The essential role of the hydrophobic effect in keeping membranes intact is illustrated by the ease with which artificial or natural bilayers are disintegrated by the addition of organic solvents; try adding a few drops of acetone to a blood sample, for example. The hydrophobic effect's contribution to the stability of a bilayer in water can be imagined as equivalent to a lateral pressure squeezing the phospholipids together and thus preventing the hydrocarbon chains from coming into contact with water.

LIPID MIXTURES

Natural membranes invariably contain more than one phospholipid, and since part of the membrane may be in the gel state at physiologic temperatures, the phase diagrams of lipid mixtures are of more than passing interest. The degree of nonideality of lipid mixtures in the liquid crystal state is also a subject that is amenable to thermodynamic analysis and of importance in determining the lateral organization of bilayers. In this book we confine ourselves to binary mixtures because far more, and more accurate, work has been carried out on these systems than on multilipid preparations. It is possible, as we will demonstrate, to derive theoretic expressions for the curves delineating the borders of the phases in the phase diagram and thereby help to confirm or demolish hypotheses as to the nature of phase transitions. No attempt is made to detail the experimental determination of phase diagrams; suffice it to say that the development of highly sensitive calorimeters has placed other techniques at a disadvantage when it comes to the accurate simultaneous determination of both transition temperatures and transition enthalpies. We start by constructing the theoretic phase diagram for a binary mixture. We treat the case of an ideal system and then describe two types of nonideal systems that appear to be excellent models for many real mixtures.

In general, for a binary system the mole fractions obey the relations:

$$\begin{aligned} x_A^s + x_B^s &= 1 \\ x_A^l + x_B^l &= 1 \end{aligned} \tag{6.10}$$

where the subscripts A and B refer to the different species and l and s to liquid and solid. The chemical potentials, neglecting pressure–volume terms, are written as

$$\begin{aligned} \mu_A^l &= (\mu_A^l)^\circ + RT \ln \gamma_A^l x_A^l = (\mu_A^l)_{\text{ideal}} + RT \ln \gamma_A^l = (\mu_A^l)_{\text{ideal}} + (\mu_A^l)^e \\ \mu_A^s &= (\mu_A^s)^\circ + RT \ln \gamma_A^s x_A^s = (\mu_A^s)_{\text{ideal}} + RT \ln \gamma_A^s = (\mu_A^s)_{\text{ideal}} + (\mu_A^s)^e \end{aligned} \tag{6.11}$$

with two corresponding equations for component B and where $\mu_{\text{ideal}} = \mu^\circ + RT \ln x$ and the terms of the form $RT \ln \gamma$ are termed *excess chemical potentials,* μ^e. In the case of ideal systems we put $RT \ln \gamma = 0$. A commonly occurring type of binary mixture is one in which there is complete or nearly complete immiscibility in the solid phase. This considerably simplifies the theoretic treatment since in the solid phase we simply write μ_A^s and μ_B^s as $(\mu_A^s)^\circ$ and $(\mu_B^s)^\circ$, the standard chemical potentials of the pure solids. This is possible because in immiscible mixtures the components exist, at worst, as micro-domains that have such a large ratio of volume to surface that we can ignore A–B interactions at the borders of unlike domains.

An example of a binary lipid mixture that is not absurdly remote from ideality is a mixture of two phospholipids with the same head group and similar chain length, for example, DMPC (C_{14}) and DPPC (C_{16}). These molecules are so similar that we could hope that in both the liquid and solid phase they will be completely miscible and form almost ideal solutions. At thermo-dynamic equilibrium:

$$\begin{aligned} \mu_A^s &= \mu_A^l \\ \mu_B^s &= \mu_B^l \end{aligned} \tag{6.12}$$

Substituting from equations 6.11 we have, for ideal solutions,

$$\ln (x_A^l/x_A^s) = [(\mu_A^s)^\circ - (\mu_A^l)^\circ]/RT \tag{6.13}$$

with a corresponding equality for component B. Differentiating with respect to temperature:

$$\frac{\partial \ln (x_A^l/x_A^s)}{\partial T} = \left[\frac{\partial}{\partial T}\frac{(\mu_A^s)^\circ}{T} - \frac{\partial}{\partial T}\frac{(\mu_A^l)^\circ}{T}\right]/R \tag{6.14}$$

Now $\mu^\circ/T = H^\circ/T - S^\circ$,

$$\frac{\partial}{\partial T}\left(\frac{\mu^\circ}{T}\right) = -\frac{H^\circ}{T^2} \tag{6.15}$$

From (6.14) and (6.15):

$$\ln (x_A^l/x_A^s) = \int_{T_A}^{T} \frac{\Delta H_A^\circ(T)}{RT^2}\, dT \tag{6.16}$$

where $\Delta H_A^\circ(T) = (H_A^l)^\circ(T) - (H_A^s)^\circ(T)$, the enthalpies being in theory functions of temperature. T_A is the transition temperature of component A. We now rely on the slight temperature dependence of ΔH° and write

$$\ln (x_A^s/x_A^l) = \frac{(\Delta H_A^\circ)_{T_A}}{R}\left(\frac{1}{T} - \frac{1}{T_A}\right) \tag{16.17}$$

a familiar type of equation. A similar expression holds for component

B. Rearranging:

$$x_A^s/x_A^l = \exp\left[\frac{(\Delta H_A^\circ)_{T_A}}{R}\left(\frac{1}{T}-\frac{1}{T_A}\right)\right] = e^A \tag{6.18}$$

and likewise,

$$\frac{x_B^s}{x_B^l} = e^B \tag{6.19}$$

By using (6.10), (6.18), and (6.19) it is but a short step to the final relationships:

$$x_A^l = \frac{1-e^B}{e^A-e^B}; \qquad x_B^l = \frac{1-e^A}{e^B-e^A}$$
$$x_A^s = \frac{e^A-e^Ae^B}{e^A-e^B}; \quad x_B^s = \frac{e^B-e^Ae^B}{e^B-e^A} \tag{6.20}$$

These expressions can be used to construct a phase diagram since they give the mole fractions of both components in the liquid and solid phase as a function of temperature and the transition enthalpies and temperatures of the components. An example is shown in Figure 60. The agreement with experiment is poor and so we abandon ideality and

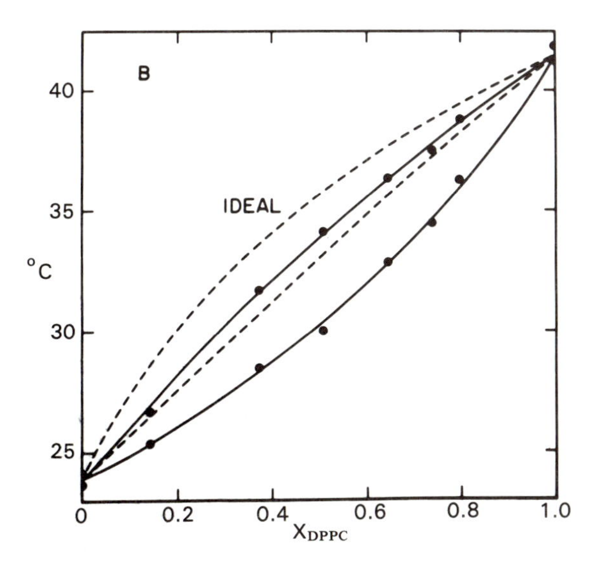

Figure 60. Phase diagram for a DPPC/DMPC mixture. The dots and solid lines are based on experiment; the dashed lines are constructed on the basis of equation (6.20). [From Mabrey, S., and Sturtevant, J. M., *Proc. Nat. Acad. Sci.* 73:3862 (1976).]

return to reality as expressed in equations (6.11). In the absence of a trustworthy means of estimating the activity coefficients, we are pushed into simplifications, two of which dominate recent theory. We note first that ideality implies that:

1. The enthalpy of mixing is zero.
2. The entropy of mixing is purely statistical:

$$\Delta S_{\mathrm{mix}} = -R(x_A \ln x_A + x_B \ln x_B) \tag{6.21}$$

3. It follows from items 1 and 2 that the free energy of mixing is given by

$$\Delta G_{\mathrm{mix}}^{\mathrm{ideal}} = -T\Delta S = RT(x_A \ln x_A + x_B \ln x_B) \tag{6.22}$$

For a real mixture, using (6.11) we have

$$\begin{aligned}\Delta G_{\mathrm{mix}} &= RT(x_A \ln x_A + x_B \ln x_B) + x_A\mu_A^e + x_B\mu_B^e \\ &= \Delta G_{\mathrm{mix}}^{\mathrm{ideal}} + \Delta G_{\mathrm{mix}}^e\end{aligned} \tag{6.23}$$

If the excess free energy is broken up into enthalpy and entropy factors, then two possible approximations are as follows:

1. Put $\Delta S_{\mathrm{mix}}^e = 0$, that is, assume that the entropy of mixing is that for an ideal solution. ΔH_{mix} is zero for an ideal solution but we now allow $\Delta H_{\mathrm{mix}}^e$ to be nonzero. These conditions define a *regular solution.*
2. Put $\Delta \mathrm{H}_{\mathrm{mix}} = \Delta H_{\mathrm{mix}}^e = 0$, but $\Delta S_{\mathrm{mix}}^e \neq 0$, which defines an *athermal solution.* We develop the regular solution equations in detail and then examine briefly the meaning of the althermal model.

The definition of a regular solution allows us to calculate the partial molar entropy of a component since it must be the same as that for an ideal solution. Note that the spatial distribution of the two components must be strictly random in this model. We have

$$\mu_i = \mu_i^{\circ}(T, P) + RT \ln x_i \qquad \text{(ideal)} \tag{6.24}$$

But

$$\mu_i^{\circ}(T, P) = H_i^{\circ} - TS_i^{\circ}$$

so that

$$\mu_i = H_i^{\circ}(T, P) - TS_i^{\circ}(T, P) + RT \ln x_i \qquad \text{(ideal)} \tag{6.25}$$

Here $\mu_i^{\circ}(T, P)$ is the standard chemical potential of i, which is a function of temperature and pressure. (Note that H_i° and S_i° are molar quantities for the pure component.) The partial molar entropy is defined by

$$\bar{S}_i = -\left[\frac{\partial \mu_i}{\partial T}\right]_{P,x} \tag{6.26}$$

From (6.25) and (6.26):

$$\bar{S}_i = -R \ln x_i + S_i^\circ \qquad \text{(regular)} \tag{6.27}$$

For a component of a *real* solution:

$$\mu_i = \bar{H}_i - T\bar{S}_i \tag{6.28}$$

where $\bar{H}_i$ and $\bar{S}_i$ are partial molar properties. From (6.27) and (6.28):

$$\begin{aligned} \mu_i &= \bar{H}_i + RT \ln x_i - TS_i^\circ \\ &= \bar{H}_i + RT \ln x_i + \mu_i^\circ - H_i^\circ \end{aligned} \tag{6.29}$$

Comparing (6.11) with (6.29) we find that

$$\mu_i^e = RT \ln \gamma_i = \bar{H}_i - H_i^\circ \qquad \text{(regular)} \tag{6.30}$$

which merely confirms formally that μ_i in a regular solution differs from its value in an ideal solution only by an energy term.

We now express the excess chemical potential in a power series of the type used by Margules[75]:

$$\mu_A^e = B_1 x_B^2 + C_1 x_B^3 + \ldots \tag{6.31}$$

The coefficients have been given their standard forms. Note that $\mu_A^e \to 0$ as $x_B \to 0$, as it should. We are not now free to choose the form for μ_B^e since the chemical potentials of all components of a solution are interrelated by the Gibbs–Duhem equation:

$$\Sigma\, n_i d\mu_i = 0 \qquad (P,\ T \text{ constant}) \tag{6.32}$$

For a binary mixture one form of the equation is

$$\text{grad } \mu_A = -(x_A/x_B) \text{ grad } \mu_B \tag{6.33}$$

which the reader should be able to show is satisfied if we write

$$\mu_B^e = (B_1 + \tfrac{3}{2}C_1)x_A^2 - C_1 x_A^3 \ldots \tag{6.34}$$

It is common practice to reduce the number of unknowns by cutting the series short and writing

$$\begin{aligned} \mu_A^e &= B_1 x_B^2 \\ \mu_B^e &= B_1 x_A^2 \end{aligned} \tag{6.35}$$

(These equations define *symmetric solutions*.)

Notice that for the enthalpy of mixing we have

$$\begin{aligned}\Delta H_{\text{mix}} &= x_A(\bar{H}_A - H_A^\circ) + x_B(\bar{H}_B - H_B^\circ) \\ &= B_1 x_A x_B^2 + B_1 x_B x_A^2 \\ &= B_1 x_A x_B(x_B + x_A) \\ &= B_1 x_A x_B\end{aligned} \tag{6.36}$$

which gives the basis for an experimental measurement of B_1. We can write ΔH_{mix} in another form using the energies U_{AA}, U_{BB}, and U_{AB}, which are the molar interaction energies of *AA, BB,* and *AB* pairs respectively. ΔH_{mix} is given by the energy loss in dissociating x_A moles of pure A and x_B moles of pure B plus the gain in energy on forming the mixture. The respective terms are given by

$$\begin{aligned}\text{Loss:}\quad & x_A Z U_{AA} + x_B Z U_{BB} \\ \text{Gain:}\quad & x_A Z(x_A U_{AA} + x_B U_{AB}) \\ & + x_B Z(x_A U_{AB} + x_B U_{BB})\end{aligned}$$

where the coordination number Z is assumed to be the same in the pure substances and the mixtures and the two terms in brackets are the average interactions of an A molecule and a B molecule with its neighbors. Rearrangement gives

$$\Delta H_{mix} = Z(2U_{AB} - U_{AA} - U_{BB})x_A x_B \tag{6.37}$$

which on comparison with (6.36) gives

$$B_1 = Z(2U_{AB} - U_{AA} - U_{BB}) \tag{6.38}$$

We stop for breath and recall our initial objective, which is to obtain expressions for the chemical potentials of A and B in a regular solution so that we can obtain expressions such as (6.20) for the mole fractions in the solid and liquid phases. Using (6.11) and (6.35) we have

$$\mu_A^l = (\mu_A^l)^\circ + RT \ln x_A^l + B_1^l(1 - x_A^l)^2 \tag{6.39}$$

where the superscript on B_1 indicates that the enthalpy of mixing need not be the same in solid and liquid. For A in the solid, assuming miscibility,

$$\mu_A^s = (\mu_A^s)^\circ + RT \ln x_A^s + B_1^s(1 - x_A^s)^2 \tag{6.40}$$

where we have eliminated x_B by using $x_A + x_B = 1$. Similar expressions hold for μ_B^l and μ_B^s. The usual equilibrium conditions (6.12) hold, and lead directly to two equations, which we write in terms of x_A. One equation reads:

$$\ln(x_A^s/x_A^l) = \frac{B_1^s(1-x_A^s)^2 - B_1^l(1-x_A^l)^2}{RT} + \frac{(\mu_A^s)^\circ - (\mu_A^l)^\circ}{RT} \tag{6.41}$$

If this equality is subjected to the manipulations that led from (6.13) to (6.17), we obtain, finally,

$$\ln(x_A^s/x_A^l) = \frac{\Delta H_A}{R}\left(\frac{1}{T} - \frac{1}{T_A}\right) + \frac{1}{RT}\left(B_1 \ldots\right) \tag{6.42}$$

The second final equation is

$$\ln\frac{(1-x_A^s)}{(1-x_A^l)} = \frac{\Delta H_B}{R}\left(\frac{1}{T} - \frac{1}{T_B}\right) + \frac{1}{RT}\left(B_1^l x_A^{l\,2} - B_1^s x_A^{s\,2}\right) \tag{6.43}$$

These equations can be solved numerically for any set of values for ΔH_A, ΔH_B, T_A, T_B, B_1^l, and B_1^s. The first four of these parameters are often known from experiments on the pure components. (Cheng[76] has suggested a method for expressing both B_1^l and B_1^s in terms of one parameter only, related to the van der Waal's forces between the molecules.) Phase diagrams constructed on the basis of the above equations are frequently close to the experimental diagrams.[77] For cases in which immiscibility in the solid phase is suspected, we write

$$\mu_A^s = (\mu_A^s)^\circ$$

and

$$\mu_B^s = (\mu_B^s)^\circ$$

and repeat the steps in the derivation of (6.42) and (6.43).[78]

An important physical implication of the athermal model is that the spatial distribution of the two components is not determined purely by chance. For random distribution we have

$$\frac{P_{AA}}{P_{AB}} = \frac{x_A}{x_B} \tag{6.44}$$

where P_{AA} and P_{AB} are the probabilities of an A species being neighbored by A or a B species respectively. This relationship is implicit in (6.37). One way of introducing non-random probabilities is to write

$$\frac{P_{AA}}{P_{AB}} = \frac{x_A}{x_B} \exp\left[-(U_{AB} - U_{AA})/RT\right] \tag{6.45}$$

which expresses the fact that it is the difference in the pair-interaction energies that determines the degree of selective association of molecules with their own kind. We do not follow up the formal consequences of this equation but instead make a number of general points.

1. Both models are internally inconsistent. The regular-solution model relies on differences in pair interactions to produce a nonzero heat

of mixing, but these differences must imply a degree of selective association that negates the premise of random mixing on which the model is based. The athermal-solution model needs different pair interactions to give a non-zero excess entropy of mixing (i.e., nonrandom mixing) but denies these differences their legitimate right to produce a heat of mixing.

2. The expansion of the excess free energy of mixing as a power series is not a consequence of either model. This means that the equations determining the form of the phase diagram are the same in both cases since they both, in a first approximation, adopt the term $B_1 x_A x_B$ for the excess free energy of mixing. In the regular solution formalism B_1 is an enthalpy; see equations (6.30), (6.35), and (6.38). In the athermal model B_1 has a different form, which, however, still relies on differences in pair-interaction energies. The truth is that we expect a binary mixture of almost any molecules to display a heat of mixing *and* nonrandom mixing, with the magnitudes of the two phenomena both being dependent on intermolecular free energies. The lateral separation of lipids has been detected in the gel state of a number of binary mixtures[79,80] but there are only sparse experimental data to support the existence of the same phenomenon in the biologically more interesting liquid crystal state.[81]

In principle the above theory should be applicable to the behavior of protein–lipid systems but the large size and comparative rigidity of protein molecules confer on them the ability to modify the configuration of nonneighboring lipid molecules in ways that depend in a complex manner on the protein/lipid ratio. These matters are best handled by methods that we postpone discussing until Chapter 9.

PHASE TRANSITIONS

We now turn to the nature of the transitions between phases as revealed by thermal measurements. For an infinite sample of a pure defectfree crystalline substance, the thermally induced transition from solid to liquid takes place over an infinitely narrow temperature range. Melting is a cooperative process involving a sudden catastrophic collapse of the whole crystal lattice. The destruction of any microscopic region of the lattice induces instability in the neighboring regions, which consequently break down and thereby create new sources of instability. Perfect infinite lattices appeal to the imagination[82] but can be only approximated in the real world. The melting of substances is in practice carried out over a finite, if sometimes very narrow, temperature range, which increases with decreasing sample size. Statistical thermodynamics can be used to show that the breadth of the transition for a single crystal can be quantitatively related to the number of molecules in the sample, a number which in this context is termed the *cooperative unit.*

The concept of cooperativity may be illustrated by the behavior of a linear assembly of entities each of which can adopt one of two thermally interconvertible states A and B. Consider the two following equilibria, both of which involve chains differing in the conformation of one (marked) unit:

$$\begin{array}{llll} (1) & \ldots\ldots AAA\underset{\uparrow}{B}BB\ldots\ldots & \ldots AA\underset{\uparrow}{B}BBB\ldots & K_1 \\ (2) & \ldots\ldots AAA\underset{\uparrow}{A}BB\ldots\ldots & \ldots AAB\underset{\uparrow}{A}BB\ldots & K_2 \equiv \sigma K_1 \end{array} \tag{6.46}$$

K_1 and K_2 ($\equiv \sigma K_1$) are the respective equilibrium constants.

Note that in (1) the marked entity is flanked by an A and a B state whereas the neighbors of the marked entity in (2) are both in state A. With regard to the magnitude of σ, there are three possibilities:

(a) $\sigma = 1$; $K_1 = K_2$. In this case the interconversions of an entity between states A and B are independent of the states of its neighbors; there is no cooperative effect.

(b) $\sigma < 1$; $K_2 < K_1$. This means that an A state is more likely to change to B if it is flanked by a B state [equilibrium (1)]. We have an example of *positive* cooperativity entirely analogous to melting. The conversion of the marked A to B will "weaken" neighboring A states.

(c) $\sigma > 1$; $K_2 > K_1$. This is a case of negative cooperativity in which states tend to assume a different state from that of their neighbors. Examples occur in spin systems but do not interest us here.

Consider an ensemble of chains each consisting of N_0 identical entities displaying positive cooperation. Each chain is allowed to be in one of two states, which we label "liquid" and "solid." Then we can define the *extent of transition* θ by

$$\begin{aligned} \theta &= \frac{\text{Fraction of chains in liquid state}}{\text{Total number of chains}} \\ &= F_L/(F_L + F_S) \end{aligned} \tag{6.47}$$

Now θ can be assumed to be a linear function of the heat absorbed by the system, and furthermore can be used to define the equilibrium constant $K = F_L/F_S$ since

$$\begin{aligned} K = F_L/F_S &= \frac{F_L}{F_L + F_S - F_L} = \frac{F_L/(F_L + F_S)}{1 - F_L/(F_L + F_S)} \\ &= \frac{\theta}{1 - \theta} \end{aligned} \tag{6.48}$$

Now for a two-state system we can apply the well-known relationship:

$$\frac{d \ln K}{dT} = \frac{\Delta H_{vH}}{RT^2} \qquad (6.49)$$

where ΔH_{vH} is the van't Hoff transition enthalpy *per mole of chains.* From (6.48) and (6.49):

$$\Delta H_{vH} = RT^2 \times \frac{1}{\theta\,(1-\theta)} \times \frac{d\theta}{dT} \qquad (6.50)$$

We now define T_t to be the temperature at the midpoint of the calorimetric transition, which corresponds to $\theta = \frac{1}{2}$. We can then write

$$\Delta H_{vH} = 4RT_t^2 \times \frac{d\theta}{dT} \qquad (6.51)$$

Since the van't Hoff transition enthalpy refers to a mole of chains, it must be divided by N_0, the number of entities in the chain, to obtain the heat of transition of a mol of entities (molecules) from the solid to the liquid phase. In symbols we have

$$N_0 = \Delta H_{vH} / \Delta H_{\mathrm{cal}} \qquad (6.52)$$

where ΔH_{cal} is the calorimetrically measured enthalpy of transition. N_0 is the number of molecules that undergo a single cooperative transition and is therefore the size of the cooperative unit in this system. Equations (6.51) and (6.52) have been used to estimate the size of the cooperative unit in the gel-to-liquid-crystal transition of single lipids,[83, 84] by calorimetric determination of $d\theta/dt$ from measurements of the kind to be seen in Figure 61. The experimental plots are not always as symmetric as those shown in the figure, perhaps because changes in heat capacity during the transition have been neglected in our derivation. The size of the cooperative unit found for lipids is generally in the range of tens to hundreds of molecules. Of course, it is not necessary that θ be determined by thermal measurements and spectroscopic methods are often more convenient in practice.[85] It is interesting that the size of the cooperative unit estimated from a heating curve does not necessarily agree with the value obtained from a cooling curve. Thus for DMPC multilayers N_0 was estimated as 200–450 from the cooling curve and 600–800 from the heating curve.[83] It could be that on cooling small domains are formed that subsequently, on a time scale much longer than that of the experiment, merge to form larger domains.

The differential scanning calorimetry curves for binary lipid mixtures are almost always asymmetric and often appear to be a superposition of two or more peaks. Their complexity, in spite of some early doubts,[86] is explicable in terms of normal macroscopic thermodynamics.[87] At a given temperature the fraction of the system in the solid phase is denoted by $f^{(s)}$. The amount of component A that is in the solid phase is thus equal to $f^{(s)}\, x_A^s$, with a similar expression for component B. For the addition of an infinitesimal quantity of thermal energy we have

$$\delta H = \delta(\Delta H_A f^{(s)} x_A^s + \Delta H_B f^{(s)} x_B^s)$$

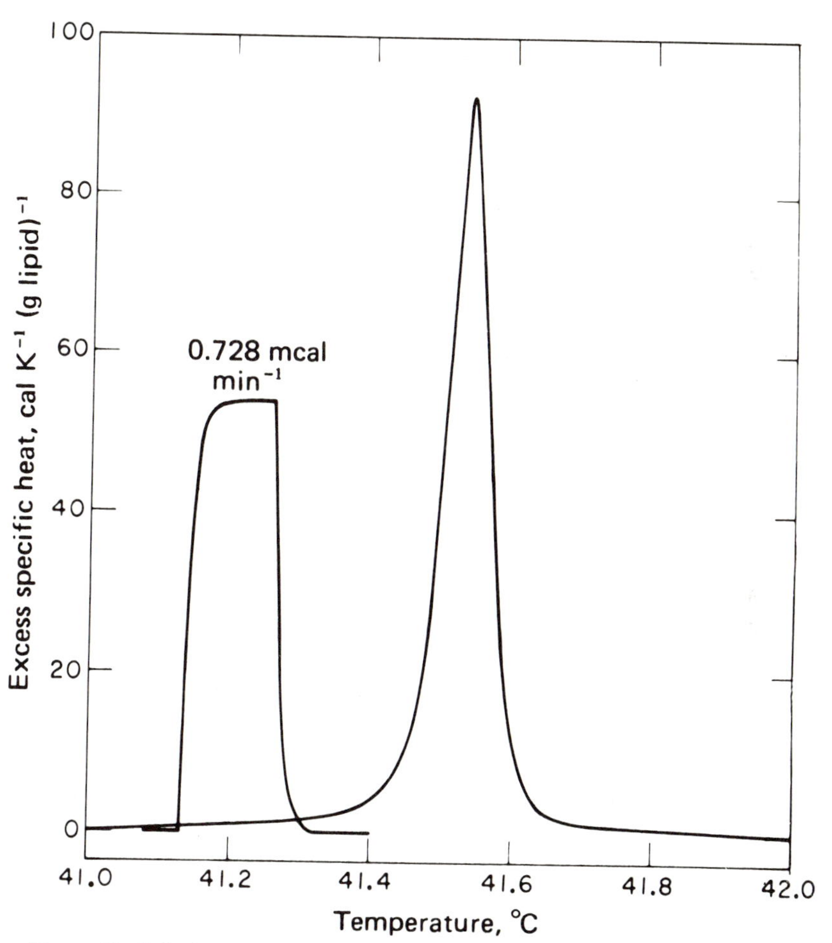

Figure 61. Calorimetric phase transition for DPPC (0.518 mg/g of suspension, scanned at 0.0232 K min^{-1}). The peak maximum is at 41.55°C. Also shown is the calibration mark made by supplying excess heat at a rate of 0.728 mcal min^{-1} to a reference cell containing water. Curves of the kind shown allow the determination of $d\theta/dT$ and hence the size of the cooperative unit involved in melting. [From Albon, N., and Sturtevant, J. M., *Proc. Nat. Acad. Sci.* 75:2258 (1978).]

or

$$\frac{dH}{dT} = C_P = \Delta H_A \frac{d}{dT} f^{(s)} x_A^s + \Delta H_B \frac{d}{dT} f^{(s)} x_B^s \qquad (6.53)$$

This equation is "ideal" in the sense that it ignores the heat that either must be supplied or released when real components are involved in a transition. The unmixing of the components in the solid phase and their mixing when they enter the liquid phase require lateral diffusion processes (Chapter 10) that are too slow to permit the measurement of mixing heats — at least at present. The differentiation in (6.53) gives an expression for C_P that contains $f^{(s)}$, ΔH_A, ΔH_B, the mole fractions of A and B in the solid and liquid phases, and the derivatives dx_B^s/dT and dx_B^l/dT. The derivatives can be obtained directly from the phase diagram, since this is a temperature–composition plot. The other quantities are also experimentally determinable and so an expression is obtained for C_P as a function of the composition of the mixture. Since composition is known as a function of temperature, we end up with the ability to construct curves of excess heat capacity against temperature. Examples are shown in Figure 62. From these and other plots[87] it can be confirmed that the main features of the experimental curves are fairly well reproduced by theory. Nonideality may account for the less than perfect agreement and pretransitional structural changes may explain the small peak seen at the left-hand side of Figure 62.

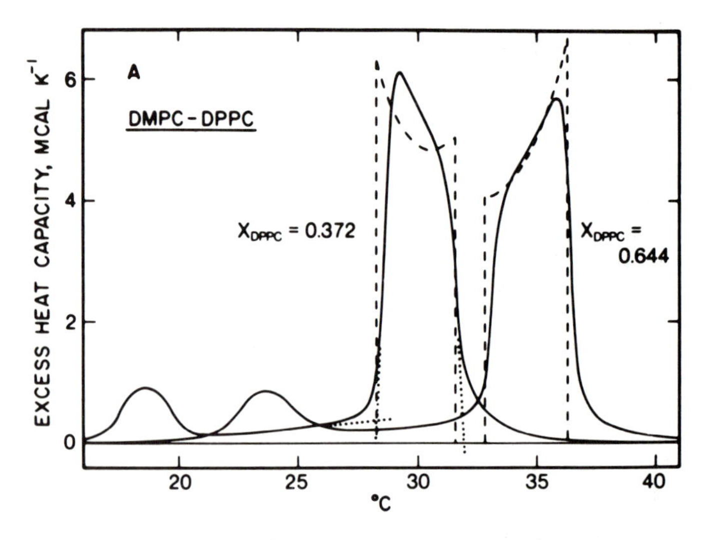

Figure 62. Excess heat capacity curves for two DPPC/DMPC mixtures. X = mole fraction. Solid lines: Experimental. Dashed lines: Theoretical curves based on equation (6.53). [From Mabrey, S., and Sturtevant, J. M., *Proc. Nat. Acad. Sci.* 73:3862 (1976).]

The experiments on which the above plots are based were carried out at very slow calorimetric scanning rates to ensure that the systems were at thermodynamic equilibrium at all times. Of course, the macroscopic equilibrium covers a microscopic dynamic equilibrium between liquid and solid domains. Clusters of molecules forming solid rafts will continually exchange molecules with surrounding liquid, that is, liquid crystal domains. The size of a given solid domain will fluctuate, having a tendency to increase because of the positive boundary free energy discussed earlier in the chapter, and a tendency to decrease so as to maximize the translational entropy. At a macroscopic level of observation, the relative amounts of solid and liquid remain constant. However, it should be realized that even at the microscopic level, each cluster is at thermodynamic equilibrium in the sense that the average chemical potential of a molecule in any cluster is the same as that of a molecule in the liquid. This is in sharp contrast to the fluctuations in density that occur in a lipid as the transition temperature is approached. In this case there are clusters of molecules with average free energies falling above or below the macroscopic average for short periods of time.

AGGREGATION

The forms adopted by aggregated lipids in aqueous media were rationalized in Chapter 4 on the basis of geometric considerations. However, the statement that a certain molecule prefers to pack in a certain way as a result of spatial factors is really a statement about inter- and intramolecular free energy and we now examine aggregation from the point of view of thermodynamics, proceeding from a specific case to a brief general discussion.

To illustrate the strategy used in studying aggregation, we consider the formation of micelles in aqueous media as treated by Ruckenstein and Nagarajan.[88] In an aqueous solution of micelle-forming amphiphiles we expect to find a distribution of aggregates ranging in size from one molecule (the monomer) upward. The number of aggregates containing g molecules each is denoted by N_g, the number of monomers by N_A, and the number of solvent molecules by N_S. The total number of "molecules" in the system is given by

$$F = N_S + N_A + \sum_{g=2}^{\infty} N_g \tag{6.54}$$

while the total number of amphiphile molecules is

$$N = N_A + \sum_{g=2}^{\infty} gN_g \tag{6.55}$$

Notice that in (6.54) we have treated each aggregate as a molecule. The standard chemical potential of a molecule of solvent is written as μ_S and that of a monomer in solution as μ_A. For an amphiphile in an aggregate, a part of its standard chemical potential is supposed to be independent of the size of the aggregate. This contribution, μ_B, can be estimated from equations 6.8, which give the difference between the chemical potential of a monomer in aqueous solution and in its own bulk phase. The hydrocarbon chains in the interior of a micelle are effectively in a bulk hydrocarbon phase but the expressions for $\mu_B - \mu_A$ have to be modified to account for the fact that in micelle formation the hydrophilic head groups remain in contact with the aqueous media. We leave the reader to consult the literature[70, 88, 89] for the details of the estimation of chemical potentials, an understanding of which is not essential to our argument.

The size-dependent part of the standard chemical potential of an aggregate of size g is denoted by μ_g, that is, μ_g/g per monomer. In micelles this term will depend on the radius, or radii, of curvature that affect both the distance between the head groups and the degree of penetration of water into the hydrophobic core. Approximate expressions for both of these contributions are given by Tanford.[89]

For an aggregate of size g the mole fraction of aggregates is defined as the number of aggregates divided by the total number of independent species in the solution, that is, by N_g/F. The free energy per mol of aggregate, assuming ideality, is

$$\mu(N_g) = \mu(N_g) + RT \ln (N_g/F) \tag{6.56}$$

so that the corresponding quantity per individual aggregate is

$$\mu(N_g)/N = \mu(N_g)/N + kT \ln (N_g/F) \tag{6.57}$$

where N is Avagadro's number. Finally, for each amphiphile in an aggregate the chemical potential is given by

$$\mu(N_g)/gN = \mu^{\circ}(N_g)/gN + (kT/g) \ln (N_g/F) \tag{6.58}$$

where $\mu^{\circ}(N_g)/gN$, the standard chemical potential per amphiphile in an aggregate of size g, is $\mu_B + \mu_g/g$. The total chemical potential of the amphiphiles in aggregates of size g is thus gN_g times (6.58). It follows that the total free energy of the system is

$$\begin{aligned} G = N_S\mu_S + kTN_S \ln (N_S/F) + N_A\mu_A + kTN_A \ln (N_A/F) \\ + \sum_{g=2}^{\infty} N_g(\mu_B g + \mu_g) + kT\sum_{g=2}^{\infty} N_g \ln (N_g/F) \end{aligned} \tag{6.59}$$

At thermodynamic equilibrium all amphiphiles have the same chemical potential, so that we can equate the potentials of a monomer in solution and a monomer in an aggregate:

$$\mu_B + \mu_g/g + (kT/g) \ln (N_g/F) = \mu_A + kT \ln (N_A/F) \qquad (6.60)$$

which can be shuffled around to give

$$N_g/F = (N_A/F)^g \exp\left[- \left(\frac{1}{kT} [(\mu_B - \mu_A)g + \mu_g]\right)\right] \qquad (6.61)$$

If we write ξ for

$$\xi = (N_A/F) \exp [- (\mu_B - \mu_A)/kT] \qquad (6.62)$$

then the size distribution function can be expressed as

$$N_g/F = \xi^g \exp[- (\mu_g/kT)] \qquad (6.63)$$

The distribution function has thus been factored into size-dependent and size-independent parts. If ξ is small then ξ^g grows too slowly with g to fight the fall in N_g/F due to the exponential function. If ξ is large enough, then there are two maxima in the distribution (Figure 63). The exact form of the function depends on the values assigned to μ_A, μ_B, and μ_g but a general result is that it requires only a small increase in N_A/F at a certain "critical" concentration to produce a huge increase in the ratio of aggregated to nonaggregated amphiphiles. Experimentally it has been known for many years that when amphiphiles are dissolved in aqueous media, there is a so-called critical micelle concentration (CMC) below which the amphiphile is overwhelmingly in the monomer form and above which added monomers appear almost exclusively in aggregates. If note is taken of the logarithmic scale in Figure 63, it will be seen that the size distribution function is strongly peaked for high values of ξ that is, of N_A/F. This is in agreement with the experimental finding that the great majority of micelles in a given solution have aggregation numbers falling within a narrow range of values. Similarly the plots for low ξ confirm the finding that at low amphiphile concentration the majority of molecules are present as monomers and the concentration of aggregates of a given g falls off monotonically with increasing g.

The above treatment of micelles emphasizes the role played by the size dependence of the standard chemical potential. We now show that this emphasis is wholly justified.

The process of aggregation relies on attractive forces between molecules. Translational entropy always operates in favor of the disintegration of molecular clusters. However, it does not follow that large aggregates are encouraged by strong forces. If the average energy of a molecule in an aggregate is independent of the aggregate size, then there is no energetic advantage in adding a molecule to a large aggregate

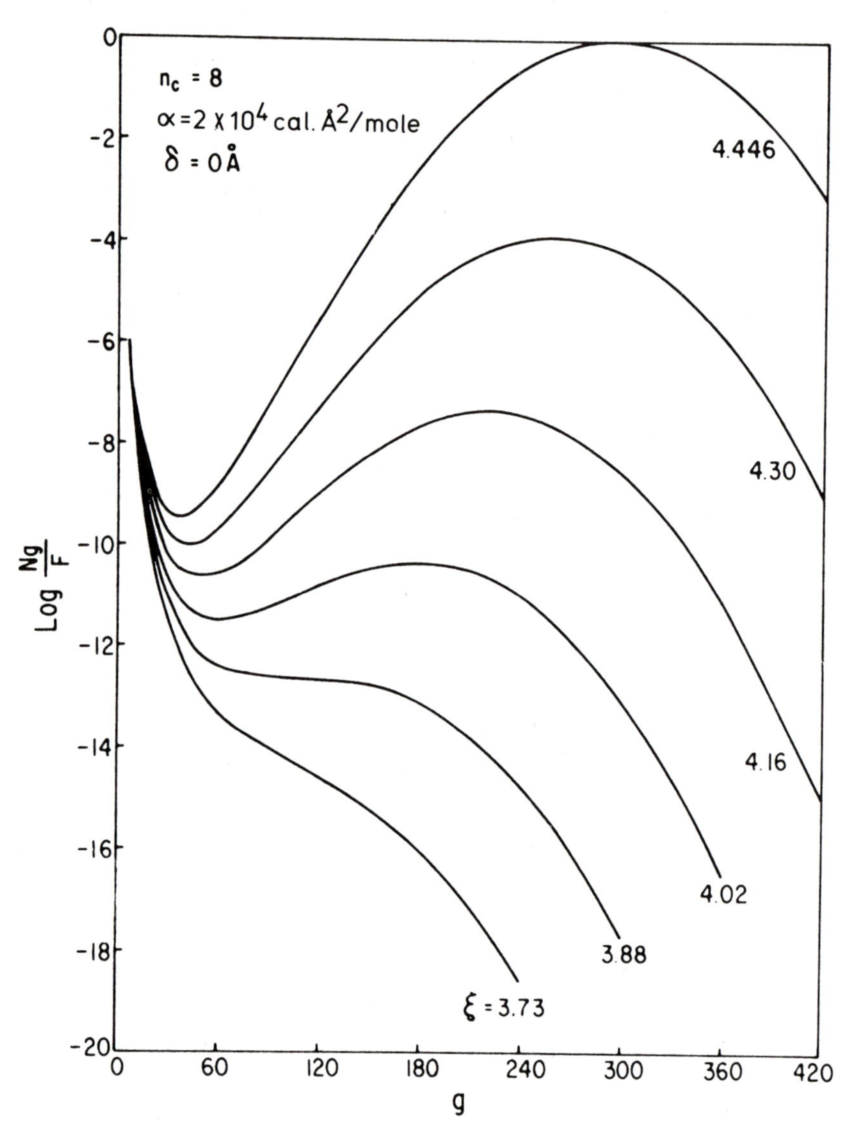

Figure 63. Dependence on g of the mole fraction of aggregates of size g for different values of the parameter ξ defined by equation (6.62). In performing the calculations, it was assumed that the surface area of the micelles is equal to that of the hydrocarbon core; that is, the "roughness" of the head groups is ignored. [From Ruckenstein, E., and Nagarayan, R., *J. Phys. Chem.* 79:2622 (1975).]

rather than to a small one, no matter how strong the forces involved, and entropy will favor small clusters. For large aggregates to predominate, we need a size-dependent average chemical potential. We consider two common general expressions for the standard chemical potential in an aggregate.[31a] We take a clue from an extreme case of aggregation, the formation of a macroscopic cube-shaped crystal in solution. The chemical potential, not including translational entropy, of the crystal depends on the potential of the bulk molecules, $\mathring{\mu}_\infty$, and the potential of the surface molecules:

$$\mathring{\mu}_{\text{crystal}} = N\mathring{\mu}_\infty + A\gamma \tag{6.64}$$

where A is the area of the crystal surface and γ the interfacial energy per unit area. Edge effects are neglected. The average potential per molecule thus is

$$\mathring{\mu}_N = \mathring{\mu}_\infty + A\gamma/N \tag{6.65}$$

But A is proportional to $N^{2/3}$, so that

$$\mathring{\mu}_N = \mathring{\mu}_\infty + C\gamma/N^{1/3} \tag{6.66}$$

which is usually written as

$$\mathring{\mu}_N = \mathring{\mu}_\infty + \alpha kT/N^{1/3} \tag{6.67}$$

and obviously applies equally well to the growth of a sphere. The size dependence of the average molecular potential here lies in the $N^{-1/3}$ factor. Similarly, and closer to our interests, the average chemical potential for a crystal growing in two dimensions can be written

$$\mathring{\mu}_N = \mathring{\mu}_\infty + \alpha kT/N^{1/2} \tag{6.68}$$

These two examples suggest the general form[90]:

$$\mathring{\mu}_N = \mathring{\mu}_\infty + \alpha kT/N^P \tag{6.69}$$

which gives an asymptotic decrease of the average potential toward its value in an infinitely large sample. A special case of (6.69) is that of linear growth. The addition of a molecule to the end of an existing linear chain leaves the amount of end surface unchanged so that the total potential is given by

$$\begin{aligned} N\mu^\circ &= (N-2)\mathring{\mu}_\infty + C' \\ &= N\mathring{\mu}_\infty + C \\ \mathring{\mu}_N &= \mathring{\mu}_\infty + C/N = \mathring{\mu}_\infty + \alpha kT/N \end{aligned} \tag{6.70}$$

Here $P = 1$ in (6.69). For $P > 1$, small aggregates are favored,[31a] as can be demonstrated formally and appreciated intuitively from the following physical argument. If $P > 1$, $\mathring{\mu}_N$ decreases rapidly with increasing N and at small values of N, $\mathring{\mu}_N$ is so near to $\mathring{\mu}_\infty$ that further additions of monomers to the aggregate are accompanied by a negligible lowering of $\mathring{\mu}_N$. Now the translational entropy of an aggregate molecule is given by the product of kT/N and a logarithmic factor that changes far more slowly with N. Thus the addition of a molecule to an aggregate results in an entropy decrease that behaves approximately as kT/N, and this decrease can outweigh the small increase in $\mathring{\mu}_N$.

While the above formalism accommodates the growth of layered structures, it fails to account for the formation of solutions of vesicles or micelles having the narrow range of aggregation number almost universally observed in experimental studies. The peaked distribution function for aggregate numbers, however, can be simulated if the average standard chemical potential of a molecule in an aggregate reaches a minimum at a finite value of N, as it does in the simple quadratic dependence given by

$$\mathring{\mu}_N - \mathring{\mu}_M = \Lambda(N - M)^2 \tag{6.71}$$

where $\mathring{\mu}_M$ is the minimum value of $\mathring{\mu}_N$, occurring at $N = M$. The greater the value of Λ, the sharper is the parabola described by (6.71) and the narrower is the size distribution of aggregates. There is more to recommend (6.71) than mathematical suitability. For those lipids with "shapes" compatible with the formation of closed structures, there are physical factors that set rough upper and lower limits on radii of curvature. With regard to lower limits, a strong bending of a bilayer will open up the hydrocarbon core to penetration by water, a process resisted by the misnamed hydrophobic force. Perhaps a more acceptable way of dealing with this resistance to bending is to treat it as the result of a surface tension acting at the water–hydrocarbon interface, it being assumed that the chains are liquid. This contribution to $\mathring{\mu}_N$ therefore can be written as an area-dependent term, γA. There are, of course, intermolecular forces, — electrostatic, dispersion, and repulsion (see the following chapter). A very general way of treating these is to write a two-dimensional analog of the familiar three-dimensional virial expansion for the energy of a real gas, $PV = 1 + B_1/V + B_2/V^2 + \ldots$. Confining ourselves to the first virial coefficient, we add a term C/A to the expression for $\mathring{\mu}_N$;

$$\mathring{\mu}_N = \gamma A + C/A \tag{6.72}$$

This function has a minimum when $A = \sqrt{C/\gamma} \equiv A_0$, the equilibrium value of the area of a lipid molecule in the plane defined by the hydrocarbon–water interface. Rewriting (6.72) in terms of A_0 gives

$$\mathring{\mu}_N = 2\gamma A_0 + \frac{\gamma}{A}(A - A_0)^2 \tag{6.73}$$

Thus a very general consideration of the forces within a bilayer lead to a parabolic dependence of standard chemical potential on surface area. The bending of a bilayer involves the stretching of one monolayer and the compression of the other, with both processes involving an increase in energy. In vesicles the situation is complicated by the fact that the area per head group already is probably smaller for the inner than for the outer surface. Nevertheless, we are encouraged to look for a parabolic dependence of potential on aggregation number by the comparative size homogeneity of vesicle preparations. We do not press this search here. Any realistic analysis must take into account the fact that the planes in which the attractive and repulsive forces in the bilayer act are probably not coincident. The resultant development is straightforward but a little too specialized for us.[91]

We have seen how some of the more obvious aspects of aggregation can be decked out in thermodynamic finery. As usual, thermodynamics helps us to select feasible forms for chemical potentials but gives us little help in defining the origin of forces. This is the subject of the following chapter.

One final word is in order on the subject of vesicles. Preparations of vesicles appear to have limited stability, even in solutions from which lipid-loving bacteria have been rigorously excluded. This casts doubts on our right to treat vesicles in terms of normal equilibrium thermodynamics. Thus comparisons of, say, enthalpies of transition for the melting of chains in bilayers and vesicles prepared from the same lipid may not be particularly meaningful if vesicles have to be regarded as metastable forms with short half-lives. It also has been pointed out that the preparation of vesicles normally involves the input of large amounts of energy into the system, for example, by sonication. Only in this way are highly stressed small vesicles formed "spontaneously" in solutions of lipids that otherwise, would contain other aggregated forms. The strong and nonuniform stress is considered[92] to throw a shadow on the suitability of vesicles as model systems for protein–lipid interactions.

7 Forces

Since membranes are constructed of molecules and belief in vital forces is unfashionable, it can be safely assumed that the forces within membranes are no different from the familiar electrostatic and van der Waals forces used to account for the properties of all molecular ensembles. If this is so, the devotion of a complete chapter to intermolecular forces appears to be an indulgence. The author's justification is twofold. First, the literature on membranes contains concepts that are unfamiliar to the average student — hydrophobic forces, image forces, structural forces, and so forth. Second, the full implications of some familiar terms (e.g., the van der Waals force) are widely misunderstood. In this chapter we will attempt to present the current viewpoint on these and other forces. Those fed on the neat certainties of undergraduate textbooks may be pleasantly surprised, and possibly challenged by the dynamic nature of the theory of intermolecular interactions. A comforting note: Principles will be stressed; mathematics will be kept in the wings.

BORN ENERGY

The passage of a charged molecule from an aqueous environment into a membrane can be modeled, in a hugely oversimplified approach, as the transfer of a charged sphere from a homogeneous medium of dielectric constant ($\epsilon_1 \sim 80$) to a homogeneous slab of hydrocarbon with dielectric constant $\epsilon_2 \sim 2$. This transfer involves an expenditure of energy that is electrostatic in origin and can be estimated as follows. Consider a sphere of radius r in a vacuum and carrying a charge of q. The potential at the surface of the sphere is q/r and the work needed to increase the charge by an infinitesimal amount dq is thus qdq/r, assuming that dq is brought from infinity. The total work involved in charging an initially uncharged sphere is, therefore,

$$w_{\text{charge}} = \int_0^Q \frac{qdq}{r} = \frac{Q^2}{2r} \qquad \text{(vacuum)} \qquad (7.1)$$

If the sphere is in a homogeneous dielectric of dielectric constant ϵ, the potential due to the charged sphere is reduced at all points in space by a factor of ϵ so that we have

$$w_{\text{charge}} = \frac{Q^2}{2\epsilon r} \qquad \text{(dielectric)} \qquad (7.2)$$

The work required for charging the sphere may be regarded as a kind of self-energy of which the value depends on the dielectric constant of the medium. Although we brought dq from infinity, we can compromise with reality and reduce the range of integration to tens of angstroms without much affecting the final result. The difference in w_{charge} for $\epsilon = 80$ and $\epsilon = 2$ is the *Born energy* of the sphere in our hydrocarbon slab. For a sodium ion, using $r = 0.95$ we find $0.35\ \text{J mol}^{-1}$, a result that is easily obtained and highly unreliable. There are several objections to our simple treatment. For the specific case of the sodium ion we have forgotten, in choosing the ionic radius, that in aqueous solution the charge of the ion is partly smeared over its hydration shell. The ion inside the membrane either is enclosed by a large organic carrier molecule (Chapter 14) or resides in a transmembrane pore. In either case the effective radius of the ion is not that obtaining in water. A more general fault is our complete neglect of the molecular structure of the aqueous and hydrocarbon media. In the simplest case the medium will be polarized by the charge and it will be necessary to take into account charge-induced dipole interactions. Furthermore, in aqueous solution the distribution of ions is affected by the presence of a charge and, therefore the dielectric constant of the solvent is a function of the distance from the charge. Closer inspection reveals that we have neglected charge–quadrupole interactions and the effect of a charge on the local structure of water. It is at present not practically possible to determine all these contributions and thus, although Born energies certainly contribute to the barrier facing ions attempting to penetrate a membrane, their calculated values are not to be taken too seriously.

For those troubled by the quantitative reliability of oversimplified models there are a variety of more complex systems that have been tackled by the theoreticians. Chan et al.[93] have considered a solvent of dielectric constant ϵ composed of spheres of radius R_2 and dipole moment μ, containing a solute of spheres of radius R_1 and charge ze. They obtain

$$F_{\text{Born}} = -\frac{(ze)^2\left(1-\frac{1}{\epsilon}\right)}{2\left(\frac{1}{2}R_1 + R_s\right)} \tag{7.3}$$

where $R_s = \left(\frac{1}{2} - \frac{3\xi}{1+4\xi}\right) R_2$ and $\xi\left(0 < \xi < \frac{1}{2}\right)$

is a calculated constant depending on temperature, the density of the solvent, and μ. This form for F_{Born} has been used empirically in the past, R_s being chosen to fit experimental data. Note that for $R_2 \ll R_1$, i.e., a small solvent) and for $\epsilon \gg 1$, (e.g., for water) $F_{\text{Born}} \longrightarrow -(ze)^2/(R_1)$, the classic result. Also, for later reference, we note that a charged ion, say Na^+, which wraps itself round with a large organic chelate molecule, increases its effective radius and reduces its Born energy, thus easing its entry into the hydrophobic core of a bilayer.

IMAGE FORCES

A better heading for this section would be "The Method of Images," but the term *image force* is probably too well established to topple. The method has been used extensively in the calculation of forces between molecules and surfaces and the forces between macroscopic bodies. We go straight to a simple example, the interaction between a positive charge, $+Q$, in a vacuum and an infinite metal plane separated from each other by a perpendicular distance D. An equal and opposite negative charge is induced on the plate (as Faraday knew) and attracts the point charge. The surface of the metal is an equipotential as it must be, otherwise the induced charges would move so that the lines of force emanating from the surface are normal to it everywhere. From the point of view of the point charge, the electric field that it "sees" might just as well emanate not from the surface, but from a charge $-Q$ located at a distance D behind the surface, and the force on the point charge is thus $-Q^2/(2D)^2$. In place of our qualitative argument the reader can look at reference 94. We accept the result and the idea of an image charge and without proof present another result, that is of greater relevance to us. A point charge in a medium of dielectric constant ϵ_3 and located at a distance D from the planar surface of a medium of dielectric constant ϵ_2 "sees" an image charge given by

$$rQ = -\frac{(\epsilon_2 - \epsilon_3)}{(\epsilon_2 + \epsilon_3)} Q \qquad (7.4)$$

where the image charge is situated, as in our first example, at a distance D behind the boundary between the media. The fact that we have specified our medium to be a dielectric means that the source of the image charge is the *polarization* induced by the point charge. (Note that the electric field outside a polarized medium is exactly the same as that produced by a surface charge numerically equal to the volume polarization.[95]) Returning to the expression for r, we consider three possible cases.

1. $\epsilon_2 > \epsilon_3$, making r negative and resulting in an attractive force between the point charge and the boundary. Thus for an ion located on the hydrocarbon side of a planar hydrocarbon–water boundary, $\epsilon_2 \simeq 80$, $\epsilon_3 \simeq 2$, and $r = -78/82$. If we wish to use this result for the case of an ion in the hydrocarbon core of a membrane, we must first realize that our model neglects the polar head groups of the membrane and the possible existence of ions in the aqueous milieu. Moreover, the value of r is not correct for a charge situated between two boundaries. The polarization induced by the charge at a given boundary creates fields felt at the opposite boundary. This problem also can be handled neatly by the method of images, but we leave the mathematical details to the curious reader and go straight to an interesting numeric result given by Parsegian,[96] who estimates that for a single electronic charge 10 Å away from one interface of a hydrocarbon slab 50 Å wide the pull on the ion is about 3×10^{-6} dynes. Newton then assures us that the pull on the surface is equal and opposite, and it is estimated that the stress on the interface is over 200 atmospheres (!) at the point nearest the ion. Furthermore, the local electric field in the same region reaches an impressive 14×10^6 volts cm^{-1}. For a symmetric system the force on a point charge must vanish at the center of the slab so that the potential energy of the charge due to its "image" has a maximum at this point. *The passage of a charged ion across our model system involves the surmounting of a potential barrier,* and in bilayers the same barrier exists, its height influenced by the detailed molecular structure of the membrane and the aqueous solution.

It is apparent that polarization has an important role in the passage of ions across membranes, and though the method of images has given its name to these effects, the jargon should not be allowed to obscure the physics.

2. $\epsilon_2 < \epsilon_3$, making r positive, resulting in an image charge of the same sign as the point charge and a repulsive force between the point charge and the surface. This case applies to an ion in water outside a membrane and suggests that the image force will strongly discourage the approach of ions. Since the Born energy also acts against the presence of ions in the hydrocarbon core of membranes, things do not look too good for ion transport, and indeed doubly charged ions that have large Born energies and image charges have great difficulty in crossing lipid bilayers. Transport through membranes, however, often is mediated by carriers (Chapter 14) or transport proteins with narrow water-filled channels.

The idea of a charge inducing an image charge of the same sign worries some people, who feel that it contradicts Faraday. The physical rationale is roughly as follows. A charge, say positive, in an infinite dielectric induces dipoles about it, and if the medium contains polar molecules, also orders these molecules. A time-average picture of this situation, smoothing out thermally caused fluctuations, is shown in Figure 64. The forces exerted on the ion by the dipoles must be symmetric if the medium, as we suppose, is isotropic. Thus the net forces along the positive and negative x-axes (Figure 64) are equal so that $F_x = 0$. If we now slice through this polarized cake and remove the semiinfinite chunk indicated in the figure, leaving a vacuum, we are in effect removing part of the force acting along the positive x-axis. The net force on the ion along this direction is no longer zero but acts away from

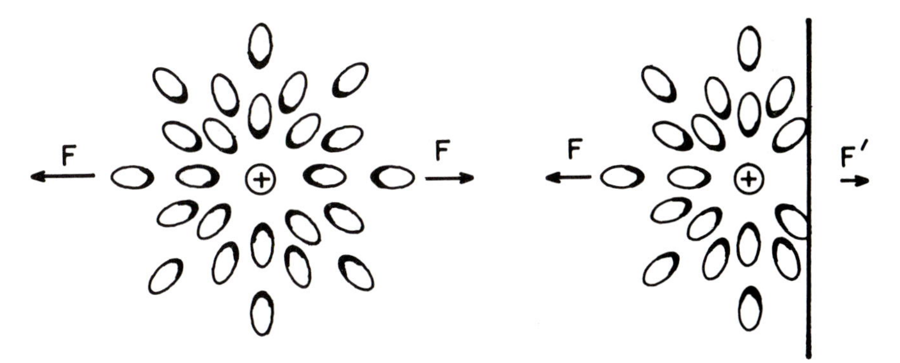

Figure 64. Charge polarizes the surrounding medium by polarizing and orienting neighboring molecules. In an infinite isotropic medium, this interaction produces no net force on the charge. If part of the medium is removed, a net force acts on the charge in a direction normal to and away from the cutting plane. If the lost piece of cake is substituted for by a medium of greater polarizability than that removed, the force will act toward the border.

the boundary. [The reader may care to work out, at least qualitatively, the result of replacing the lost chunk by another of greater dielectric constant, recalling that a larger dielectric constant implies a larger electric susceptibility (polarizability) since

$$\epsilon = 1 + 4\pi\chi_e \tag{7.5}$$

where χ_e, the susceptibility, is given by P/E, the ratio of the polarization to the field in the medium.]

3. $\epsilon_2 = \epsilon_3$, gives $r = 0$ and zero-image force. This should be obvious from the reasoning of the previous paragraph.

Dipoles, since they are nothing but charge distributions, can also induce "image dipoles," by which we mean that the forces between dipoles and boundaries can be treated by the method of images, and the force between a dipole and an interface can be attractive, repulsive, or zero, depending on the dielectric constants of the two media.

The method of images has its limitations. It assumes perfectly smooth, structureless boundaries that are mechanically undisturbed by the perturbing charge. It would be unwise to apply the method to an ion situated within a handful of angstroms from the head groups of a bilayer.

STRUCTURAL FORCES

Many of the equations used in the physical sciences are approximations. In particular, we often treat matter either as a continuum or as a collection of disembodied molecules having no volume. To account for the behavior of real molecules, it is often sufficient to add small correction terms of the kind in the van der Waals equation and more generally typified by the virial coefficients that appear, for example, in the equations for the properties of gases and dilute solutions. These corrections implicitly contain a recognition of the forces that operate between slightly crowded molecules. In liquids, where average intermolecular distances are less than molecular dimensions, intermolecular forces cannot be handled by such simplifications as point charges and dipoles, and new concepts have to be accepted. Thus the movement of one molecule of water affects not only its neighbors but also, by transmitted momentum, more distant molecules. Very-short-range repulsive forces can dominate molecular interactions, giving the behavior of liquids fascinating and comparatively unfamiliar aspects that are of immediate concern to us and that can be classified under the general title of "structural forces."

Consider two parallel bilayers separated by water or an aqueous solution. The forces that might be expected to play a role are of several kinds, the most obvious being the Coulombic interaction between electrostatic charges on the membrane surfaces due to ionized groups or adsorbed ions. This force, which is strongly modified by the presence of ions in the intervening aqueous solution, can be reliably estimated, as we will see. We also expect van der Waals forces to draw the surfaces together, and the modern theory of van der Waals forces will occupy much of this chapter. What does experiment have to say? The way in which the force between bilayers depends on distance has been determined for both neutral and charged bilayers.[97] Neutral bilayers such as DPPC, DLPC, and DMPC separated by *small* distances repel each other with an enormous force that obeys an exponential law given by

$$P = P_0 \exp\left(\frac{-d_w}{\lambda}\right)$$

where $P_0 \simeq 7 \times 10^9$ dynes/cm^2 and $\lambda \simeq 2.56$ Å. The relationship holds for separations d_w of ~3 to 25 Å, and for a wide variety of phospholipids. It has been suggested that this *hydration force* may take the same quantitative form for all hydrophilic surfaces in water. The distance between neutral bilayers in water increases with the proportion of water present but reaches a maximum above which addition of further water has no effect on the separation (Figure 27A).

The existence of such an equilibrium implies a balance between repulsive and attractive forces, the latter presumbly being van der Waals forces. By using the equilibrium distance and the exponential repulsion expression given above, we can estimate the van der Waals force at that distance. The answer is not consistent with the classic theory of van der Waals forces based on the summation of the calculated van der Waals forces between all pairs of atoms chosen, one each, from the two bilayers. We deal with this problem soon, merely commenting here that it has been solved. The problem that cannot be said to have received a convincing quantitative explanation as yet is the hydration force. The fact that the force is effectively independent of the type of lipid rules out electrostatic repulsion between dipolar head groups. In any case, this force can be estimated, and it is far too small to account for the experimentally measured repulsion, even in the case of highly charged lipids (Figure 65). The present explanation is that the structure of water in the interlayer space allows mechanical forces to be transmitted directly between the layers. Our knowledge of the structure of liquids is poor in comparison with what we know of gases and solids. Radial distribution functions available for many liquids show that molecular positions are correlated over distances of several molecular diameters. Surfaces undoubtedly perturb and often partially order neighboring solvent molecules. The head groups of lipids, charged or dipolar, are certainly solvated, as we saw from neutron diffraction.

The existence of a relatively stable layer of adsorbed water in itself imposes limitations on the positioning of neighboring water molecules. Such a propagation of order might be combined with a propagation of polarization such that bilayers at distances of up to 20 or 30 Å could be aware of each other through the agency of a structured intervening

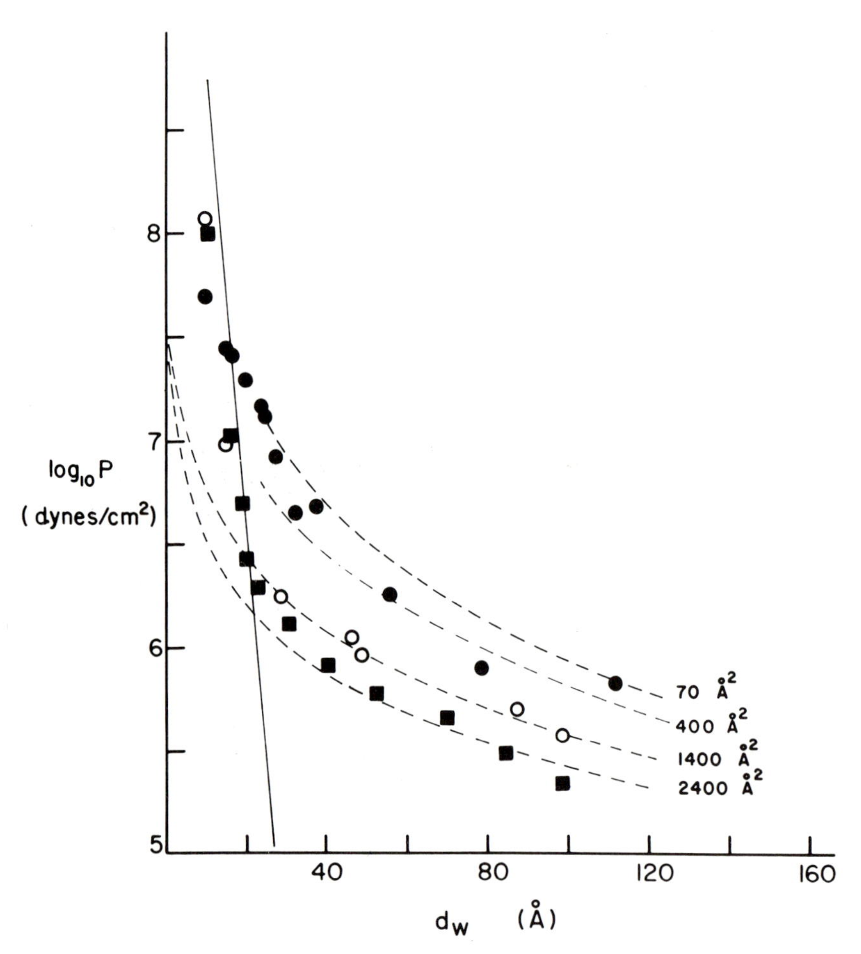

Figure 65. Pressure between two parallel bilayers of egg phosphatidylcholine containing 5, 10, and 100 mol% phosphatidylglycerol. Solid lines: The experimental hydration repulsion. Dashed lines: Theoretical electrostatic repulsion. The figures indicate the surface area per charge. [From Rand, R. P., *Ann. Rev. Biophys. Bioeng.* 10:277 (1981).]

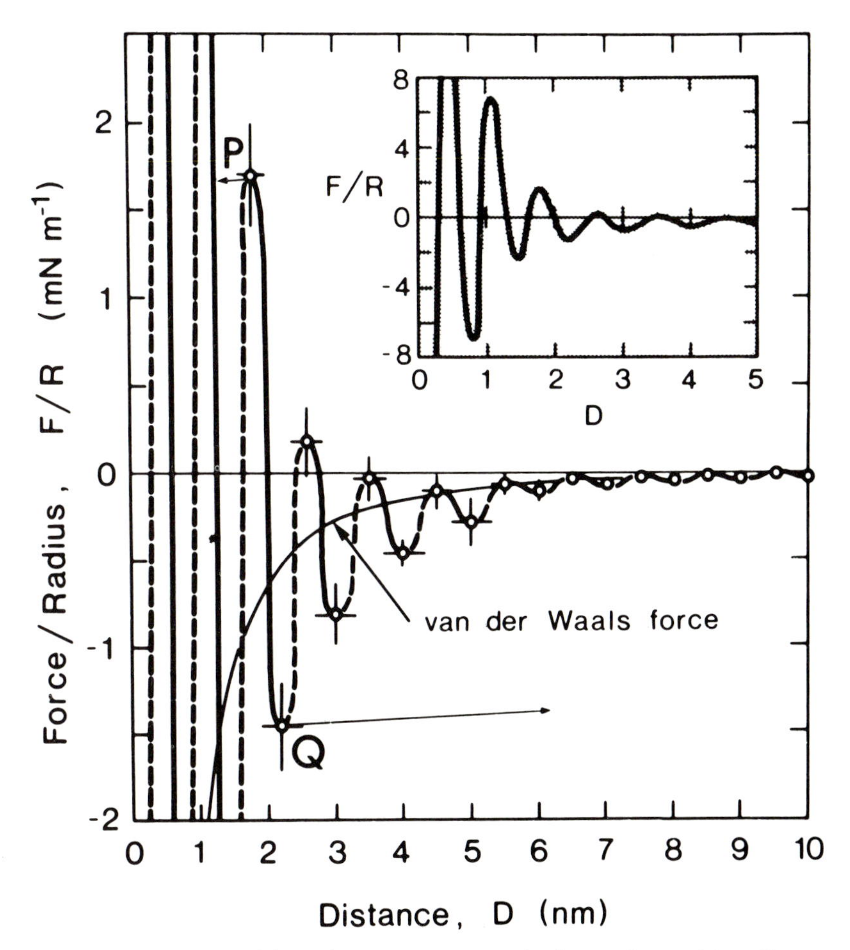

Figure 66. Measured force between two curved mica surfaces with radius of curvature ≈ 1 cm and separated by octamethylcyclotetrasilane. Temperature, 22°C. The calculated van der Waals attraction for an intervening *continuum* is indicated. The arrows at P and Q indicate jumps that occur experimentally from force maxima or minima to stable positions. Thus the arrow from Q ends on the $F = 0$ axis. [From Horn, R. G., and Israelachvili, J. N., *Chem. Phys. Lett.* 71:192 (1980).]

medium. This "explanation" is definitely of the hand-waving variety but some extraordinary experiments on interfacial forces suggest that structural forces may be quite real. Figure 66 shows the measured repulsive force between two mica surfaces separated by nearly spherical molecules of diameter 10 Å. The scale is important, the oscillations having a "wavelength" of about 10 Å and dying out at about 70 Å. A qualitatively similar state of affairs is evident in Figure 67. At interfacial distances smaller than the dimensions of the intervening molecules, the force rises very steeply. It is believed that the phenomena typified by these two examples are due, as presumably are hydration forces, to the structure of the liquid medium. Liquids are not continua, but consist of molecules having volume, shape, polarizability, and a strong repulsion

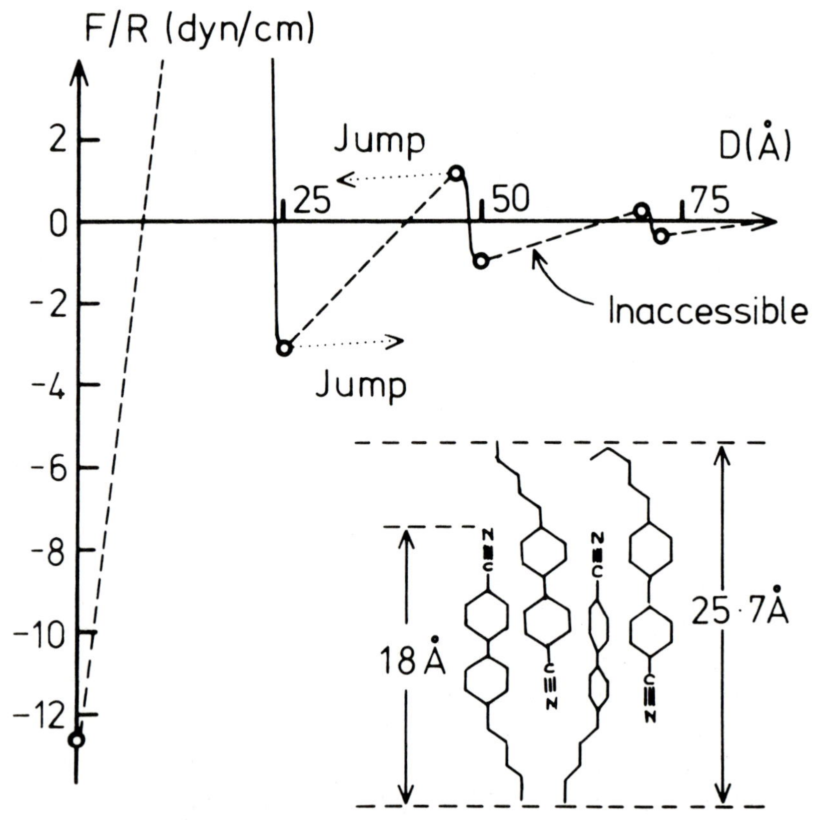

Figure 67. Measured force between two curved mica surfaces separated by a nematic liquid crystal. The whole range of distance is not experimentally accessible as indicated by the dashed lines and the occurrence of jumps—compare Figure 66. [From Ninham, B. W., *J. Phys. Chem.* 84:1423 (1980).]

at close distances. A calculation of the free energy per unit area of two hard walls interacting across a Lennard–Jones fluid gives a force that decays with distance in an oscillatory manner. A neat general explanation of the jumpy nature of the forces in the above examples has been given by Ninham and is presented in Figure 68.

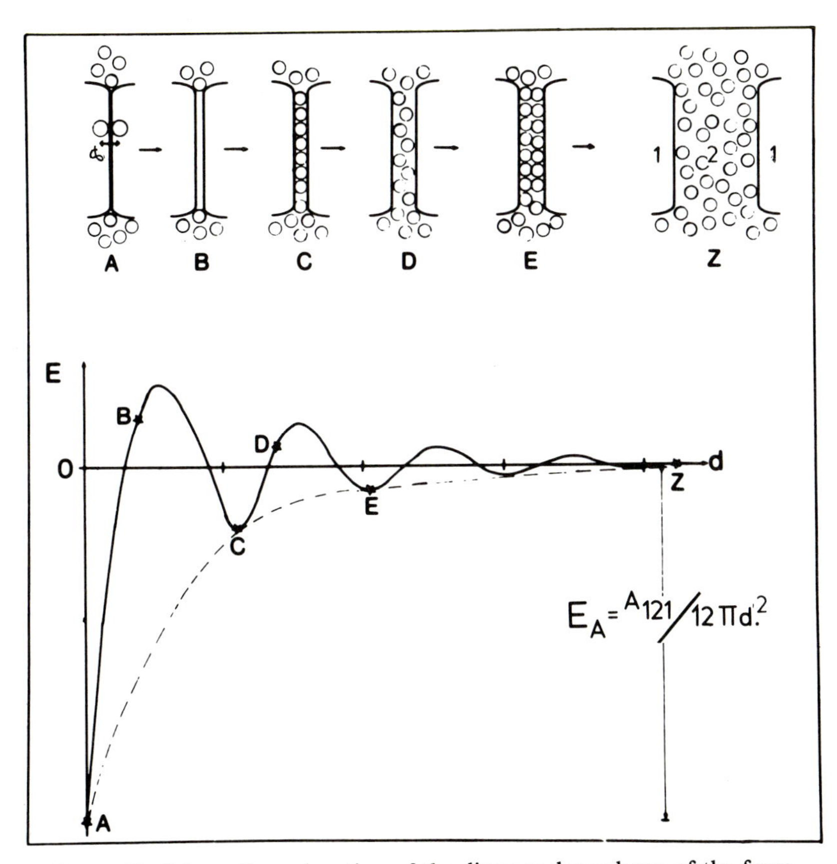

Figure 68. Schematic explanation of the distance dependence of the force between two closely spaced surfaces separated by a liquid. **A.** The surfaces are in contact and the potential energy of the system is at a minimum. **B.** The gap is too small to permit the entrance of liquid molecules and the vaccuum does not screen the interaction. **C.** The gap exactly accommodates a layer of molecules that reduce the force by an amount determined by the Hamaker constant A_{121}. **D.** The intervening liquid is imperfectly packed so that the screening is less than that expected from the liquid at its normal, bulk, density. **E.** The packing is again optimal and the screening likewise. [From Ninham, B. W., *J. Phys. Chem.* 84:1423 (1980).]

The moral of the story is that solvent-mediated forces can be of overriding importance at distances of less than about ten molecular diameters. For surfaces less microscopically planar than mica, and for liquids such as water with its dipole moment and hydrogen bonds, it is not likely that oscillatory behavior will be observed, but both theory and experiment strongly support the hypothesis of structural forces between bilayers and membranes. We do not expect oscillations to be evidenced at interfacial distances of over several molecular diameters, but it is interesting to ask what the discrete nature of molecules contributes to interfacial forces at large distances. A provocative calculation has been made by Chan et al.,[98] their model being two parallel attractive hard walls separated by a liquid consisting of spherical molecules with a hard-core diameter, interacting via a two-body London potential. We can calculate the force between the walls using modern van der Waals theory and assuming the liquid is a dielectric continuum, and the result can be compared with that obtained by taking into account the intermolecular forces within the liquid. A general result is that the force between the walls was modified significantly by the liquid at interwall distances ranging from about 10 to 100 molecular diameters, depending on the values assumed for the molecular parameters. Although the numbers coming out of these calculations are to be taken with a pinch of salt, the central conclusion is clear: The discrete nature of the liquid cannot be ignored.

VAN DER WAALS FORCES

An attractive force between molecules, physically necessary to account for condensation, made an implicit appearance long ago in the van der Waals equation. The nature of this force, which exists even for neutral nonpolar molecules, was explained by London in terms of the ever-present fluctuations of the charge cloud, expected for all molecules and manifested as fluctuating dipoles and higher electric multipoles. At distances for which quadrupole and higher moments can be neglected, the mutual interactions of the time-varying dipoles can, in the case of nonpolar molecules, quantitatively account for the van der Waals forces in gases. Molecules with permanent electric dipoles interact by additional mechanisms, including a direct interaction between dipoles, treated by Keesom, and an interaction between a dipole and the induced dipole it creates in neighboring neutral molecules, the so-called Debye contribution to intermolecular attraction. The early theory dealt with pairs of molecules interacting through a vacuum, a model system suited to gases but not to the large molecular assemblies and liquid media typical of biologic material. In an attempt to handle the forces between large bodies, it was suggested[99] that an appropriate procedure would be to sum the van der Waals forces between all pairs of atoms, chosen one from each body. This seemingly reasonable procedure does not work, basically because it fails to take into account the fact that fluctuations in macroscopic bodies are correlated; they cannot be treated as the sum of independent fluctuations of constituent molecules. In simple terms, the time-varying electric field created by the twitching of one molecule influences the charge fluctuations of the neighboring molecules in the same body. Further, the fields due to one body must affect the fluctuations in the other. We face a many-body problem usually complicated by the presence of a liquid medium between the macroscopic bodies, a medium itself composed of molecules, fluctuating and being subjected to time-dependent fields. Many-body physics is not a trivial subject, and it might seem that there would be little hope of obtaining even a semiquantitative estimate of the van der Waals force between two bilayers, a force crucial in explaining the behavior of mutually approaching cells. In fact, the problem was solved in 1956 by Lifshitz,[100] using quantum electrodynamics — which is not every biophysicist's cup of tea. The great attraction of Lifshitz's theory was that its application required only a knowledge of the macroscopic properties of the interacting bodies and the intervening medium. Nevertheless, the theory was considered for some time to be elegant but unusable since it seemed to require physical data that were impractical to obtain. Subsequently it became clear that values of the physical parameters needed to apply the Lifshitz theory were more accessible than had at first appeared and many calculations have been made of van der Waals forces between macroscopic bodies. Where is has been possible to check the results against experiment, the agreement almost always has been satisfactory. Our object in this section is to hint at the physical basis of the theory, to show and briefly analyze the central equations, and to present the important qualitative conclusions. At the level of this book it is certainly not appropriate to attempt more than this, but it would be negligent to do less since the theory is unfamiliar to most physical chemists and yet the forces it explains are of major importance. We stress that the predictions of the theory, verified in the laboratory, are sometimes at complete variance with naive expectations based on classic, pre-Lifshitz, van der Waals theory. Thus bodies may repel each other because of van der Waals forces, the energy of interaction almost never follows an r^{-6} law, the forces can be of extraordinary strength at long range, and so forth.

The theory is upheld by two pillars — the interaction of fluctuating dipoles and the fluctuation–dissipation theorem. We take them in that order.

To expose the essential point of our argument, we consider the interaction between two isotropic nonpolar molecules in a vaccuum. By "isotropic" we imply that the measured properties of the molecule are independent of the mutual orientation of instrument to molecule. By "nonpolar" we mean that the time-average dipole moment is zero. However, both the electron and nuclear charge distributions in a molecule are continually fluctuating at the demand of the uncertainty principle and thermal excitation. At any given moment, even a nonpolar molecule is almost certain to have a finite dipole moment. The origin of the van der Waals force is that the fluctuations in the dipoles of different molecules are not independent. The time-varying dipole of a given molecule produces a time-varying electromagnetic field at the position of the second molecule, which does its best to adjust its own fluctuations to those of the field in such a way as to lower the energy of interaction of its dipole with the field. Since both molecules in our simple system are fluctuating and each is bathed in the field of the other, it is best to regard them as comprising a single system, coupled by the radiation field in such a way as to minimize the energy of the system. In recalling the case of two static dipoles, one will remember that, provided they are free to orient themselves, their interaction is always finally attractive. In a similar way the charge distributions of fluctuating molecules will show a preference for those configurations that result in an attraction across a vaccuum. In the accepted jargon we say that the fluctuations are correlated. A molecular spectroscopist might prefer to say that we are dealing with coupled oscillators, and in fact one approximate method of developing the modern theory of van der Waals forces is to set up a model comprised of a set of coupled oscillators and then to derive the normal modes of the system.[101] It is clear that two vital factors in determining the coupling between molecules, and therefore the force between them, are the strength and frequency distribution of their emitted radiation and their fluctuating dipoles. A molecule that undergoes large fluctuations in charge distribution will produce a stronger electromagnetic field and will exhibit a larger time-varying dipole with which to interact with the fields of other molecules. However, the essential point of the model is the *coupling* of dipoles, which we can regard as due to the influence of the field of one molecule on the direction, magnitude, and frequency of the dipole in another molecule. This is where frequency enters, since the response of any system, whether tuning fork, electric circuit, or molecule, depends not only on the strength of a perturbation, but also on its frequency. Oscillations in a tuning fork are most effectively produced over a rather small range of sound frequencies. Electric circuits can be tuned to respond over narrow frequency bands. The magnitude of the induced charge fluctuations in a molecule depend strongly on the frequency of the electromagnetic radiation. We say that the polarizability is a function of frequency and we correctly expect that the polarizabilities of molecules will appear in the expression for the van der Waals force between them. But molecules, as we have seen, are transmitters as well as receivers, and just as a tuning fork transmits best at the frequency to which it resonates best, it is easy to believe that molecules fluctuate spontaneously with a frequency distribution exactly paralleling their talents as receivers. The fluctuations of a molecule, therefore, are determined by both their internal structure and the persuasion of external fields. Polarizability controls the fluctuations, and through them the frequency spectrum of transmitted radiation.

Our argument has been shamefully qualitative but the reader is asked to accept the following conclusions:

1. Fluctuating molecular dipoles interact across a vaccuum in such a way as to reduce the total energy of the system. The van der Waals, or "dispersion," force is attractive.
2. The strength of the coupling between molecules depends, among other things, on their polarizabilities.

If we are to attempt quantitative force calculations, we need a means of either measuring or estimating polarizabilities. For the two-molecule model we require a knowledge of the polarizabilities of the molecules in a vaccuum, and this is obtainable from spectroscopic data since, as we saw, it is the polarizability that determines the strength of the molecule's response to radiation of a given frequency. In practice, interest usually centers on the forces between bodies containing very large numbers of molecules that are internally coupled to each other to give a radiation field that is a product of all the molecules in the body. The problem of determining the spontaneous fluctuations of a macroscopic body might seem overwhelming but we have already hinted at the solution by our remark on the parallel receiving and transmitting characteristics of a tuning fork. The fluctuation-dissipation theory tells us that the response of a system to an external disturbance, in the limit of a linear response, can be predicted completely from a knowledge of the random fluctuations occurring in the system at macroscopic equilibrium.[102] Conversely, we can learn about the spontaneous microscopic fluctuations in an assembly of molecules at thermodynamic equilibrium by measuring its response to an external disturbance. We will have more to say about this in Chapter 17. Here we state without proof that the response of a system to an electromagnetic field allows us to determine the nature of the electrical (and magnetic) fluctuations within the system in the absence of the field. The response is manifested by the conventional spectral characteristics, or at low frequencies by the dielectric permeability. At frequencies at which the system absorbs strongly, we can be sure that there are strong spontaneous microscopic fluctuations and a corresponding peak in the plot of polarizability against frequency. We thus have at our disposal, at least in principle, a means of aquiring the information that we believe to be necessary for the calculation of the dispersion forces between macroscopic bodies.

It is instructive to return to the case of two isotropic molecules interacting across a vaccuum. If the molecules are, in fact, atoms, there is an approximate and convenient expression for the electronic polarizability;

$$\alpha_e(\omega) = fe^2/m(\omega_e^2 - \omega^2) \tag{7.6}$$

where ω is the frequency of the radiation, e and m the charge and mass of the electron, and f the effective number of oscillating electrons, an oscillator strength. At $\omega = \omega_e$ there is a maximum in the polarizability and, therefore, in the electronic absorption spectrum. London,[103] and others have derived expressions for the interaction of two atoms or molecules. We will discuss this result, but before doing so, some general points need to be made.

We show no derivations, as they would not be suitable in a book of this kind. The formula that we give will be for the dispersion force unless specifically stated. Molecules with permanent dipoles or higher multipoles can, of course, interact in other, well-documented, ways.

We now consider the formula for the free energy of interaction between two isotropic bodies in free space separated by a distance r large compared with their dimensions[104]:

$$G = -\frac{6kT}{r^6} \sum_{n=0}^{\infty}{}' \alpha_1(i\omega_n)\alpha_2(i\omega_n) \tag{7.7}$$

The prime on the summation sign indicates that the $n = 0$ term is multiplied by 1/2. $\omega_n = n(2\pi kT/\hbar)$ and $\alpha_1(i\omega_n)$ can be obtained by taking the expression for the frequency-dependent polarizability and replacing ω by $i\omega$, where $i = (-1)^{1/2}$. For example, using this procedure with 7.6, we obtain for an atom:

$$\alpha_e(i\omega_n) = fe^2/m(\omega_e^2 + \omega_n^2) \tag{7.8}$$

At 300 K $\omega_n = n(2.5 \times 10^{14})$ rad sec^{-1}. For $n = 0$ we have the static polarizability $\alpha(0)$, for $n = 1$ ω_1 is well above rotational frequencies for molecules, and $\omega_{10} = 2.5 \times 10^{15}$ rad sec^{-1} is well into the ultraviolet range. Since for most atoms ω_e is around 10^{16} rad sec^{-1}, we see from (7.8) that for atoms the values of $\alpha_e(i\omega_n)$ that contribute most to G will be in the ultraviolet, and the static polarizability plays a negligible role.

The term in $n = 0$ does not contribute to the *dispersion* energy since it contains only the static polarization. The sum over the terms for $n > 0$ can be shown under normal physical conditions to give

$$G_{n>0} = -\frac{3\hbar}{2r^6} \frac{\omega_{1e}\,\omega_{2e}}{(\omega_{1e} + \omega_{2e})} \alpha_{1e}(0)\,\alpha_{2e}(0) \tag{7.9}$$

This is London's formula for the dispersion energy of two atoms in vacuum. We do not press the analysis of (7.9) since we are concerned with forces between multimolecular bodies, but there are two important physical situations that require equation (7.7) to be modified, and the nature of the modifications is of direct relevance to the forces between membranes. We consider in turn the consequences of large values of r and of the presence of a medium between the particles.

We have assumed that messages between molecules are delivered instantaneously, but radiation, of course, takes time to travel through space. The expression for the dispersion energy given in (7.7) is valid provided the time taken for radiation to reach the receiving particle and be "reflected" back to the transmitting particle is small compared with the period of the oscillation of the transmitter — that is, provided that $\omega_e \ll c/r$ where c is the velocity of light in vacuum. If this condition is not satisfied, the 'reflected' radiation may not be in phase with the oscillations of the transmitter. This mismatching prevents the system from minimizing its energy, and at values of r large enough to give $\omega_e \gg c/r$, expressions (7.7) and (7.9) no longer hold and have to be replaced by the following expression for the so-called *retarded* dispersion energy, that is, the energy when signals arrive late,[105]

$$G = -\frac{23\hbar c}{4\pi r^7} \alpha_{1e}(0)\,\alpha_{2e}(0) \qquad \text{for } \omega_e \gg c/r \tag{7.10}$$

Notice the r^{-7} dependence. $\alpha_{1e}(0)$ is the polarizability of particle 1 due to electronic displacements, that is, excluding polarization attributable to the orienting of permanent dipoles. Notice that the distance at which the onset of retardation becomes significant depends on ω_e. Since a particle may have appreciable spontaneous fluctuations at a variety of frequencies, it follows that for each frequency there is a nonretarded and a retarded range of distances for the dispersion force and that the larger the frequency, the shorter is the distance at which one range passes into the other. Under these circumstances we can hardly expect a simple dependence on distance for the van der Waals energy for two particles since the nonretarded and retarded terms depend on $1/r^6$ and $1/r^7$ respectively.

If two small isotropic particles are separated by a medium rather than a vacuum, expression (7.7) again needs modification and the result for the nonretarded free energy is[104]

$$G = -\frac{6kT}{r^6} \sum_{n=0}^{\infty}{}' \alpha_1^*(i\omega_n)\alpha_2^*(i\omega_n)/\epsilon_3^2(i\omega_n) \qquad \text{for dielectric medium} \tag{7.11a}$$

The retarded energy is given by

$$G = -23\hbar c\alpha_1^*(0)\alpha_2^*(0)/4\pi\epsilon_3^{5/2}(0)r^7 \tag{7.11b}$$

In these expressions $\epsilon_3(i\omega)$ is the dielectric permeability of the *medium* at imaginary frequencies. At $\omega = 0$, $\epsilon_3(0)$ is simply the familiar static dielectric constant. For oscillating fields, $\epsilon(\omega) = 1 + P(\omega)/E(\omega)$ where $P(\omega)$ is the polarization that is produced by and oscillates at the same frequency as the field $E(\omega)$. We enter neither into the analysis of the polarization into in-phase and out-of-phase component nor into the physical significance of $e_3(i\omega)$, which we treat as a function for which good approximate formulas exists. Four points are of importance to us. First, the medium reduces the intensity of the electric field of the radiation by $1/\epsilon_3$ and since both the transmitted and reflected fields are affected, a factor of $1/\epsilon_3^2(i\omega_n)$ appears in (7.11a). Second, the polarizability of a particle depends on its environment since polarization involves a distortion of charge distribution. In (7.11) α^* is an excess polarizability of the particle over that of the medium. Note that α^* can be negative, or zero. Third, since light is slowed down by a factor of $\sqrt{\epsilon_3}$ by the medium, the journey between the particles takes longer and retardation sets in at smaller distances than in vacuum. Last, we see in (7.11) the first appearance of *bulk* rather than molecular properties. The medium can be liquid or solid dielectric, but all that we need to know is its macroscopic permeability; the theory thus bypasses the need to consider microscopic fluctuations directly.

The case of two small isotropic particles, glanced at above, is of far less interest to us than the case of a particle and an interface, or two interfaces, for which we now present formulas.

For a small isotropic particle 1 interacting with a single interface from which it separated by a medium 3 (Figure 69), the nonretarded free energy is given by[106,107]

$$G = -\frac{\mathrm{kT}}{2r^3} \sum_{n=0}^{\infty}{}' \alpha_1^*(i\omega_n)\Delta_{23}/\epsilon_3(i\omega_n)$$

where (7.12)

$$\Delta_{23} = \frac{\epsilon_2(i\omega_n) - \epsilon_3(i\omega_n)}{\epsilon_2(i\omega_n) + \epsilon_3(i\omega_n)}$$

and r is the distance between the particle and the interface.

This expression, which applies to a body with small dimensions compared with r, contains a term in $n = 0$ that is very small for nonpolar molecules, and which for reasonably polar molecules depends overwhelmingly on permanent dipole interactions. The terms for $n > 0$ give the dispersion energy since they are the terms that arise from $\omega_n \neq 0$, that is, from fluctuations. The application of the formula requires a determination of α_1^*, the excess permeability of the particle in the medium. For small spherical particles of radius R, which can be approximated as continuous medium of permeability ϵ_1, expressions for α_1^* have been derived[108] and lead to the following formula for the nonretarded energies of interaction with an interface (see Figure 69):

$$G = -\frac{\hbar R^3}{4\pi r^3} \int_0^\infty \frac{\epsilon_1 - \epsilon_3}{\epsilon_1 + 2\epsilon_3} \times \frac{\epsilon_2 - \epsilon_3}{\epsilon_2 + \epsilon_3} \, d\omega \tag{7.13}$$

for particle 1 in medium 3, and

$$G = -\frac{\hbar R^3}{4\pi r^3} \int_0^\infty \frac{\epsilon_1 - \epsilon_2}{\epsilon_1 + 2\epsilon_2} \times \frac{\epsilon_3 - \epsilon_2}{\epsilon_3 + \epsilon_2} \, d\omega \tag{7.14}$$

for particle 1 in medium 2. In these formulas $\epsilon_n = \epsilon_n(i\omega)$. Notice that the free energy is proportional to r^{-3} as compared with the r^{-6} dependence for the interaction between two particles. Since, for the geometry shown in Figure 69, the force on the particle is given by $F(r) = -\partial G/\partial r$ we see that if the integral in (7.13) is positive (negative) and the particle is attracted to (repelled by) the interface.

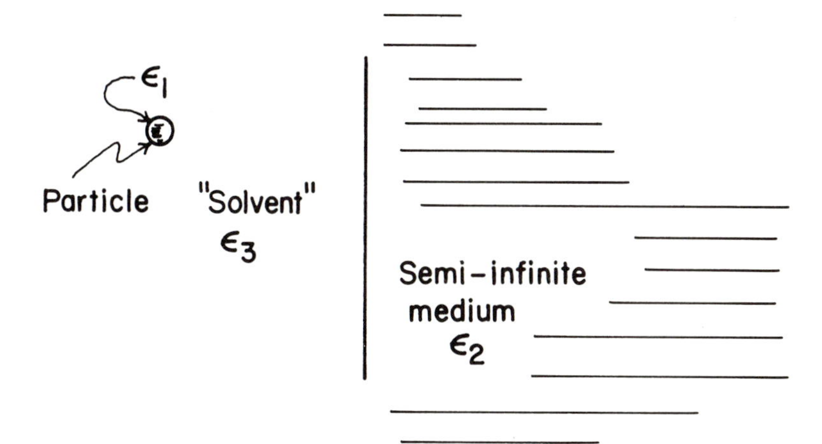

Figure 69. Isotropic particle in a "solvent" interacting with a semi-infinite medium. ϵ_1, ϵ_2, and ϵ_3 are the frequency-dependent *macroscopic* permeabilities appearing in equation (7.12).

The free energy given by (7.13) will be zero if either $\epsilon_1 = \epsilon_3$ or $\epsilon_2 = \epsilon_3$ at all frequencies. The vanishing of the force in these rather unique cases can be readily understood in terms of the underlying physics. If $\epsilon_2 = \epsilon_3$, the interface "disappears" as far as polarizability properties are concerned and the model reduces to a particle in an infinite homogeneous medium with no expected resultant force. If $\epsilon_1 = \epsilon_3$, we have a dielectric analog of a body of the same specific gravity as the liquid in which it is immersed; Archimedes tells us that no resultant force acts on the body. Similar arguments can be applied to (7.14). Of course, the integrals in (7.13) and (6.14) are not likely to be identically zero, but at least they could be very small for suitable combinations of materials. Note that if, say, $\epsilon_1 = \epsilon_3$, then a particle in medium 3 will feel no force, but if placed in medium 2, it will feel a force: There is no unique particle–interface interaction.

In general, it is not to be expected that the integrals in (7.13) and (7.14) will vanish. For models closer to reality we can distinguish two special cases.

1. If $\epsilon_2 > \epsilon_1 > \epsilon_3$ for all ω, then G is negative and a particle in either medium 2 or 3 will be attracted to the interface. [The inequalities need not hold over the whole range of ω, provided the integrals (7.13) and (7.14) are positive.]

2. If $\epsilon_1 > \epsilon_3 > \epsilon_2$ over the whole range of ω, then in (7.14) G is positive and in (7.13) G is negative. This means that a particle originally placed in medium 2 will be attracted toward the interface, and if its momentum carries it through to medium 3, it will be repelled from the interface. Van der Waals dispersion forces can thus drive a particle through an interface. The at-first-sight suspicious result that van der Waals forces can be repulsive is physically explicable in the same terms as our discussion of repulsive image forces.

On looking at (7.4) one can see that for the arrangement in Figure 69 the *image* force on a charge Q residing on particle 1 will be given by

$$\frac{Q^2}{(2r)^2}\left[\frac{\epsilon_2(0) - \epsilon_3(0)}{\epsilon_2(0) + \epsilon_3(0)}\right] \times \frac{1}{\epsilon_3(0)} \tag{7.15}$$

where the factor $1/\epsilon_3(0)$ accounts for the diminution of the image field by medium 3. Compare (7.15) with (7.13). For the image force it is the *static* permeabilities that are needed in contrast to the frequency-dependent permeabilities that control dispersion forces. Both phenomena are attributable to polarization. As we saw earlier, image force can be either repulsive or attractive and thus a charge can be driven across an interface by these forces, just as a neutral molecule can be driven by dispersion forces.

We now consider the van der Waals force between two parallel interfaces separated by a dielectric medium. This is the system to which Lifshitz first applied himself and for which he found the following expression for the non-retarded free energy per unit surface[109]:

$$G = -\frac{kT}{8\pi r^2} \sum_{n=0}^{\infty}{}' \sum_{m=1}^{\infty} \frac{1}{m^3} \Delta_{13}^m \Delta_{23}^m \tag{7.16}$$

where r is the distance between the semi-infinite dielectric media 1 and 2 and Δ_{ij} is given by (7.12). A good approximation for G is to ignore quadratic and higher terms in Δ_{ij}, that is, all terms having $m > 1$.

$$G \simeq -\frac{kT}{8\pi r^2} \sum_{n=0}^{\infty}{}' \Delta_{13}\Delta_{23} \tag{7.17}$$

Again, the $n = 0$ term is a zero-frequency term not to be included in the dispersion force. The contribution of the terms for which $n > 0$ can be expressed as an integral:

$$G_{n>0} \simeq \frac{-\hbar}{16\pi^2 r^2} \int_0^{\infty} \Delta_{13}\Delta_{23}\, d\omega \tag{7.18}$$

The similarity between (7.18) and (7.13) allows us to shorten our discussion and merely state that, depending on the values of the permeabilities that enter into Δ_{ij}, the nonretarded force between two semi-infinite media can be attractive, repulsive, or zero. As usual, the nature of the medium is critical. In biologic systems the medium separating membranes often contains electrolytes and the expressions for the intermembrane dispersion force are complicated by the presence of terms accounting for the screening of the dispersion force by the electrolytes.[110,111] Without entering into details, we can say that for high-frequency electric fields the ions in solution cannot respond fast enough to affect the permeability. For low frequency fields the ions sway back and forth in response to the field and the net effect is to increase the polarization, and thus the permeability of the solution. For two identical semi-infinite media we see from (7.12) and (7.16) that an increase in ϵ_3 results in a decrease in G. In practice the numbers show that only ω_0 is low enough to be taken into account, which means that dispersion forces are not affected, but the "static" interactions between polar molecules can be heavily screened. Moreover, the change in ϵ_3 will shorten the distance at which retardation commences.

The nonretarded free energy (7.16) has an inverse square dependence on r but this simple relationship is destroyed by the contribution of the retarded free energy. Since, as we saw above, the onset of retardation occurs at distances that decrease as the fluctuation frequency increases, we cannot expect a simple dependence of free energy on distance, except at small values of r. We do not give the exact expression for the retarded energy but consider an approximate form later.

Van der Waals energies are often discussed in terms of the Hamaker coefficient. In 1937 Hamaker[99] proposed that dispersion forces between two bodies could be evaluated by summing pairwise the interactions between molecules in the different bodies using the London r^{-6} formula to obtain the individual pair energies. For two semi-infinite media separated by a distance l the energy calculated in this way was written:

$$G_{LmR}(l) = -\frac{A_{LmR}}{12\pi l^2} \tag{7.19}$$

A_{LmR} is the Hamaker *constant* where L, m, and R, indicate left, medium, and right; that is, they label the three components of the system. This expression implies that G depends on l^{-2}, which we know to be untrue. Nevertheless, it is customary to express the results of modern theory in terms of a Hamaker *coefficient* by writing

$$G_{LmR}(l) = -\frac{A_{LmR}(l)}{12\pi l^2} \tag{7.20}$$

This equality recognizes the fact that G is not a function of l^{-2} and, therefore, that A_{LmR} is not a constant but a function of l. Using the theoretical expressions for the dispersion energy, A_{LmR} can be calculated and the results are often presented as plots of $A(l)$ against l. Such a plot is shown in Figure 70, which illustrates a number of points. We see that in water $A(l)$ is reasonably constant up to $l \sim 50$ Å, which means that an l^{-2} law holds to a good approximation as expected from (7.16). At larger distances the higher frequency terms in the dispersion energy are successively attenuated by retardation until at around 50,000 Å the energy is completely dominated by the zero-frequency term, which is not retarded because there is no question of mismatching of a reflected wave. In salt solutions the zero-frequency term is heavily shielded so that the energy drops more rapidly with distance and the higher permeability as compared with water also reduces the energy at all separations.

A model system closer to our interests would be two slabs of *finite* thickness separated by a dielectric. Because we have removed material from the semi-infinite slabs of our previous model, we lose interactions. However, at close separations the energy is dominated by the interaction between parts of the slabs immediately neighboring the medium (because of the l^{-2} law) and the energy is not much different from that for semi-infinite slabs. At distances l comparable to the thickness of the slabs, deviations from the previous model become marked.

Calculations have been performed on many models, including interacting spheres, cylinders, and slabs coated with a layer of a second

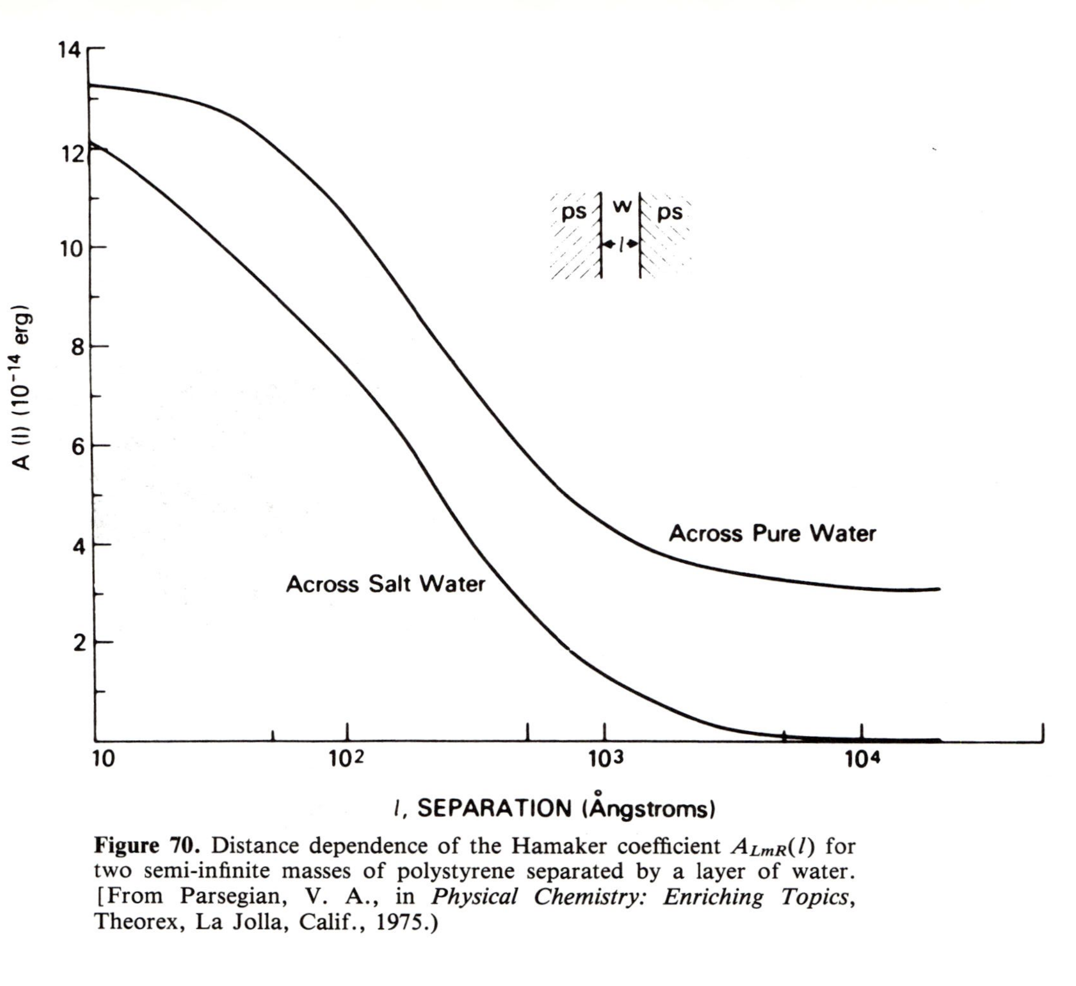

Figure 70. Distance dependence of the Hamaker coefficient $A_{LmR}(l)$ for two semi-infinite masses of polystyrene separated by a layer of water. [From Parsegian, V. A., in *Physical Chemistry: Enriching Topics*, Theorex, La Jolla, Calif., 1975.)

substance. The attempt to use an essentially continuum theory for the interaction of discrete complex structures such as membranes requires approximations and faith, but there can be no doubt that for these systems the theory is capable of providing qualitative answers and, at long distances, reasonably good quantitative results. The experimental results for simpler systems such as the force between two mica surfaces show excellent agreement with theory.[112]

To apply equations (7.7) through (7.18), we need to know the functions $\epsilon(i\omega)$, which, as mentioned above, can be obtained by substituting $i\omega$ for ω in the expressions for $\epsilon(\omega)$, the permeability. We refer the reader to the literature[113] for details of the determination of $\epsilon(\omega)$ from the experimental data, which are basically the absorption spectra in the ultraviolet, infrared, and microwave regions. Of the many systems for which calculations have been made, we consider two that are relevant to membrane–membrane interactions. First, we return to the case of parallel interfaces separated by a dielectric, it being understood that the separation is small compared with the facial dimensions. Gingell and Parsegian[114] have expressed the van der Waals energy in an approximate but convenient form:

$$G_{LmR}(l) \simeq -\frac{kT}{8\pi l^2}\left[\frac{\Delta_{Lm}^{(o)}\Delta_{Rm}^{(o)}}{2} + \sum_{n=1}^{\infty} \Delta_{Lm}^{(n)}\Delta_{Rm}^{(n)}(1+r_n)e^{-r_n}\right] \quad (7.21)$$

where $\Delta_{ij}^{(n)}$ is given by

$$\Delta_{ij}^{(n)} = \frac{\epsilon_i(i\omega_n) - \epsilon_j(i\omega_n)}{\epsilon_i(i\omega_n) + \epsilon_j(i\omega_n)} \quad (7.22)$$

and as usual $\omega_n = n(2\pi kT/h)$. (Note that the superscripts o and n are *not* exponents.) r_n is the ratio of the time taken for a message to travel twice across the gap to the period of the oscillation associated with frequency ω_n:

$$r_n = 2l\epsilon_m^{1/2}\omega_n/c \quad (7.23)$$

The $n=0$ term is the static contribution. The terms with $n>0$, the dispersion terms, are multiplied by a term $(1+r_n)e^{-r_n}$ that accounts for retardation. When r_n is very small, this term goes to unity; when r_n is very large, the term goes to zero. For $r_n = 1$ the term is equal to 0.74; for $r_n = 10$, it is 0.0005.

One of the interesting tests of equation (7.21) is provided by the measurement of the mutal attraction of two bodies of water separated by a film of glycerol mono-oleate[115]. Writing $G(l) = -A(l)/12\pi l^2$, it was found experimentally that $A(l)$ for $l \sim 50$ Å was approximately 4.7×10^{14} erg. Using equation (7.21), a theoretic estimate of $A(50$ Å$)$ gives an answer in the range[113] $3.4 \leq A(50$ Å$) \leq 6.8 \times 10^{14}$ erg. The static contribution ($\omega = 0$) can be estimated from the dielectric constants of water and hydrocarbon, which are approximately 80 and 2, respectively, giving $\Delta^{(o)} \simeq 0.95$. At room temperature $kT \simeq 4 \times 10^{-14}$ erg and the contribution of the static terms to A is thus $\sim 3 \times 10^{-14}$ erg. For a medium of 50 Å in thickness, retardation is negligible since for $\omega = 10^{17}$ rad sec^{-1} (the high-frequency edge of the absorption spectra of water and hydrocarbon) $r_n = 2 \times 50 \times 10^{-8} \times 2^{1/2} \times 2\pi \times 10^{17}/3 \times 10^{10} = 29.6$, which means that the retardation term $(1 + r_n)e^{-r_n} = 4 \times 10^{-12}$.

A second example, closer to membrane interactions, is the attraction between two slabs of hydrocarbon in water. Again the low-frequency (microwave) contribution dominates, accounting for $\sim 50\%$ of the free energy while frequencies in the infrared and ultraviolet account for $\sim 10\%$ and $\sim 40\%$ respectively. Taking the slabs to be 50 Å wide, the calculated values of A (expressed as a percentage of the value at zero separation) are shown in Figure 71. The infrared and ultraviolet terms are subject to retardation effects, which set in first for the ultraviolet frequencies since these are higher. The low-frequency terms, although effectively free from retardation at the distances involved, are subject to the effect of electrolyte concentration, which gives a factor of approximately $(1 + 2Kl)e^{-2K}$, where K is the Debye constant (see next section)

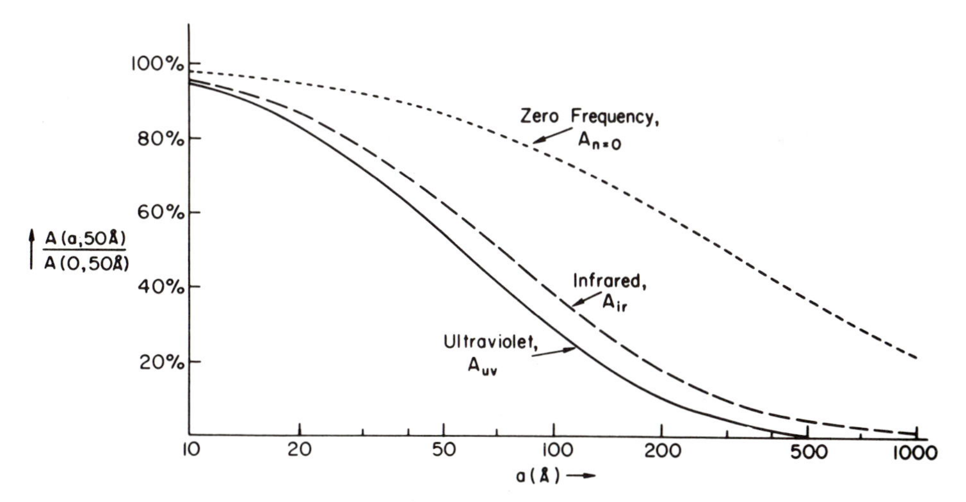

Figure 71. Relative distance dependences of three contributions to the van der Waals force between two interacting polar slabs of thickness b and separation a. For two hydrocarbon slabs in pure water, the microwave, infrared, and ultraviolet contributions are roughly in the ratio 50% : 10% : 40% when the slabs are very close ($b \gg a$). [From Parsegian, V. A., *Ann. Rev. Biophys. Bioeng.* 2:221 (1973).]

and l the separation of the slabs.[116] The average physiological environment can be approximated as 0.14 M (molar) 1:1 electrolyte solution for which $K \simeq 8$ Å^{-1} so that for $l = 10$, 20, and 50 Å, the values of this factor are 0.29, 0.04, and 5×10^{-5} respectively. Thus, at large values of l, retardation kills the high-frequency terms and the electrolyte suppresses the low-frequency terms.

A more realistic model for the interaction of two bilayers is one in which a slab of hydrocarbon is coated with layers of a polar material (Figure 72) and approximate expressions have been derived for the force per unit area between bilayers of this kind[117]:

$$-F_A = \frac{A_1}{6\pi}\left[\frac{1}{d_a^3} - \frac{2}{(d_a + 2d_p + d_h)^3} + \frac{1}{(d_a + 4d_p + 2d_h)^3}\cdots\right]$$
$$+\frac{2A_2}{6\pi}\left[\frac{1}{(d_a + d_p)} - \frac{1}{(d_a + d_p + d_h)^3}\cdots\right] \quad (7.24)$$
$$+\frac{A_3}{6\pi}\left[\frac{1}{(d_a + 2d_p)^3} - \frac{1}{(d_a + 2d_p + d_h)^3}\cdots\right]$$

This approximation has itself been pared away to give useful simpler forms. Calculation shows that the terms in A_2 and A_3 can be neglected,[118] and by truncating the remaining series the following expression is obtained:

$$F_A = \frac{H}{6}\left[\frac{1}{d_w^3} - \frac{2}{(d_w + d_1)^3} + \frac{1}{(d_w + 2d_1)^3}\right] \quad (7.25)$$

where d_w, the thickness of the water layer, and d_1, the bilayer thickness, are equivalent to d_a and $d_h + 2d_p$ in (7.24). For $d_w \ll d_1$ a final simplification yields:

$$F_A = \frac{H}{6} \cdot \frac{1}{d_w^3} \quad (7.26)$$

Equation (7.25) has been used to interpret experimental data on force–distance relationships in multibilayers,[119] a subject to which we return shortly.

We have concentrated on the dispersion force between large bodies but forces between individual molecules are responsible for the structure of the lipid aggregates that are our main topic. In particular, the nonpolar hydrophobic core of bilayers consists of hydrocarbon chains interacting through repulsive close-range forces and attractive van der Waals forces. A primitive model for the calculation of the dispersive force is a system comprised of two thin, dielectrically anisotropic rods in a vaccuum. It has been shown that the interaction between the rods will tend to align them so that they are adjacent and parallel. The result is almost self-evident and has been used as the basis for one theoretic approach to the behavior of the lipid chains in the liquid crystal state of hydrated bilayers.[120]

An interesting consequence of the theory of dispersion forces is the prediction that in systems containing two types of bodies, A and B, the formation of like pairs AA and BB, is favored over the formation of unlike pairs AB. This follows from the form of (7.21) that depends on the product of terms $\Delta_{Lm}^{(n)}\ \Delta_{Rm}^{(n)}$, which we write as xy for short. The difference in energy between the system AA, BB and the system $2AB$ depends on the difference between $x^2 + y^2$ and $2xy$. Now $(x-y)^2 = x^2 + y^2 - 2xy > 0$ for $x \neq y$. Thus $x^2 + y^2 > 2xy$ and, from

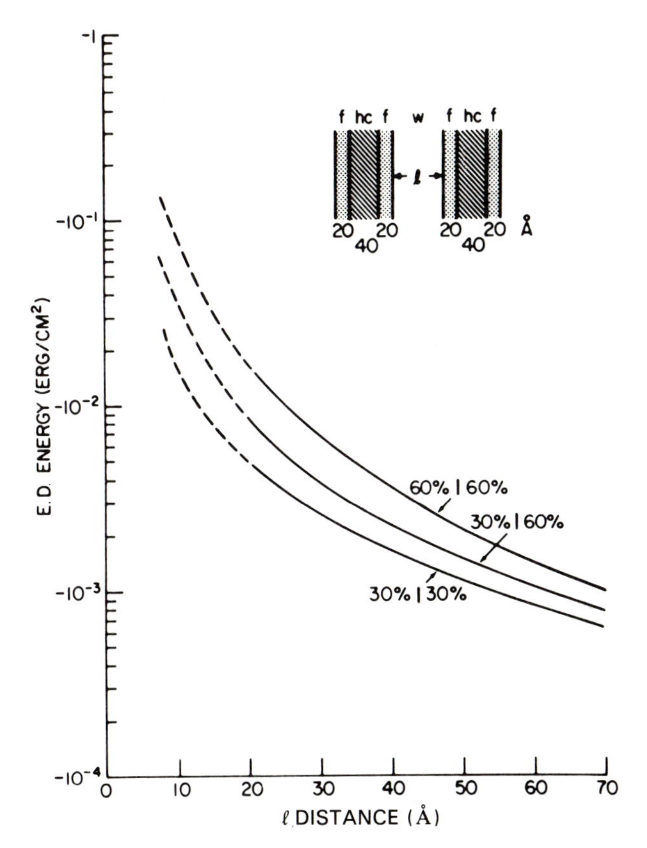

Figure 72. Approximate dispersion attraction energy for parallel planar slabs. *hc*: Hydrocarbon; *f*: Polar surface "fuzz" with the polarizability properties of sucrose solutions of the weight percentage composition indicated near the curves. [From Parsegian, V. A., *Ann. Rev. Biophys, Bioeng.* 2:221 (1973).]

(7.21), $G_{AA}(l) + G_{BB}(l)$ is lower than $2G_{AB}(l)$. This inequality may be relevant to the phenomenon of specific aggregation of biological cells.[121]

The foregoing account of dispersion forces has omitted serious reference to numerous studies on the interaction between nonplanar surfaces and between small finite bodies, but the principles involved in dealing with such systems are no different from those leading to the equations we have presented. Our bias toward planar surfaces is dictated by the fact that they provide a suitable model for the interaction of closely spaced bilayers, a system which has received much experimental study and which approximates the encounter of two cell membranes. That dispersion forces should play an important role in such systems can be predicted from Parsegian's[122] conclusion that in the absence of other forces, dispersion forces are capable of holding together bodies separated by distances much less than their size. Thus, while cells probably can be pulled together by van der Waals interactions, the force between two molecules is not likely to exceed kT, except at such small distances that the continuum model of Lifshitz is inapplicable.

For the reader wishing to delve deeper into the theory, three authoritative reviews are recommended.[122–124]

Structural and dispersion forces operate between all surfaces in aqueous solution. For bilayers comprised of uncharged lipids such as DPPC, these are the only forces that need to be taken into account. For lipids with charged heads, and for cell membranes, the presence of surface charge both affect the distribution of ions near the surface and produces intersurface forces. For this reason we now turn to electrostatic forces.

ELECTROSTATIC FORCES

Initially we tread relatively familiar ground and request the reader's patience while we discuss some well-known equations. All natural membranes are charged due to the ionization of head groups and proteins and in vivo all membranes are bathed in electrolyte solutions. We therefore need to understand the electrostatic interaction between charged surfaces and ionic solutions. We begin by looking at a single planar charged surface in a solution of strong electrolyte; image forces are not taken into account. We work in centimeter-gram-second units, which seem more natural in this field than SI units.

The basic law in electrostatics in Coulomb's law, which states that the force between two charges q_1, q_2 separated by a distance r is given by

$$\hat{F} = q_1 q_2 / \epsilon r^2 \qquad (7.27)$$

where ϵ is the dielectric constant of the medium. The force due to q_1 on a unit charge q_2 is given by $q_1/\epsilon r^2$, and since a force is expressible as the gradient of a potential, we can write

$$q_1/\epsilon r^2 = \frac{d\phi(\hat{r})}{dr} \tag{7.28}$$

which gives the potential due to a point charge:

$$\phi(\hat{r}) = q_1/\epsilon r \tag{7.29}$$

From (7.27) we can also derive the electric field at q_2 because $\hat{F}(\hat{r}) = q_2\hat{E}(\hat{r})$, giving

$$\hat{E}(\hat{r}) = q_1/\epsilon r^2 \tag{7.30}$$

from which we can recover the potential since

$$\hat{E}(\hat{r}) = -\operatorname{grad}\phi(\hat{r}) = -\frac{d\phi(\hat{r})}{dr}$$

For an infinite plane uniformly smeared with charge at a density of σ per unit area and immersed in a vacuum, an isotropic dielectric, or an electrolyte solution, the field is everywhere perpendicular to the plane and the potential is a function of only one coordinate, the distance from the plane. If the plane is in a dielectric of static dielectric constant ϵ, then the field at a distance x from the plane is given by $4\pi\sigma/\epsilon$ and the potential at that point by

$$\phi(x) = 4\pi\sigma x/\epsilon \tag{7.31}$$

Thus the field is independent of x and the potential falls (or rises) linearly with distance. A collection of static charges placed in the dielectric complicate the electric field but there is no problem in calculating the potential at all points in space from our previous equations; potentials are additive. If the charges are endowed with thermal kinetic energy, the potential at any point is time varying. The problem facing us if we have Avagadro's number of mobile charges is susceptible to the methods of statistical thermodynamics, but not susceptible enough to have yet found a completely satisfactory answer. We take a simple approach and subsequently criticize our basic assumptions. The model is a uniformly charged planar surface immersed in a solution of electrolyte in which ions are represented by point charges and the solvent by a structureless continuum. At a given distance from the surface there will be a time-varying potential resulting from the static charge on the plane and the mobile electrolyte ions. We define a time-averaged potential $\phi(x)$, which we put equal to zero at an infinite distance from the plane. This potential is a function of the average spatial distribution of ions, which is in turn partially determined by the potential. To find a self-consistent solution, we need Poisson's equation — which relates the electric potential at a given point to the charge density $\varrho(x)$ at that point:

$$\frac{\partial^2\phi(x)}{\partial x^2} = -\frac{4\pi\varrho(x)}{\epsilon} \tag{7.32}$$

where we assume the dielectric constant to be independent of x and have written the one-dimensional form of the equation. For an electrolyte solution, the charge density can be written in terms of $n_i(x)$, the number density of ions of type i at the point x:

$$\varrho(x) = \sum_i n_i(x)z_i e \tag{7.33}$$

where the summation is over all types of ion in the unit volume around point x and $z_i e$ is the charge on the ion of type i. We need another relationship in order to solve for $\varrho(x)$, and this can be obtained by remembering that the electrochemical potential of a certain ion must, at equilibrium, be independent of its position in the solution,

$$\tilde{\mu}_i(x) = \tilde{\mu}_i(\infty) = \text{constant for all } x \tag{7.34}$$

where ∞ is a point where $\phi(x) = 0$. *If we now assume that the solution is ideal,* we can write for a single ion:

$$\tilde{\mu}_i(x) = (\mu_i^0)_{T,P} + kT \ln\,[n_i(x)] + z_i e\phi(x) \tag{7.35}$$

Writing the equivalent expression for $\mu_i(\infty)$, we find from (7.34):

$$n_i(x) = n_i(\infty)e^{-z_i e\phi(x)/kT} \tag{7.36}$$

or, defining a reduced potential, $f(x) = e\phi(x)/kT$,

$$n_i(x) = n_i(\infty)e^{-z_i f(x)} \tag{7.37}$$

Equation (7.37) can be obtained more straightforwardly by realizing that the distribution of an ion in a potential field will be governed by the Boltzmann distribution. However, although this is the standard approach, the derivation via the electrochemical potential is perhaps more revealing since it shows that the Boltzmann distribution *as written* implies that the solution is ideal, which no electrolyte solution ever is. We later press this point further. For the moment we proceed with the mathematics and substitute (7.37) into (7.33) to obtain an expression for $\varrho(x)$, which we put into Poisson's equation to give

$$\frac{\partial^2 f(x)}{\partial x^2} = -\frac{4\pi e^2}{\epsilon kT}\sum_i z_i n_i(\infty)e^{-z_i f(x)} \tag{7.38}$$

This *Poisson-Boltzmann (P–B) equation* in one dimension has an analytical solution. The boundary conditions are

$$\begin{aligned} \phi(x) &= \phi_0 \qquad \text{at } x = 0 \\ \phi(x) &= \frac{d\phi(x)}{dx} = 0 \qquad \text{at} \qquad x = \infty \end{aligned} \tag{7.39}$$

The solution for a 1:1 electrolyte is

$$f(x) = e\phi(x)/kT = \frac{2}{z} \ln \left[\frac{1 + \alpha \exp(-Kx)}{1 - \alpha \exp(-Kx)}\right] \tag{7.40}$$

where

$$\alpha = \frac{\exp(ze\phi_0/2kT) - 1}{\exp(ze\phi_0/2kT) + 1} \tag{7.41}$$

$$K = (8\pi ne^2/\epsilon kT)^{1/2} \tag{7.42}$$

and $n = 1/2 \sum_i n_i(\infty)z^2$ is the ionic strength.

ϕ_0, the potential at $x = 0$, is referred to as the surface potential. K is the Debye constant, which has the units of reciprocal length. The Debye length, $1/K$, is a basic quantity in the theory of electrolyte solutions. Equation (7.40) defines a *very* roughly exponential decay of potential from the surface value ϕ_0 to zero at $x = \infty$.

Under certain circumstances the P-B equation and its solution can be considerably simplified, namely, when $z_i f(x) \ll 1$; that is, when $z_i e\phi(x) \ll kT$. (For $z_i = 1$ and $T = 300$ K, $e\phi = kT$ for $\phi \simeq 25$ mV.) This condition allows us to write

$$e^{-z_i f(x)} \simeq 1 - z_i f(x) \tag{7.43}$$

an approximation that converts (7.38) into a linear equation:

$$\frac{\partial^2 f(x)}{\partial x^2} = K^2 f(x) \tag{7.44}$$

with the immediate solution

$$\begin{aligned} f(x) &= f_0 e^{-Kx} \\ \phi(x) &= \phi_0 e^{-Kx} \end{aligned} \tag{7.45}$$

This equation predicts that the potential drops exponentially from its value on the surface, reaching ϕ_0/e at the Debye length $1/K$ (here e is not to be confused with the electronic charge). In going back to (7.42) we see that $1/K \propto [n(\infty)]^{-1/2}$, that is, the Debye length decreases with increasing electrolyte concentration.

The expression for the potential can be used to derive the spatial distribution of the ions from (7.36). Taking the simplest case, we put $z_i = \pm 1$ and obtain plots of the kind shown in Figure 73, from which it is clear that negative ions have a higher concentration near the positive surface than in the bulk and that positive ions are repelled. In contrast to the bulk solution far from the plane, there is a net charge density in the solution close to the charged surface. Note that the distribution curves of cations and anions in Figure 73 are not reflections of one another. It can be shown that the center of gravity of the space charge (for the case $ze\phi(x)/kT \ll 1$) falls on the plane $x = 1/K$ that is, at the Debye length. This is what is meant by saying that $1/K$ is the "thickness of the double layer." For ionic concentrations of 0.1 and 0.001 M monovalent 1:1 electrolyte, $1/K$ at room temperature are approximately 10 Å and 100 Å respectively. The Debye length is a measure of how fast the net space charge near the surface, and induced by the surface charge, dies away. Beyond two or three Debye lengths, the solution is effectively electroneutral.

We can relate the surface potential ϕ_0 and the surface charge σ by realizing that the total system of surface plus solution must be effectively electroneutral. From Figure 74 we see that the charge σ on unit area of surface must thus be equal and opposite to the net charge in a cylinder perpendicular to the area and stretching to infinity:

$$\sigma = -\int_0^\infty \sigma(x)dx = \frac{\epsilon}{4\pi}\int_0^\infty \frac{d^2\phi(x)}{dx^2}\,dx \tag{7.46}$$

The second equality follows from Poisson. The solution is the Gouy–Chapman equation

$$\sigma = [2n(\infty)\epsilon kT/\pi]^{1/2}\sinh(ze\phi_0/2kT) \tag{7.47}$$

where $\sinh(ze\phi_0/2kT) = \frac{1}{2}[\exp(ze\phi_0/2kT) - \exp(-ze\phi/2kT)]$. Two limiting cases are of interest. For ϕ_0 that is very large, $\sinh(ze\phi_0/2kT) \to \frac{1}{2}\exp(ze\phi_0/2kT)$ and the surface potential ϕ_0 is proportional to $\ln\sigma$. At small values of ϕ_0, $\sinh(ze\sigma_0/2kT) \to ze\phi_0/2kT$ and ϕ_0 is proportional to σ. Another system for which voltage is proportional to charge is a parallel-plate capacitor with spacing d for which the capacitance per unit area is given by $C = 4\pi\epsilon/d$. It follows that a charged surface, together with its associated double layer, can be replaced, for the present purpose, by a parallel-plate capacitor with separation $1/K$ and capacitance $4\pi K\epsilon$. Thus K can be derived from measurements of capacitance in the limit of small surface potential.

Before pointing out the defects of the theory, we make some general comments. The physical basis for the theory, which remains effectively unchanged by subsequent improvements, is the unceasing war between

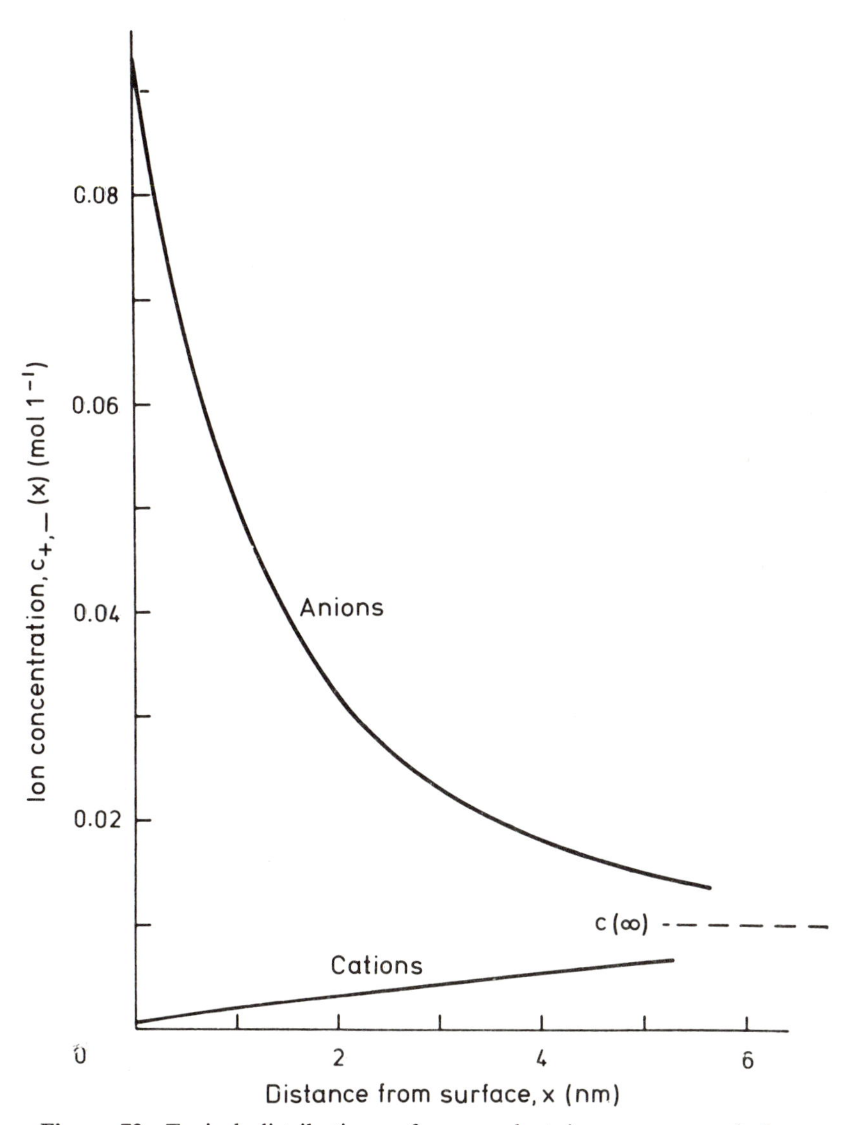

Figure 73. Typical distributions of monovalent ions near negatively charged surface. The bulk concentration at large distances from the surface is 0.01 $\mathrm{mol\,l^{-1}}$. $\phi(0) - \phi(\infty)$ is taken as 56.4 mV, ϵ as 80, and $T = 20°\mathrm{C}$. [From Aveyard, R., and Haydon, D. A., *An Introduction to the Principles of Surface Chemistry*, Cambridge University Press, Cambridge, England, 1973.]

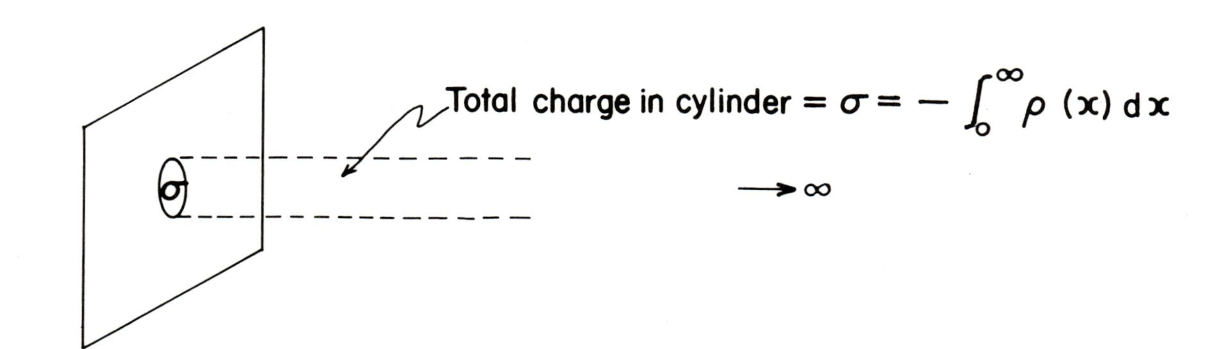

Figure 74.

energy and entropy. Consider a membrane placed in pure water. Dissociated ions produced by ionization of the head groups will tend to diffuse away from the surface and become homogeneously distributed through the available volume of water. This is entropy at work. The electrostatic repulsion between the dissociated ions is an energetic fifth column working on the side of entropy. The attraction between the ionized head groups and their counter-ions is the energetic factor tending to drag the counter-ions back to the surface. The final ionic distribution is a compromise that minimizes the free energy. For solutions containing both positive and negative ions, the details are complicated but the principles remain the same. The electrochemical potential, which was our starting point, contains a pure energetic factor, $z_i e\phi(x)$ and a pure entropy term, $kT \ln[n_i(x)]$. For real, nonideal solutions, these terms remain, but they need modification.

Experimental tests show that the theory based on the linearized P-B equation (7.44) gives the right answers only for very dilute electrolytes. Use of the full P-B equation (7.38) improves matters, but not that much. Without showing evidence of the discrepancy, we go straight to the causes.

The expression we used for the electrochemical potential is appropriate for an ideal solution. In real solutions the concentration of the ions must be multiplied by an activity coefficient that appears as a consequence of intermolecular interactions. In (7.32) we have already tried to take into account one intermolecular force since $\phi(x)$ is supposed to be an average potential at an ion due to the surface charge *and all the other charged ions.* However, this is not enough, as it treats molecules as dimensionless points and the medium — water — as a structureless continuum. The intermolecular forces between real ions include close-range repulsive forces that can be modeled as a Lennard–Jones 6–12 potential or a hard-sphere model or some other approximation. At low electrolyte concentrations, the effects of this potential in bulk solution are negligible, but the fact that ions occupy space means that an ion cannot approach nearer to the surface than its effective radius r. Thus the expressions (7.36) and (7.37) for the distribution of ions in the solution have no meaning for $x < r$. The absurd consequences of (7.36) can be appreciated by putting $\phi_0 = 300$ mV and $n(\infty) = 10^{-2}$ molar, which gives a concentration of over 1000 molar at the surface. This difficulty was realized by Stern,[125] who modified the Gouy–Chapman model by placing a layer of adsorbed ions on the charged surface. The theory has been discussed in too many places[126] to warrant further dissection, but we note that the idea of an adsorbed layer links up with our previous discussion of structural forces. Stern limited the effect of finite ionic size to the adsorbed layer, but we know that it is in general impermissible to ignore molecular volumes in the liquid phase. In technical terms, the pair-correlation function in liquids oscillates for distances up to a few molecular diameters. This fact has been literally smoothed over by writing the energy terms in the electrochemical potential as a function only of the electrostatic potential and excluding short-range repulsive forces.

There are other things wrong with the Gouy–Chapman theory, but this is not the place to discuss them. To be fair, the theory is a good qualitative guide to reality, and even for physiological saline (~0.14 M) agreement between theory and experiment for lipid bilayers is quite reasonable for many systems. In deciding whether to use the approximate form (7.45) or the exact form (7.40), we should bear in mind that surface potentials of both artificial and natural membranes usually range from 50 to 100 mV, which means that $e\phi/kT > 1$, invalidating the condition for the approximation (7.43). For distances exceeding a few Debye lengths (7.40) reduces (for a 1:1 electrolyte and $e\phi_0/kT < 1$) to

$$\phi(x) = \frac{4kT}{ze}\alpha e^{-Kx} \qquad (7.48)$$

(Use $(1+y)/(1-y) \simeq 1 + 2y$ *for* $y \ll 1$.)

Two specific features of membranes require us to modify our physical model: The head groups are neither completely ionized nor can they be considered to provide uniformly smeared surface charge. The second point is illustrated by Figure 75, which shows a map of the electrostatic potential near the surface of a DPPE bilayer calculated[127] on the basis of the crystal structure shown in Chapter 1. It is doubtful whether it is worth trying to incorporate maps of this kind into a quantitative double-layer theory, especially in view of the complex chemical constitution of real membranes with their mixture of lipids. However, the consequences of partial ionization can be predicted. We follow the reasoning of Ninham and Parsegian[128] and consider a surface carrying acidic groups, *HA*, with a surface density of $1/S$, where S is the average area per acid group — which in practice falls in the range 100 $Å^2$ to 400 $Å^2$. Depending on pH, temperature, ionic strength, and so on, a certain fraction α of the groups will be ionized, resulting in a negative surface surface charge density due to the anionic A^- group:

$$\sigma = -e\alpha/S \qquad (7.49)$$

At the surface a familiar chemical relationship must be obeyed:

$$\frac{[H^+][A^-]}{[HA]} = K_a, \quad \text{the acid dissociation constant} \qquad (7.50)$$

Since $\alpha = [A^-]/([A^-] + [HA])$, we can rewrite (7.50):

$$[H^+]_s = K_a \frac{1-\alpha}{\alpha} \tag{7.51}$$

where the subscript s is a reminder that we are concerned with the concentration of protons immediately in the vicinity of the surface. Now $[H^+]_s$ also obeys the Boltzmann distribution,

$$[H^+]_s = [H^+]_{\text{bulk}} e^{-(e\phi_0/kT)} \tag{7.52}$$

Of course $[H^+]_s$ in (7.51) and (7.52) must have the same value. Furthermore, ϕ_0 must be partially determined by α. Substituting (7.52) into (7.51), we have

$$K_a = [H^+]_{\text{bulk}} \frac{\alpha}{1-\alpha} e^{-(e\phi_0/kT)} \tag{7.53}$$

which, if we assume K_a to be known, is an equation containing two unknowns. There is another relationship between ϕ and α:

$$\left(\frac{d\sigma}{dx}\right)_0 = -\frac{4\pi\sigma}{\epsilon} = \frac{4\pi e\alpha}{\epsilon S} \tag{7.54}$$

The second equality follows from (7.49), and the first from the theorem in electrostatics that states that the electric field $\hat{E} = -d\phi/dx$ due to a charged surface is given by $4\pi\sigma/\epsilon$. From (7.53) we can derive an expression for α in terms of ϕ_0, which can be substituted into the boundary condition (7.54). The remaining basic equations needed to define the system are the Poisson equation, the charge density $\varrho(x)$ in terms of ionic concentrations, and the boundary condition $d\phi(x)/dx = 0$ for $x = \infty$. Ninham and Parsegian, who first attacked this problem, addressed themselves to the interesting problem of the interaction between *two* identical parallel ionizable planar surfaces. All we need to do to obtain the basic equations for this case is to change the final boundary condition $d\phi(x)/dx = 0$ at $x = \infty$ to

$$\frac{d\phi(x)}{dx} = 0 \qquad \text{at } x = b \tag{7.55}$$

where the distance between the plates is $2b$ (Figure 76.) The mathematics is too lengthy to be repeated here; instead we summarize the physics. The authors included the presence of both mono and divalent cations and plots of ion concentration against distance are shown in Figure 76. The plots refer to the concentrations between the plates that sandwich a solution in contact with an external reservoir at pH 7, for $n_+^{(r)}/n_{++}^{(r)} = 38.8$ and for total concentration $n^{(r)} = 0.972 \times 10^{20}$ ions cm^{-3} (i.e., 0.16 M). These conditions are reasonably physiological. The main conclusions to

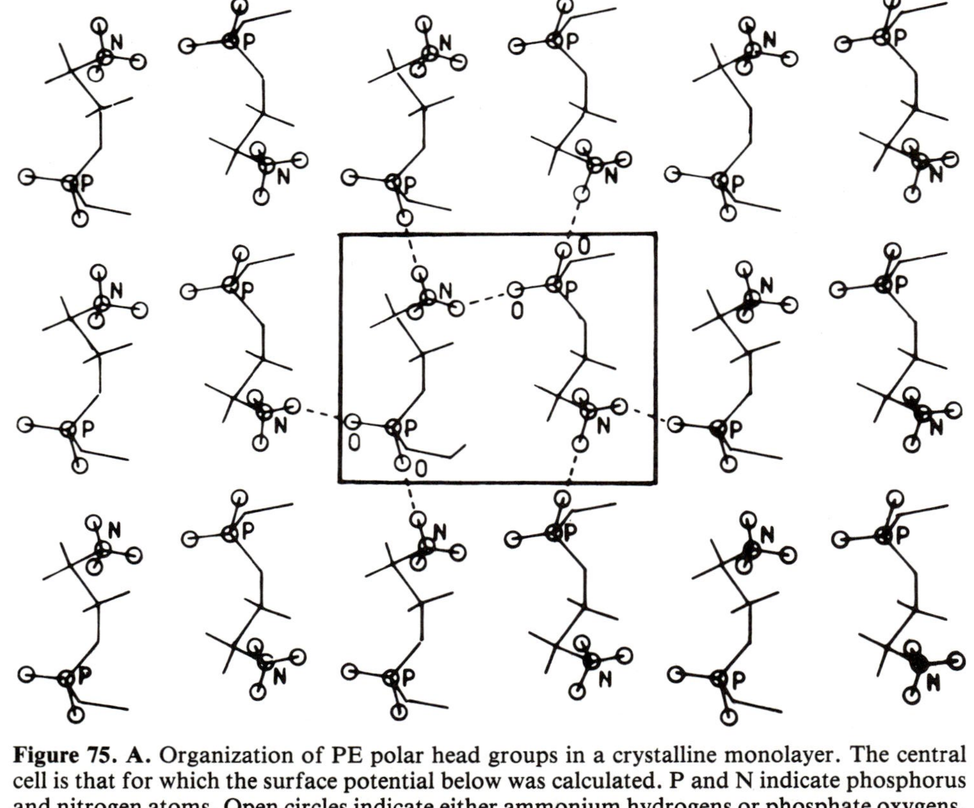

Figure 75. A. Organization of PE polar head groups in a crystalline monolayer. The central cell is that for which the surface potential below was calculated. P and N indicate phosphorus and nitrogen atoms. Open circles indicate either ammonium hydrogens or phosphate oxygens.

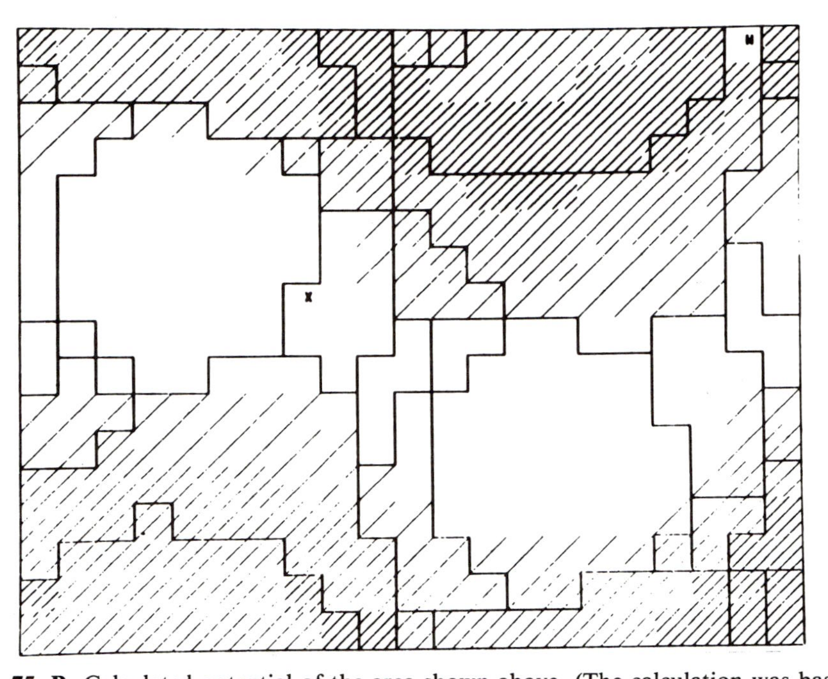

Figure 75. B. Calculated potential of the area shown above. (The calculation was based on 98 molecules, that is, on 49 cells.) The potential becomes increasingly negative with increasing density of shading. The white areas (high positive potential) are associated with nitrogen and the dark areas (high negative potential) with the phosphate groups. [Both from Zakrzewska, K., and Pullman, B., *FEBS Lett.* 131:77 (1981).]

be drawn from the plots are:

1. The ionic concentrations at the midpoint are not very different from the reservoir for spacings of 20 and 60 Å.
2. Anions are strongly repelled from the neighborhood of the surfaces.
3. Divalent cations have concentrations near the surfaces com-

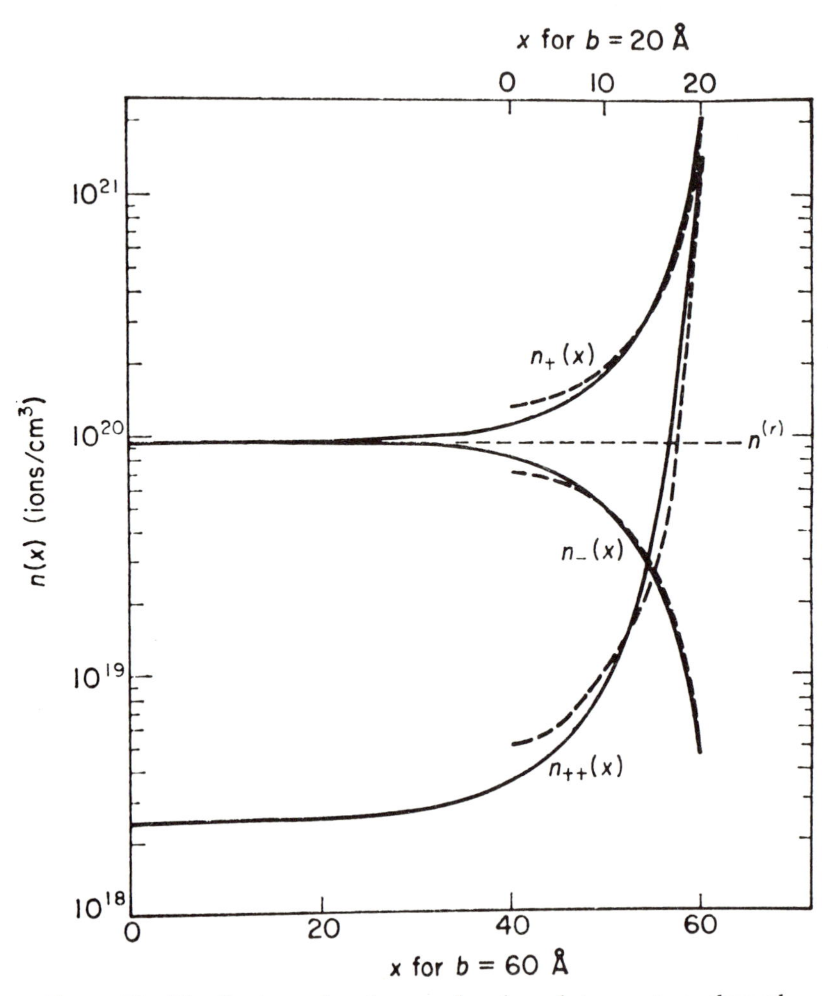

Figure 76. Distribution of cations and anions between two charged surfaces separated by 40 Å (dashed lines) and 120 Å (solid lines). $n^{(r)}$ is the concentration of the reservoir. Notice that the ratio of surface to bulk cation concentration is far higher for doubly charged ions. [From Ninham, B. W., and Parsegian, V. A., *J. Theor. Biol.* 31:405 (1971).]

parable to those of monovalent cations even though in the reservoir $n_{+}^{(r)}/n_{++}^{(r)} \simeq 40$. This may explain the strong effects of small concentrations of divalent cations in biologic systems.

4. For the conditions stated, the Debye length is 7.8 Å. The plots show that deviations from the reservoir concentrations only become significant at distances of less than about two Debye lengths from the surface.

Another important finding was that the pH of the surfaces is very insensitive to surface separation, even to distances well below the Debye length. In looking at equations (7.51) and (7.52), we see that relative constancy of pH implies that both surface charge and surface potential are insensitive to surface separation.

We turn to a question of central importance in weighing the forces between approaching cells: What is the *electrostatic force* between two ionizable surfaces separated by an electrolyte?

The force acting between two parallel planar surfaces carrying *fixed* charges has been derived in a number of ways.[129] We give one method here and then treat the effects of replacing fixed charges by ionizable groups.

We consider two very large charged plates separated by a dielectric medium containing, for simplicity, a 1:1 electrolyte. The sandwiched solution is continuous with the solution outside the plates containing electrolyte at a concentration of $n(\infty)$ moles liter^{-1}. Ions cannot penetrate the surfaces. The electric potential far from the plates is put at zero and the pressure at $P(\infty)$. We make the likely assumption that the charged plates exert a force on one another and are held in position by an externally applied pressure, which we now attempt to find. If the system is at equilibrium, then every infinitesimal volume of solution between the plates is at *mechanical* equilibrium. The forces acting on such a volume are of two kinds, a purely mechanical force due to the hydrostatic pressure gradient $dP(x)/(dx)$ and a force that is electrostatic in origin, $\varrho(x)d\phi(x)/(dx)$. These two forces must sum to zero:

$$\frac{dP(x)}{dx} + \varrho(x)\,\frac{d\phi(x)}{dx} = 0 \qquad (7.56)$$

Note that the origin of the coordinates lies midway between the surfaces. We now substitute for ϱ using Poisson's equation:

$$\frac{dP(x)}{dx} - \frac{\epsilon}{4\pi}\cdot\frac{d^2\phi(x)}{dx^2}\cdot\frac{d\phi(x)}{dx} = 0 \qquad (7.57)$$

or

$$\frac{d}{dx}\left[P(x) - \frac{\epsilon}{8\pi}\left(\frac{d\phi(x)}{dx}\right)^2\right] = 0 \qquad (7.58)$$

which means that at all points:

$$P(x) - \frac{\epsilon}{8\pi}\left(\frac{d\phi(x)}{dx}\right)^2 = \text{constant} \tag{7.59}$$

If the plates carry equal charge density, then the system is symmetric and at the middle point $d\phi(x)/dx = 0$ (Figure 77). Substituting in (7.59), the constant is given by $P(0)$. The force acting on unit area of a plate is given by $P = P(0) - P(\infty)$, which can be found by integrating (7.56) from the point midway between the plates to a point in the external solution:

$$P = P(0) - P(\infty) = \int_{\infty}^{0} dP(x) = -\int_{0}^{\phi(0)} \varrho(x)d\phi \tag{7.60}$$

Now for a 1:1 electrolyte we can write

$$n_+(x) = n(\infty)\exp(-e\phi(x)/kT)$$
$$n_-(x) = n(\infty)\exp(e\phi(x)/kT)$$

So that

$$\varrho(x) = e[n_+(x) - n_-(x)] = -2\,ne\sinh(e\phi(x)/kT) \tag{7.61}$$

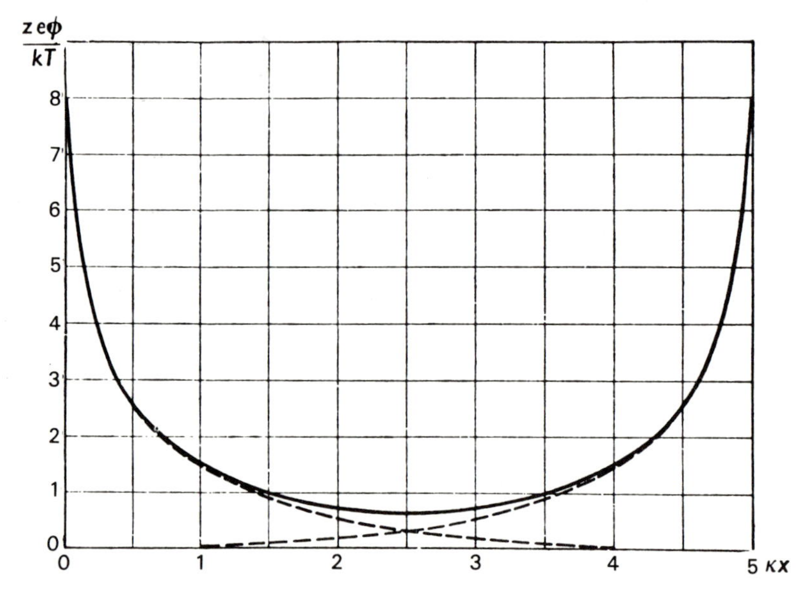

Figure 77. Illustrating the construction of an approximate potential function for two slightly overlapping double layers by the superposition of the individual potentials.

Therefore:

$$-\int_0^{\phi(0)} \varrho(x)d\phi = \int_0^{\phi(0)} 2ne \sinh(e\phi(x)/kT)d\phi =$$

$$2nkT\left[\cosh\left(\frac{e\phi(0)}{KT}\right) - 1\right]$$

And thus:

$$P = 2nkT\left[\cosh\left(\frac{e\phi(0)}{kT}\right) - 1\right] \qquad (7.62)$$

which solves the problem if we know $\phi(0)$, the potential at the midpoint.

A generalization of (7.62) that takes into account the presence of divalent ions is

$$P = nkT[\xi - 1 + (1/\xi - 1)(1 - \eta/2) + (1/\xi^2 - 2)\eta/2] \qquad (7.63)$$

where the pressure is in dynes cm^{-2}, $\xi = \exp(e\phi(0)/kT)$, and η is defined by

$$n_+^{(r)} = n(1 - \eta)$$
$$n_{++}^{(r)} = n\eta/2 \qquad (7.64)$$

For $\eta = 0$ (7.63) reduces to (7.62). The above expressions for pressure require a knowledge of $\phi(0)$ before they can yield quantitative answers. Since $\phi(0)$ must be a function of the distance between the opposing surfaces, calculations can be lengthy. A convenient and quantititatively reliable approximation can be obtained, however, in the case of large distances, in which we can neglect the effect of one surface's double layer on the other surface. The potential at any point between the surfaces can then be taken to be a simple superposition of the potentials due to each plate. Furthermore, at the midpoint, which is presumed to be several Debye lengths from the surfaces, we can use (7.48) for the potential due to each surface resulting in $\phi(0) = (8kT/ze)\alpha \exp(-Kd)$ (Figure 77). This potential is substituted into (7.62) and the resulting expression is simplified by expanding the exponentials up to the quadratic terms. The result is

$$P = 64nkT\alpha^2 e^{-2Kd} \qquad (7.65)$$

where α is given by (7.41). For $e\phi_0/2kT \ll 1$, a further approximation yields

$$P = 16nkT(e\phi_0/kT)^2 e^{-2Kd} \qquad (7.66)$$

If we assume that the surface potential is independent of the surface separation, then both (7.65) and (7.66) give a simple exponential dependence of pressure on distance. The assumption of constant potential is a feature of the famous DLVO (Deryaguin–Landau–Vervey–Overbeek) theory of the forces between colloidal particles. For two surfaces having different and constant potentials, Parsegian and Gingell[130] give the pressure as

$$P_\phi(d) = \frac{4e^2n}{kT} \cdot \frac{-(\phi_2^2 + \phi_1^2) + \sigma_1\sigma_2(e^{2Kd} + e^{-2Kd})}{(e^{2Kd} - e^{-2Kd})^2} \tag{7.67}$$

If it is assumed that surface *charge* is independent of separation, then

$$P_\sigma(d) = \frac{8\pi}{\epsilon} \cdot \frac{(\sigma_2^2 + \sigma_1^2) + \sigma_1\sigma_2(e^{2Kd} + e^{-2Kd})}{(e^{2Kd} - e^{-2Kd})^2} \tag{7.68}$$

For $\sigma_1 = \sigma_2$ and $Kd \gg 1$, this reduces to

$$P = \frac{8\pi}{\epsilon}\,\sigma^2 e^{-2Kd} \tag{7.69}$$

[We can now substitute for σ using (7.47), and again putting $e\phi_0/kT \ll 1$, we recover (7.66)].

A more realistic model for our purposes is that of two opposing ionizable surfaces. The ionic concentrations between the surfaces were discussed earlier in the chapter (see Figure 76) and the preferential attraction of divalent cations to the (negatively charged) surface is again evident in the strong effect that these ions have in reducing the pressure between ionizable surfaces. Notice that the Debye length, which is a measure of the extent of the influence of the double layer, decreases if the charge on the cations is increased; see (7.42). The calculated pressures shown in Figure 77 display near-exponential dependence on separation down to the Debye length. This means that the simple forms (7.66) and (7.69) are reliable for most purposes.

While this chapter is primarily concerned with theory rather than experiment, the laboratory consequences of the double layer are of interest, particularly those that provide a more or less direct means of measuring potentials. We will glance at two phenomena that ultimately depend on surface charge.

The movement of a solution of ions past a charged surface provides a means of estimating the potential near the surface. We consider a planar charged surface and focus on a parallelopiped in the neighboring solution having area A and depth δx (Figure 78). If this volume element is within the double layer, there will be a finite charge density — which we label $\varrho(x)$. An electric field applied parallel to the surface will cause the element to move by exerting a force $E\varrho(x)A\delta x$. When a steady-state velocity is reached, this force is balanced by the frictional force opposing motion:

$$F(x+\delta x) - F(x) = E\varrho(x)A\delta x \qquad (7.70)$$

where the frictional forces are given by

$$F(x) = \eta \frac{dv(x)}{dx} A \qquad (7.71)$$

with v being the velocity. From (7.71) we have

$$\frac{dF(x)}{dx} = \eta \frac{d^2v(x)}{dx^2} A \qquad (7.72)$$

and from (7.70):

$$\underset{x\to 0}{Lt} \frac{F(x+\delta x) - F(x)}{\delta x} = \frac{dF(x)}{dx} = E\varrho(x)A \qquad (7.73)$$

If this result is substituted into (7.72) and $\varrho(x)$ is replaced by the use of the Poisson equation (7.32), we find that

$$-\epsilon E \cdot \frac{d^2\phi(x)}{dx^2} = 4\pi\eta \frac{d^2v(x)}{dx^2} \qquad (7.74)$$

A first integration, using the boundary conditions $d\phi(x)/dx \to 0$ and $dv(x)/dx \to 0$ as $x \to \infty$, gives

$$-\epsilon E \cdot \frac{d\phi(x)}{dx} = 4\pi\eta \frac{dv(x)}{dx} \qquad (7.75)$$

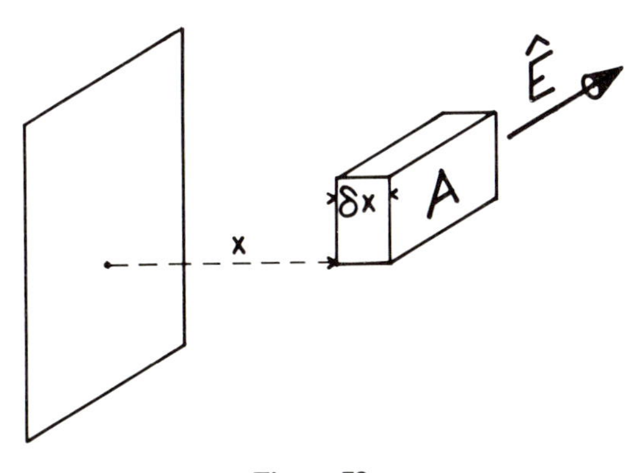

Figure 78.

A second integration yields

$$-\epsilon E\phi(x) = 4\pi\eta v(x) + \text{constant}$$

The normal boundary condition on the velocity in the case of a liquid continuum would be to take the flow velocity at the solid surface as zero. In the present case, being conscious of the probable reality of the Stern layer of strongly adsorbed ions, we take the so-called "slip plane" to be situated at a small distance from the surface. At this slip, or shear, plane the velocity is zero and the potential is termed the ζ potential. With these conditions, (7.76) becomes

$$-\epsilon E\phi(x) = 4\pi\eta v(x) - \epsilon E\zeta \tag{7.77}$$

However, at large distances from the surface $\phi(x) \to 0$, and $v(x)$ approaches a maximum value, which we denote by v_{bulk}. These physical constraints convert (7.77) into:

$$\zeta = \frac{4\pi\eta v_{\text{bulk}}}{\epsilon E} \tag{7.78}$$

This is the Helmholtz–Smoluchowski equation and provides the basis of a practical method for measuring the potential at a rather ill-defined point within the double layer. For a spherical particle the corresponding expression is

$$\zeta = \frac{6\pi\eta v}{\epsilon f(Kr)} \tag{7.79}$$

In this expression v is the *electrophoretic mobility* of the body (which could be a vesicle), that is, velocity per unit field. The function $f(Kr)$ is not simple, but clearly must approach a value of 1.5 as r becomes very large so as to agree with the result (7.78) for a planar surface. For very small r, the function $f(Kr)$ approaches unity. If we are prepared to equate the experimental ζ potential with the surface potential, we can use (7.47) to deduce the surface charge density.

A very different strategy for the determination of surface charge and potential is the use of spectroscopic probes. As an example, we look at the measurement of surface potential by means of spin labels. Charged spin labels of the general type

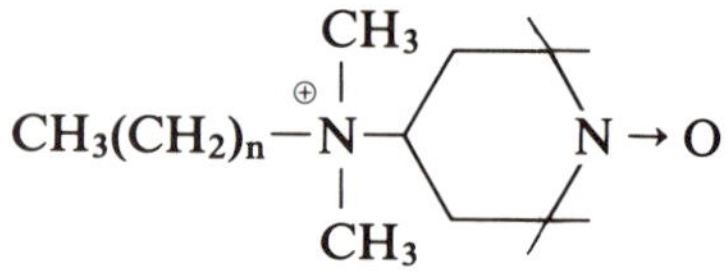

$$CH_3(CH_2)_n-\overset{\oplus}{N}(CH_3)_2-C_5H_6(CH_3)_4N\rightarrow O$$

bind to bilayers and biological membranes by penetrating into the hydrophobic core. In the absence of a surface potential, a partition coefficient can be defined by:

$$\lambda_0 = \frac{\text{Number of bound labels}}{\text{Number of free labels}} \tag{7.80}$$

In the presence of a surface potential, the partition coefficient becomes

$$\lambda = \lambda_0 \exp(-\phi_0 zF/RT) \tag{7.81}$$

For bilayers it is possible to vary the surface potential by changing the ionic strength of the medium. The resulting changes in the partition coefficient of the spin label are easy to follow since the bound and free spin labels are characterized by slightly different spectra. The variation of surface potential is in excellent agreement with that expected from the simple double-layer theory given earlier in this chapter. Spin-label techniques have also been applied to the estimation of potential differences across membranes and to the measurement of pH gradients.[131] Fluorescent probes have also been employed for similar purposes, a particularly interesting type of experiment being based on the Stark effect. Strong electric fields produce readily observable changes in the electronic spectra of molecules. If $\Delta\hat{\mu}$ is the change in the dipole moment of a chromophore due to electronic excitation and $\Delta\alpha$ is the corresponding change in polarizability, then an electric field of strength E will result in a frequency shift of the transition given by

$$\Delta\nu = -\frac{1}{h}\Delta\hat{\mu}\cdot\hat{E} - \frac{1}{2h}\Delta\alpha E^2 \tag{7.82}$$

While not all the changes in the spectra of fluorescent probes adsorbed on membranes can be attributed solely to the Stark effect, the very existence of changes due to variation of the surface potential is proving to be a sensitive means of examining membrane surface properties, particularly for excitable membranes.[132] This type of experiment and others too numerous and varied to describe here are rapidly strengthening our belief in our ability to predict the electrical properties of bilayers with satisfying accuracy. Real membranes are a tougher problem theoretically since they contain many kinds of molecules arranged in unknown patterns and having ill-defined charge distributions. It is highly probable that some components of membranes have conformations that are sensitive to ionic strength, pH, specific cations, and other factors. There is no a priori reason to rule out the existence of membrane electric fields having lateral components. Exotic jungles await our exploration.

SUMMARY OF INTERMEMBRANE FORCES

One of our main objects in this chapter has been to examine the distance dependence of the various forces acting between two bilayers. These forces are measurable[119] and we now review and compare them. Subsequently we consider briefly the nature of the internal forces that are responsible for the structural integrity of the bilayer.

The hydration force can be ignored at interlayer distances of over ~20 Å. At shorter distances the hydration force easily dominates electrostatic repulsion and attractive dispersion force, even for membranes carrying high surface charge.

Electrostatic forces dominate the *repulsive* interaction between lipid bilayers and between vesicles at distances over ~20–30 Å. For strongly charged bilayers, electrostatic repulsion can overcome the attraction due to dispersion forces at all distances so that stable multilamellar assemblies never form. The same situation can arise for weakly charged bilayers in media of low ionic strength, a fact that can be appreciated by examining the consequence of a decrease in the Debye constant in (7.69). Electrostatic repulsion may originate in the adsorbtion of divalent cations on neutral bilayers, in which case the rise in surface potential accompanying the mutual approach of the bilayers can result in desorbtion.

Dispersion forces are, for neutral or weakly charged surfaces, of longer range than other forces, and we can picture the formation of stable stacks of bilayers or of clusters of cells as resulting from an initial long-range attraction drawing the bodies together until repulsive forces grow strong enough to balance the attractive force. At this equilibrium distance, the system reaches a potential minimum (Figure. 79). For neutral bilayers a measurement of the bilayer separation at equilibrium and of the bilayer thickness allows an estimation of both the hydration force (using $P = P_0 \exp(-d_w/\lambda)$ with values of the parameters taken from experiment) and of the dispersion force using (7.25). The only unknown is the Hamaker coefficient H, which can be obtained by equating the two forces. Examples of such an analysis are given in Table 5 based on the data of Lis et al.[119]

Up to now we have implicitly assumed that bilayers have an internal structure that is independent of their separation. This is certainly not likely to be true in mixed lipid systems in which, for example, the approach of two vesicles could result in the migration of more highly charged lipids away from the directly opposing areas. However, even for planar one-lipid systems it can be expected that the imposition of force on a bilayer can affect the molecular packing. Direct experimental evidence on this point is contained in the work of Lis et al.[133] on the compressibility of bilayers. The main qualitative conclusions of their study

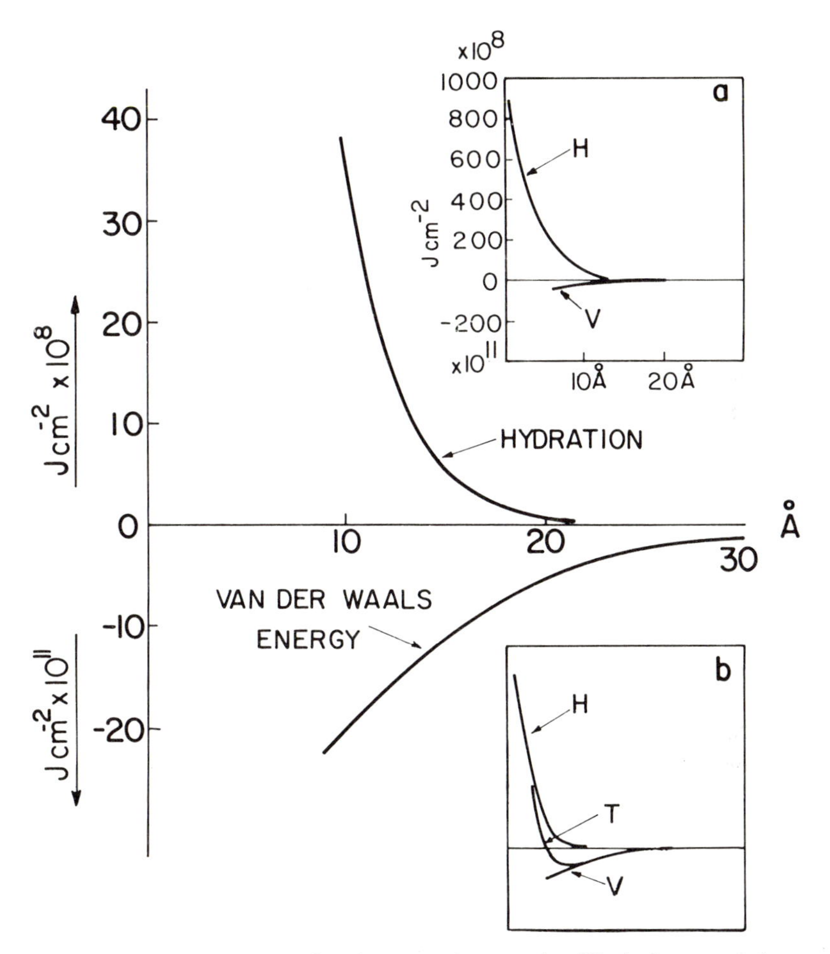

Figure 79. Typical curves for the attractive van der Waals force and the repulsive hydration force between planar bilayers. Note the different energy scale above and below the zero energy line. **A**. Extended vertical scale. **B**. Schematic illustration of the decisive role played by hydration energy in determining the position of the potential energy minimum for the total energy *T*.

are that (1) lateral pressure decreases the surface area of the lipid molecules; (2) at high pressures, phase changes often occur, such as crystallization of the chains; (3) the compressibility curves are not explicable in terms of a single intramembrane interaction, and (4) chain melting does not result in a large increase in compressibility, in contrast to the effect of added cholesterol or the introduction of unsaturated chains.

The modest effect of chain melting is perhaps consistent with the small change in volume accompanying the gel–liquid crystal transition. The nonlinear variation of a molecular cross-sectional area with force is not too surprising, considering that there are several interactions within the bilayer that have to be taken into account: repulsion between head groups, repulsion between chains, and the interfacial surface tension between the fluid hydrocarbon chains and the hydophilic water-containing head-group zone. It has been found that the pressure–area curves for bilayers bear little resemblance to those for the corresponding monolayers. Studies of this kind are important in elucidating not only forces within membranes, but also the fluctuations in lateral packing of lipids reflected in the compressibility data.[134] Note the variety of forces within the membrane: Chain repulsion is an excluded volume effect, head-group repulsion is a mixture of excluded volume and electrostatic forces, and the surface tension term is attributed to the "hydrophobic force" discussed in the previous chapter. The balance between the effective lateral pressure, due to the surface tension, and the repulsive forces maintain the membrane structure and can provide a framework within

Table 5
The Distance Dependence of the Pressure Between Neutral Phospholipid Bilayers

Values are listed for P_o and λ in the expression $P_o \exp(-d_w/\lambda)$.

Lipid	$\log_{10} P_o$	λ
DLPC	9.72	2.6
DMPC(27°C)	9.94	2.6
DPPC	9.83	2.0
DPPC(50°C)	10.99	2.2
DOPC	9.6	2.9
Egg PC	9.76	2.6
Egg PE	10.57	2.1

Temperature is 25°C unless indicated otherwise. Units of pressure, dynes cm^{-2}. [From Le Neveu et al., *Biophys, J.* 37:657 (1982).]

which to treat the mechanical properties of membranes, such as force–deformation relationships.[31] Modulation of any of the forces will affect membrane geometry, and consequently intermembrane forces. A simple example is the effect of divalent cations, which can influence not only the intermembrane electrostatic force, but also, by modifying the charge state of the surface, can induce changes in chain packing, and even phase, as we saw in Chapter 4. In the following chapter we will have a closer look at the interior of the bilayer.

We end with a problem. The aggregation of bilayers and of cells (both dead and living) seems to be explicable in terms of the forces described in this chapter. However, in the laboratory the approach of two phospholipid vesicles is often followed by their fusion. Fusion can frequently be induced or accelerated by divalent cations such as Ca^{++} or Mn^{++}, which may well displace water from the vicinity of the head groups and also reduce the electrostatic repulsion between negatively charged bilayers. The mechanism of fusion is not known. The hydration force, in the absence of structural rearrangement of the bilayer, would seem to present an almost insuperable barrier to the close approach of bilayers, even in the absence of electrostatic forces. Despite an enormous experimental effort,[29] we are a long way from understanding the details of fusion at a level that would satisfy a physical organic chemist.

8 Theory and the Lipid Bilayer

The structural and dynamic complexity of phospholipid aggregates has not deterred theoreticians. In Chapter 4 we saw how steric factors have provided a conceptual basis for predicting organizational variations in single and mixed phospholipid systems but the model employs structureless geometric shapes to represent molecules and cannot accommodate properties such as order parameters that depend on detailed molecular structure. In this chapter we delve deeper and review some attempts at a theoretical description of the dynamics of lipid chains in bilayers. This does not imply that theoreticians have not addressed themselves to other properties of bilayers, such as their mechanical and electrical characteristics. Any but a cursory explanation of mechanical properties is beyond our scope (see Chapter 7). Electrical properties were dealt with at an elementary level in our discussion of forces in Chapter 7 and the conductivity of membranes will concern us in Chapter 13. Attempts have been made to calculate the fields due to membranes[127] but the difficulty of finding simple and credible models for the charge distribution in the polar head groups is one reason for the comparative theoretical neglect of the surface of the membrane. The dynamic behavior of lipid chains in bilayers is, on the other hand, understood reasonably well, as we shall see. The nature of bilayers containing proteins is the subject of a feud of Guelph–Ghibelline ferocity, which we summarize in Chapter IX. Here we concentrate on what theory has to say about the chains in simple bilayers.

Theoretical chemistry might be defined as the art of approximating but the specific method used to distort molecular reality and the choice of facts to be quietly buried at night are matters as personal as religious faith. In this chapter we will outline the basic tenets of three major creeds: the excluded volume approach (Catholic in its rigidity), the mean field theory (a softer orthodoxy), and the molecular dynamics method (fundamentalist). Our starting point must be a consideration of forces.

The dependence of a given type of force on molecular identity and intermolecular distance can usually be described by approximate expressions that have the backing of both theory and experiment (see Chapter 7). Occasionally we can avoid dealing directly with forces, as when we use a mass of experimental data to put a value on the energy difference between two conformations, say trans and gauche bonds in an alkane. Since one of the principal objectives of theoretical work on bilayers has been an attempt to account for or predict the behavior of lipid chains, it has been inter- and intrachain forces that have received the most attention rather than forces between head groups. In fact many model makers have avoided head-group interactions by regarding the polar surface of bilayers as a structure onto which lipid chains are anchored and whose properties are simulated by a simple surface tension type of term in the Hamiltonian such as γA, where A is a unit area and the value of γ is guessed at.

In building a model for the hydrophobic core, we need to take into account both intramolecular and intermolecular forces. For a lipid chain in a vacuum, an important determinant of conformation is the collection of internal forces that favor trans over gauche conformations by ~ 0.5 kcal/mol. Strong short-range repulsions also have to be implicitly taken into account since they prevent conformations in which different sections of the alkane chain approach each other too closely. The intermolecular forces also include this repulsive force, together with longer range attractive dispersion (van der Waals) forces. Different authors have treated intermolecular forces in rather different ways. One starting point is the realization that short-range repulsive forces can have a decisive influence on molecular orientation in condensed phases. The reader may care to try the following exercise: Throw the contents of a box of matches onto a large sheet of paper, using a helping finger to ensure that no match rests on another. Measure the angle θ that each match makes with one side of the sheet and calculate the average value of $\frac{1}{2}(3 \cos^2 \theta - 1)$. The result will be close to one quarter, which is the expected result for a collection of lines arranged randomly in a plane.

(Note the difference between this result and that for three dimensions, — $S_{ii} = 0$.) Repeat the procedure using progressively smaller pieces of paper. As the area shrinks, the matches are forced into an increasing degree of alignment. Quantitative measurements confirm the visual impression. Results of such an experiment are shown in Figure 80. The system goes from complete disorder to nearly perfect ordering, but the change is not gradual. The order parameter rises rapidly over a small range of diminution of area — we have a two-dimensional "phase transition" between two states that could be labeled "liquid" and "liquid crystal." There are no attractive forces between our molecules: The sole reason for the transition is the inability of one match to penetrate the volume occupied by another. We speak of an *excluded volume* effect. For neutral molecules or moieties this effect is a direct consequence of powerful repulsive forces, but these are of such short range that they can be well approximated by a simple, infinitely hard potential defined as being zero for any distance above a certain minimum and infinitely large below that distance.

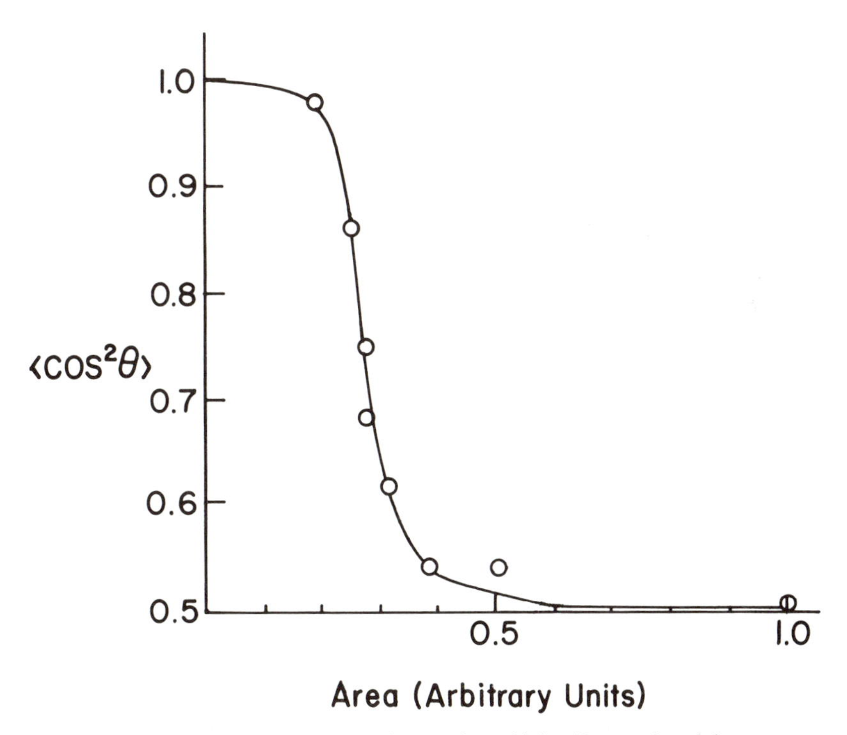

Figure 80. Results of an experiment in which 40 match sticks were distributed at random on rectangular planes of varying size. See text for details.

The match model should not be carelessly taken as an analog of the *thermally induced* isotropic-to-nematic phase change of a liquid crystal. It is obvious that temperature changes would not induce ordering changes in the assembly of matches: we need an *attractive force* to induce molecules or matches to come together in a constant-volume system. For lipid chains in bilayers, attractive forces manifest themselves directly in the experimental fact that T_t rises with increasing chain length, which is a consequence of the increasing dispersion forces between the chains. For a given head group, increasing chain length diminishes the relative importance of the head-group interactions, as emphasized by the fact that a plot of T_t against $1/(n-3)$, where n is the number of chain carbons, extrapolates to 137°C for $n = \infty$,[21] while polyethylene melts at 138°C. Dispersion forces between chains are obviously not to be ignored in setting up a model for the bilayer and its order–disorder transition. In the excluded volume theories, the onus of ordering is placed on repulsive forces and the "soft" attractive forces are denied the right to induce order directly, their role is limited to compression of the system, exactly as we diminished the area available to the matches.

One alternative to the excluded volume approach is provided by mean field theory in which ordering is usually attributed to attractive rather than repulsive forces, and in which we forgo any attempt to define the instantaneous forces acting between specific molecules. We regard the force acting on a given molecule to be an *average* taken over the time-varying interactions with the rest of the system. An analogy is a man being jostled in a dense crowd. If the crowd as a whole is static, then the average force on the man is zero even though he may come home bruised by countless digs in the ribs. Furthermore, if the man attempts to move away from his average position, he will meet a restoring force which, on the average, has cylindrical symmetry. In the case of human beings this force is an excluded volume effect, and in a dense crowd it would act on an individual to maintain both his location and upright position. He is one molecule in a monolayer of oriented humans and is partly responsible for the restoring force acting on other crowd members. In the mean field theory of nematic crystals the restoring force on each molecule in the nematic phase is approximated as a time average over the molecule's interactions with the rest of the system. This force — sometimes supposed to be a dispersion force, sometimes unspecified in origin — is responsible for the intermolecular contribution to the molecule's energy and affects the molecule's average conformation and orientation. The individual molecules have a range of allowable conformations and the mean field determines the probability distribution of these conformations at a given temperature, just as an electric or magnetic field determines the orientational distribution of electric or magnetic dipoles.

Now since the mean field force originates on molecules that are themselves subject to the same field, we have a self-consistent problem whose solution, for a bilayer, gives the distribution function for the conformations of the chains from which thermodynamic properties and order parameters can be derived.

We are now at a crossroads. Our aim is to construct a theory that rationalizes the structure and properties of bilayers. We are aware of the forces involved but know that we cannot hope to avoid approximations and simplifications. Thankfully the major paths followed by theoreticians — the excluded volume and mean field theories — have in common some basic steps dictated by the precepts of statistical thermodynamics. We must first define the allowable microstates of the system. In the present case this entails specification of the different acceptable chain conformations. The next step is to assign energies to the microstates, using whatever approximations our consciences permit. The energies are the raw material for the partition function, which in turn can be processed to yield the thermodynamic properties of the system.

We consider first the excluded volume method following the variation formulated by Nagle.[135] The reader should realize that the mathematical apparatus is well above the level of this book; our object is to give a feeling for the model and its consequences. In enumerating the microstates, a basic assumption is that C—C bonds in lipid chains are allowed only three states: t, g^+, and g^-, the degenerate gauche states lying ~0.5 kcal/mol above the trans. The time spent between these conformations is effectively placed at zero and the states have long enough average lifetimes to ignore uncertainty in their energies. Even with these restrictions, the number of allowed microstates in a macroscopic system is enormous.

Nagle's strategy is to avoid the seemingly inevitable approximate solution by setting up a simple model that is exactly soluble, at least as far as the configurational problem is concerned. He reduces the problem to two dimensions while retaining a critical view of the concurrent dangers. The different chain configurations are represented by paths drawn on a two-dimensional lattice (Figure 81). A trans bond is represented by a vertical line, and the two gauche conformations by diagonal lines. The paths are infinitely long — a seemingly suicidal choice made in order to obtain the infinitely sharp phase transition that mathematics guarantees only for infinitely large systems. Small systems show broad transitions (see Chapter 4). A given chain cannot appear twice at the same level of the lattice, a rule which precludes doubling back of chains. *No point on the lattice can be crossed by more than one chain.*

We see that the excluded volume method as used here implies an infinitely hard repulsive potential, which, when translated into topologic restrictions, results in the complete exclusion of certain conformations; that is, it effectively assigns infinite energy to these conformations. This automatically results in a limited range of microstates, favouring in particular, those with roughly aligned chains. Thus the model does what is demanded of it in that it attributes ordering to repulsive forces. (There are those who will be reminded of the self-avoiding random walk.) Any one *set* of lines represents a possible microstate of our two-dimensional model and the set of all diagrams thus represents the set of all microstates of the system. The unoccupied sites correspond to states of less than maximum density. Apparently we have to construct all the possible diagrams and list their energies. This is hardly a practical proposition but the whole point in choosing the model is that we can sidestep the construction of diagrams. This is so because there is a well-known problem in statistical thermodynamics that is both exactly soluble and isomathematical with our model of lipid chains. Each diagram can be converted,

Figure 81. Set of possible states for chains confined to a triangular lattice. The solid lines are hydrocarbon chains; the dotted lines indicate the lattice. The chain on the left is in the all-trans state for which E_{rot} is zero. Next to it is an all-gauche configuration with $E_{rot} = 8\epsilon$. The chain on the right has $E_{rot} = 3\epsilon$ and the remaining chain $E_{rot} = 2\epsilon$. [From Nagle, J. F., *J. Chem. Phys.* 58:252 (1973).]

by a straightforward procedure, into a diagram representing a set of two-atom fragments (dimers) on a hexagonal lattice. Some correspondences of this kind are shown in Figure 82 and the reader can perhaps deduce the rules that take a chain diagram into a dimer diagram (or she can read Nagle's paper). The correspondence is unique. Now if the forces between dimers are defined, each dimer diagram can be assigned an energy so that the complete set of dimer diagrams and associated energies represent the possible microstates of the system. (A physical realization could be the adsorbtion of a homonuclear diatomic gas on a crystalline surface.)

We now have the information necessary to evaluate the partition function of the dimer model but appear to face the same problem as we face for the chain diagrams: There is too much counting to be done, especially if the diagrams are infinite. The partition function for the dimer model, however, can be obtained by mathematical trickery

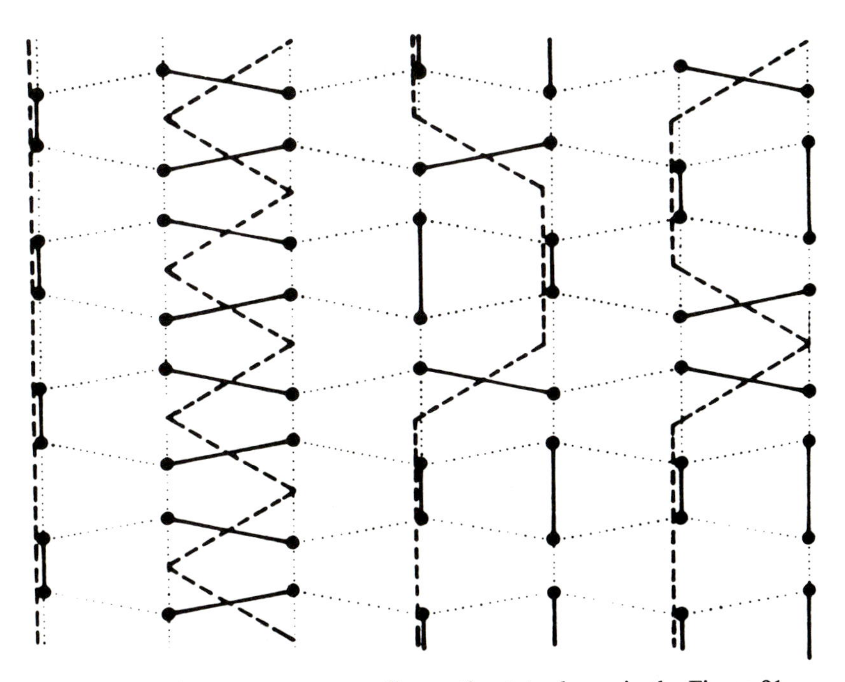

Figure 82. Dimer state corresponding to the state shown in the Figure 81. The dimers are represented by solid lines joining pairs of vertices on the new lattice, which is indicated by dotted lines. Dashed lines show the original chains. The triangular lattice has been repressed. [From Nagle, J. F., *J. Chem. Phys.* 58:252 (1973).]

without the need to draw the diagrams. Leaving much mathematics aside, we ask the reader to believe that because of the one-to-one correspondence of diagrams it is also possible to evaluate the partition function for the chains-on-a-lattice problem. It must be realized that the information fed into the partition function up to now includes only the energies of the different configurations of the total system; the attractive forces maintaining the integrity of the bilayer have not yet appeared. Thus it is proper to speak of the *configurational partition function* from which we can derive the configurational free energy:

$$F_{\text{conf}} = U - TS = -kT \ln(p.f.)_{\text{conf}} \qquad (8.1)$$

The energy arising from the internal conformation of each chain and the excluded volume effect on the allowable system configurations have been treated exactly.

An elementary example may help the reader to believe in the connection between the diagrams and the thermodynamic parameters of the system: An all-trans chain is represented by a straight vertical line. If we bend the line at one point (see the two right-hand lines in Figure 82), we add one gauche bond to the system. This adds 0.5 kcal to U in (8.1) because of the gauche–trans energy difference. It also adds $-k \ln 2$ to the configurational entropy because there are two energetically equivalent ways of bending the chain and Boltzmann told us long ago that $S = k \ln W$. The introduction of a gauche bond will affect the attractive force between chains, a factor not yet discussed.

In dealing with attractive forces, Nagle makes no attempt to define their microscopic form. The *total* attractive energy of the sample is assumed to be a function of the average density ϱ of the system, which seems reasonable since the dispersion energy is a bulk property. (Notice that the model does not include head-group interactions, which are perhaps better taken to be surface rather than bulk dependent.) The dispersion energy is expressed in the form:

$$E_{\text{disp}} = a_{vdW} \varrho^{b} \qquad (8.2)$$

where a value of around 5 kcal/mol CH_2 is given to a_{vdW} on the basis of the heat of sublimation of hydrocarbon chains. The final results of the theory are not too dependent on physically reasonable variations in b. By incorporating the dispersion energy into the internal energy of the system and adding a PV term, we can obtain the Gibbs free energy of the model. [Changes in volume with temperature are rather small (Chapter 3), even in going through the bilayer transition temperature.] If we fix the temperature and minimize the free energy, we obtain the thermodynamically stable states of the system. From the value of the free-energy turning points and the values of its derivatives with respect to, say, temperature and pressure, we then can derive the thermodynamic properties of the equilibrium states and the phase diagram. In view of the gross simplification of reality involved in the two-dimensional lattice model, we should not expect too much in the way of agreement with experiment. The predicted transition temperature of 353 K for DPPC can be compared with the experimental value of 315 K. The main value of the theory is the demonstration that the excluded volume interaction can account for a *rotameric* disordering transition, in which the number of gauche bonds is reduced on freezing. The most obvious criticism that can be leveled at the theory is that the two-dimensional lattice overestimates the excluded volume effect in that the addition of a third dimension would allow chains to avoid each other more easily and thus provide additional allowable conformations.

In comparison with excluded volume theories, mean field theory is mathematically straightforward. We roughly follow the footsteps of Marcelja,[120] whose work is based mainly on the ideas used by Maier and Saupe[136] in their theory of nematic liquid crystals. Focusing on a particular lipid molecule, we write its energy in the form:

$$E^{(i)} = E^{(i)}_{\text{int}} + E^{(i)}_{\text{disp}} + \gamma A^{(i)} \tag{8.3}$$

The superscript labels a specific conformation i of the lipid chain. The first term in the energy is the intramolecular contribution, which depends only on the number of gauche bonds. The second term is the van der Waals intermolecular force between the molecule and all other molecules in the system. It is the assumption that this force is a time-independent average over the motion of the "other" molecules that lies at the heart of the mean field method. The last term is area dependent and carries within its artfully simple form steric repulsions, electrostatic interactions, and hydrophobic effects. We now examine the nature of $E^{(i)}_{\text{disp}}$.

For a *rigid* rod-shaped molecule in the nematic phase, Maier and Saupe took E_{disp} to have the form:

$$E_{\text{disp}} = -\phi\left[\tfrac{1}{2}(3\cos^2\beta - 1)\right] \tag{8.4}$$

where β is the angle between the long axis of the molecule and the director. The strength of the interaction with the environment for different nematics is contained in the magnitude of ϕ, which is sometimes termed the molecular field. E_{disp} has cylindrical symmetry about the director and a minimum at $\beta = 0$. There is no a priori reason for taking the angular dependence to have the form given by Maier and Saupe, and more general functions can certainly be written down, but the ideas of the mean field theory are not dependent on refinements of E_{disp}. [(8.4) can be written $E_{\text{disp}} = -\phi P^2_0$, where P^2_0 is a Legendre polynomial. Expansion of E_{disp} in the even Legendre polynomials is theoretically attractive but the experimental determination of the expansion coefficients leaves much to be desired.] By accepting (8.4) for a rigid molecule, we can approximate the form of E_{disp} for a lipid chain by supposing that each segment in a chain makes an independent contribution to the dispersion energy. Marcelja writes

$$E_{disp}^{(i)} = -\phi(n_{tr}^{(i)}/n) \sum_j \left(\frac{3}{2}\cos^2 \beta_j^{(i)} - \frac{1}{2}\right) \tag{8.5}$$

where $n_{tr}^{(i)}/n$ is the fraction of trans C—C bonds in the conformation i and the sum is over all segments j in the chain. The angle $\beta_j^{(i)}$ is the angle between the director and the direction of the jth segment defined as that normal to the plane containing the two C—H bonds. We leave aside discussion of the uncertain theoretical basis for the factor $n_{tr}^{(i)}/n$. Now the molecular field ϕ depends on the average conformations of all the molecules in the bilayers, and this is expressed as follows by Marcelja:

$$\phi = V_0 \left\langle (n_{tr}/n) \sum_j \left(\frac{3}{2}\cos^2 \beta_j^{(i)} - \frac{1}{2}\right)\right\rangle \tag{8.6}$$

where the angular brackets indicate an average taken over all the chains j in the system. Note that in the crystalline state every chain is in the all-trans configuration, which implies that $n_{tr}/n = 1$ and all $\beta_j^{(i)} = 0$ so that $\phi = V_0$. This means that for the crystal $E_{disp}^{(i)} = -nV_0$. In the hypothetical completely random configurational state of the bilayer's core, the value of $(\frac{3}{2}\cos^2 \beta_j - \frac{1}{2})$ is zero when averaged over the motion of a segment so that $E_{disp}^{(i)}$ is zero. We therefore can roughly identify V_0 with the melting energy per CH_2 group of the appropriate crystalline hydrocarbon. From the data for polyethylene, this gives $V_0 = 680$ cal/mol CH_2 unit as an upper limit to V_0 in bilayers.

To perform the average in (8.6) we have to know the probability of occurrence of each configuration, and this in turn is determined, via the Boltzmann distribution, by the total energy $E^{(i)}$ of each configuration,

$$p^{(i)} = \exp(-E^{(i)}/kT)/\sum_i \exp(-E^{(i)}/kT) \tag{8.7}$$

where the denominator is, of course, the partition function. We see that to evaluate $p^{(i)}$ we need the $E^{(i)}$, and so we have come full circle: To find the molecular field ϕ, we need the $p^{(i)}$. This requires a knowledge of $E^{(i)}$, which can be evaluated only if we know ϕ. We have a self-consistent problem that can be solved by asking the computer to try a range of values of ϕ (at a given temperature) until a value is found that allows a simultaneous self-consistent solution to equations (8.5) through (8.7). Of the two constants V_0 and γ, the first is usually left as a free parameter, with its value chosen to give the best fit of theory to experiment for one property of the bilayer, say the order-parameter profile. The physical meaning and the value of γ are the subjects of much learned hand waving. An increase in the surface area of a liquid at, say, the gas–liquid interface is almost invariably associated with a proportional increase in the number of molecules of liquid at the surface. The free-energy difference between bulk and surface molecules is reflected in the magnitude of γ, the surface tension. The meaning of the term γA in equation (8.3) is different; it represents the energy required to change the area of a lipid molecule at the bilayer surface while *preserving* the total number of molecules at the surface. The term is thus more analogous to the energy of distortion of a two-dimensional elastic sheet. The physical forces contribution to A presumably include the interactions between head groups and the ill-named "hydrophobic force," the raising of the free energy when lipid chains are exposed to water. On the basis of guesswork and back-of-the-envelope types of calculation, γ is usually assigned a value of ~ 18 dynes/cm.

A good example of the above approach is found in the work of Schindler and Seelig.[137] By taking $V_0 = 590$ cal/mol and $\gamma = 18.5$ dynes/cm, they obtained an excellent fit to the order-parameter profile of DPPC at 41°C. Using the same values of V_0 and γ, a phase transition was found at 39.5°C; at this temperature there are two physically meaningful self-consistent values of ϕ, each associated with the same free energy for the system, corresponding to two phases in equilibrium. The result is pleasantly close to the experimental value of 41°C for T_t. Furthermore, the average chain length is calculated to be 13.7 Å at 41°C while the all-trans chains are 19.7 Å long. The predicted shortening on melting is thus ~ 12 Å for the bilayers as compared with 11–12 Å found by X-ray studies. If we assume that the cross-sectional area of a chain in a given conformation is expressible as

$$A^{(i)} = 20.4 \text{ Å}^2 \times (\text{length of all-trans chain})/l^{(i)} \qquad (8.8)$$

where $l^{(i)}$ is the projected length on the bilayer normal, then the average cross-sectional area of a chain can be deduced from the average length. The result at 41°C is 29.3 Å^2 or 58.6 Å^2 per phospholipid molecule, a result to be compared with the value of 66 Å^2 at 45°C as found by Janiak et al.[15] using X-ray diffraction. The theory gives excellent agreement between the theoretical and experimental values of the linear expansion coefficient.

The success of the mean field theory and the comparative ease with which it can be applied account for its current popularity and raise questions as to the profitability of the theoretically more complex excluded volume method. The mean field model certainly appears to be nearer to physical reality in the case of the liquid crystal phase, but it cannot provide a proper account of the detailed molecular nature of the phase transition. This is so because, unlike the excluded volume diagrammatic approach, the mean field model does not admit the existence of fluctuations; the essence of the mean field theory is its time-averaged view of the world. The excluded volume theory takes into account all possible conformations of the *system*, not just one chain, and it therefore includes conformations that represent states of the system well removed from spatial homogeneity. Such states are of central importance in the vicinity of phase transitions and critical points. Furthermore, the mean field theory tends to be vague as to the origin of the field. The nomenclature E_{disp} presumes that the force acting on a chain is of the van der Waals brand and it is easy to show (and fairly obvious) that attractive forces will favor parallel alignment of two rodlike molecules, thus producing an ordering force. Order-of-magnitude estimates of the force between chains made according to the methods of Chapter 7 give molecular fields far smaller than those emerging from self-consistent publications. This is not too surprising since it is certain that the field must contain a strong contribution from repulsive forces. Luckily this shortcoming does not affect the results of the theory since in practice only the functional form of ϕ is specified, the magnitude being taken care of by the empirically determined value of V_0.

Caillé et al.[138,139] have presented a theory that incorporates both the excluded volume and mean field approaches. A modification has been used by Pink, Green, and Chapman in their interpretation of the Raman spectra of bilayers (Chapter 5). We sketch the more important aspects of this interesting synthesis in its original version.

One basic simplification characterizing the theory is the limitation of chain configurations to only two: the all-trans state, labeled 0, and a fictitious state m, which is an average over all the disordered chain states. The cross-sectional areas at the surface of the two states are written A_0 and mA_0, where $m > 1$. Chains are arranged on a hexagonal lattice having unit cells of area A. State 0 chains occupy one cell; state m chains "occupy" two cells, even though m is certainly less than 2. This insistence on maintaining the regularity of the hexagonal lattice has great computational advantages but falsifies reality. One consequence is that the gain in spatial entropy accompanying melting does not appear in the theory. However, this is not too serious a drawback since it is the much larger increase in entropy attributable to rotomeric disordering that is the main driving factor for melting.

The attractive forces between chains are confined to nearest neighbors and can be written

$$W = \sum_{\alpha,\beta} K_{\alpha\beta} N_{\alpha\beta} \qquad (\alpha, \beta = 0, m) \tag{8.9}$$

where $N_{\alpha\beta}$ is the number of adajcent pairs of type (α, β) and $K_{\alpha\beta}$ is the corresponding interaction energy for a single pair. (Readers may hear echoes of the Bragg–Williams theory of binary alloys.) The internal energy, entropy, and free energy of the chains are given by E_α, S_α, and f_α, respectively, where $\alpha = 0$ or m. The partition function for a single chain is thus given by

$$Z_\alpha = \exp(-f_\alpha/kT) \qquad (\alpha = 0,\ m) \tag{8.10}$$

We can now construct the Hamiltonian and the system partition function. The Hamiltonian is written:

$$\begin{aligned} H = &-K_{\alpha\alpha}\sum_{ij}\mathcal{L}_{\alpha i}\mathcal{L}_{\alpha j} - K_{\beta\beta}\sum_{ij}\mathcal{L}_{\beta i}\mathcal{L}_{\beta j} \\ &-K_{\alpha\beta}\sum_{ij}(\mathcal{L}_{i\alpha}\mathcal{L}_{j\beta} + \mathcal{L}_{i\beta}\mathcal{L}_{j\alpha}) + \sum_{\substack{k=\alpha,\beta \\ i}}(E_k - TS_k + \pi A_k)\mathcal{L}_{ki} \end{aligned} \tag{8.11}$$

π is the surface pressure (Chapter 7). The summations $\langle ij\rangle$ are over neighboring pairs of chains. The $\mathcal{L}$'s are projection operators whose properties are very simple: When $\mathcal{L}_{i\alpha}$, for example, operates on the system wave function, it gives the wave function multiplied by unity, if the chain at site i is in state α, or zero, if the chain is in state β. Thus the term $-K_{\alpha\alpha}\Sigma_{\langle ij\rangle}\,\mathcal{L}_{\alpha i}\mathcal{L}_{\alpha j}$ gives the total energy of interaction of all neighboring chains, both of which are in the state α. H is now converted to the following mean-field Hamiltonian, an expression that gives the energy of a *single site* in terms of the average state of the whole system:

$$H_{MF} = \sum_i (h_\alpha\,\mathcal{L}_{\alpha i} + h_\beta\,\mathcal{L}_{\beta i}) \tag{8.12}$$

where

$$h_\alpha = -zK_{\alpha\alpha}\langle\mathcal{L}_\alpha\rangle - zK_{\alpha\beta}\langle\mathcal{L}_\beta\rangle + (E_\alpha - TS_\alpha + \pi A_\alpha) \tag{8.13}$$

$$h_\beta = -zK_{\beta\alpha}\langle\mathcal{L}_\beta\rangle - zK_{\beta\alpha}\langle\mathcal{L}_\alpha\rangle + (E_\beta - TS_\beta + \pi A_\beta) \tag{8.14}$$

and z is the coordination number of the site — in our case six. Notice that all the operators in (8.12), (8.13), and (8.14) are single-site operators, consistent with the mean-field philosophy. [A feeling for the significance of the terms in H_{MF} can be obtained by putting all chains in the state α. The $\mathcal{L}_{\alpha i}$ has the eigenvalue unity at all sites and $\mathcal{L}_{\beta i}$ has the eigenvalue zero, so that H_{MF} reduces to

$$H_{MF} = \sum_i h_\alpha\,\mathcal{L}_{\alpha i} \tag{8.15}$$

Now $\langle\mathcal{L}_\beta\rangle$ in h_α is the average value of the eigenvalue of $\mathcal{L}_{\beta i}$, which is clearly zero, just as the eigenvalue of $\langle\mathcal{L}_\alpha\rangle$ is unity, so that

$$H_{MF} = \sum_i\,[-zK_{\alpha\alpha} + E_\alpha - TS_\alpha + \pi A_\alpha]\,\mathcal{L}_{\alpha i} \tag{8.16}$$

which is just an expression for the internal energy of all the chains and the energy of interaction of all pairs.] The Hamiltonian allows the energy levels to be determined, and hence the partition function:

$$Z(N, \pi, T) = \sum_{N_m} \sum_{N_{\alpha\beta}} \Omega(N, N_m, N_{\alpha\beta})(Z_0^{N_0} Z_m^{N_m}) \exp\left(-\frac{W}{kT} - \frac{\pi A_t}{kT}\right) \quad (8.17)$$

A_t is the total area of the system; Z_0 and Z_m have been defined above, and they are intramolecular partition functions. The exponential function contains the intermolecular energy and the pressure–area term. $\Omega(N, N_m, N_{\alpha\beta})$ is the number of intermolecular configurations for a given number of chains in state m and a given number of nearest neighbor pairs. Configurations with fixed N_m and $N_{\alpha\beta}$ are degenerate and it is clear that Ω can reach enormous values. The evaluation of Ω is simplified by the assumption of a hexagonal lattice — the combinatorial problem has been solved for this and more complex cases.[140] (If we replace our two chain states by the two states α and β of an electron and the intermolecular dispersion force by the spin-exchange interaction, we find we have the famous Ising model for planar magnetic systems.) As is usual in problems involving this type of degeneracy in systems containing many molecules, one set of values of N_m and $N_{\alpha\beta}$ gives much the greatest contribution to the partition function, and this single maximum term is used in place of Z. The free energy and other properties of the system follow. Where is the excluded-volume interaction in all this? It is hiding in Ω, for this function of N_m and $N_{\alpha\beta}$ is determined by the fact that we have restricted molecules to lattice sites and this restriction is, as in Nagle's treatment, the topologic expression of the molecules' mutual impenetrability — that is, their hard-core repulsions. (A similar model can be used to give a fair account of our match experiment.) If the fraction of molecules in the disordered state is given by $y_m = N_m/N$, then the free energy per molecule obtained from the partition function is given by

$$G/N = kT[y_m \ln y_m + (1 - y_m)\ln(1 - y_m)] + \tfrac{1}{2}zK_{00}(1 - y_m)^2 + y_m(f_m - f_0) + \pi A_0(1 + my_m) \quad (8.18)$$

At thermodynamic equilibrium this is at a minimum. Notice that only K_{00} is retained, thus reducing to a minimum the parameters while leaving the form of the equation unchanged. Putting $(\partial G/\partial y_m)_{T,\Pi} = 0$, we find:

$$\mathfrak{M} = \tanh(1/2kT)\left(\frac{z\,|K_{00}|\,\mathfrak{M}}{2} + H(T)\right) \quad (8.19)$$

where

$$\mathfrak{M} = 1 - 2y_m \quad (8.20)$$

and

$$H(T) = T(S_m - S_0) - (E_m - E_0) - \pi m A_0 - z\,|K_{00}|/2 \qquad (8.21)$$

$\mathfrak{M}$, which takes the values 1 and -1 for an all-trans and a completely disordered layer respectively, is a measure of order — that is, an order parameter. It is easy to rewrite $H(T)$ as the difference between the free energy for the 0 and m states. At equilibrium between two phases this difference must vanish so that the condition for a first-order transition is $H(T) = 0$, and we can write

$$0 = T(S_m - S_0) - (E_m - E_0) - \pi_{eq} m A_0 - z\,|K_{00}|/2 \qquad (8.22)$$

(at equilibrium)

Similarly at the critical temperature, where the distinction between the phases vanishes, we have

$$0 = T^*(S_m - S_0) - (E_m - E_0) - \pi^* m A_0 - z\,|K_{00}|/2 \qquad (8.23)$$

It follows that

$$(\pi^* - \pi_{eq}) = (T^* - T)\,\Delta S/mA_0 \qquad (8.24)$$

where

$$\Delta S = S_m - S_0$$

Equation (8.24) can be compared with the experimental pressure versus area curves for a phospholipid monolayer spread on water. Furthermore, a result in the mean field treatment of the Ising model in magnetism can be interpreted in our terms as

$$T^* = z\,|K_{00}|/4k \qquad (8.25)$$

which means that the experimental critical temperature T^* gives us a value for K_{00}. The area A_0 is known and m can be estimated at roughly 0.6. From (8.24) and the experimental data for π^* and T^*, we can now estimate ΔS and, finally, from (8.23), ΔE. It is true that we have appealed to experiment but the achievements of the theory should not be disdained. A first-order transition is predicted and the curve giving the locus of points of coexistence of two phases is not too far from experiment. The value of the order parameter $\mathfrak{M}$ can be obtained by putting $H(T) = 0$ in (8.19), inserting the value of K_{00}, and solving the resulting equation, but $\mathfrak{M}$ does not have too much connection with the order parameters normally measured in bilayers. For that matter, the theory is for *monolayers*, although we can pretend that it is for bilayers by fixing the lateral pressure at ~ 50 dynes/cm (see Chapter 7).

The above theory has been expanded considerably by using ten allowable chain conformations, introducing segment order parameters, and treating the dispersion force in more detail. The modified theory has been used to discuss the Raman spectra of lipids and other properties of bilayers, including protein–lipid systems. In general, it can be said that the theory is an attempt to give both excluded volume and dispersion forces their due while retaining a physically quite acceptable model of the phospholipid system.

The theories sketched above have their deficiences and leave one with a guilty feeling of compromise. A priori calculations on a structure of the complexity of a bilayer are not feasible at present but physical models with considerably more credibility than those described earlier in the chapter have been handled by the power of the computer. As an example of computer-simulated dynamics we describe a study of alkane chains by van der Ploeg and Berendsen.[141] They studied an ensemble of decane molecules by the *molecular dynamics* technique, which has its philosophic roots in the 18th century idea that, given the position and velocity of every particle in the universe at a certain time, the whole future and past course of history could be predicted by the application of Newton's laws. The starting point of the molecular dynamics method is the allocation of a position to each atom or group of atoms in the model. Next the forces operating on the atoms are specified and the system is then "released." The atoms move in response to the forces defined by the initial configuration. As they move, interatomic distances and forces change and continually determine the subsequent motion of the system. The configuration of the system can be computed as a function of time, the size of the experimental model being limited by the size of the computer. Since the initial state of the system is arbitrary, and, except by coincidence, will not represent a state near equilibrium, sufficient time must be allowed at the beginning of the "experiment" for the system to equilibrate. This relaxation time is naturally chosen to be long in comparison with the time scale of molecular motion, which, in a bilayer, is determined by molecular vibrations and rotations. A period of several tens of picoseconds is sufficient.

We now consider the details of the model treated by van der Ploeg and Berendsen. Their bilayer consisted of two opposing layers, each containing 16 decane molecules arranged in a square and constrained to be periodic in two dimensions. The alkane chains were assigned an area at the surface of 25 Å^2. The heads of the decane molecules were confined to the vicinity of the bilayer surface by a harmonic potential

$$V(z) = \tfrac{1}{2}k_h(z - \bar{z})^2 \tag{8.26}$$

where $k_h = 300 \text{ kJ mol}^{-1} \text{nm}^{-2}$ and the z-axis is normal to the bilayer.

The potential is a surrogate for those forces that maintain the integrity of the head group sheet in bilayers built from amphiphiles. In writing down the inter- and intramolecular forces, the CH_2 and CH_3 groups are taken as united atoms interacting via Lennard–Jones-type potentials. All intermolecular pairs interact; only those intramolecular pairs separated by more than three carbon atoms interact, a limitation corresponding to that on the closest distance of approach of two carbon atoms in the same chain. The potential has the familiar form:

$$V(r_{ij}) = 4\epsilon\left[\left(\frac{\sigma}{r_{ij}}\right)^{12} - \left(\frac{\sigma}{r_{ij}}\right)^{6}\right] \tag{8.27}$$

The values for ϵ and σ were taken as 0.4301 and 0.374 for methylene groups, as 0.6423 and 0.274 for methyl groups, and as 0.9203 and 0.422 for head groups. The units are kJ mol^{-1} and nm. In the excluded volume and mean field methods, the rotational isomerism is handled by specifying the gauche–trans energy difference. In the molecular dynamics method, the complete dihedral potential function, giving the energy for all angles, is essential to obtain the time-varying configuration of the system. The function, based on work on liquid alkanes, is written as an expansion in the dihedral angle ϕ,

$$V(\phi) = \sum C_i(\cos\phi)^2 \quad \text{kJ mol}^{-1} \tag{8.28}$$

with $C_0 = 9.2789$, $C_1 = 12.1557$, $C_2 = -13.1201$, $C_3 = -3.0597$, $C_4 = 26.2403$, $C_5 = -31.4950$. These parameters give $V(\phi) = 0$ for $\phi = 0$; 12.3428 kJ mol^{-1} for $\phi = 60$ degrees, and 2.9288 kJ mol^{-1} for $\phi = 120$ degrees. The bond-bending potential for the C—C bond was assumed to be harmonic:

$$V(\alpha) = \tfrac{1}{2}k_b(\alpha - \alpha_0)^2 \tag{8.29}$$

where $k_b = 520$ kJ mol^{-1} rad.$^{-2}$

The configuration of the bilayer chains was computed at intervals of 8×10^{-3} picoseconds (ps) over a period of 80 ps. The initial equilibration time was 100 ps. Order parameters are obtained by averaging over the time-dependent orientations of the chain segments. The results are in excellent agreement with those measured experimentally for S_{CD}. However, additional information emerges from the simulation. Concentrating on one methylene group, we define the x- and y-axes as the H—H vector and the bisector of the H—C—H angle respectively. Order parameters S_{xx} and S_{yy} for these vectors can be defined in the same way as S_{zz}. In our discussion of order parameters in Chapter 5, we assumed isotropic rotation of the chains about the molecular long axis. If this is true, then we can write $S_{xx} = S_{yy} = -\tfrac{1}{2}S_{zz}$, where the second equality follows from the fact that the order-parameter tensor is traceless. The molecular dynamics treatment of the decane bilayer shows that S_{xx} is not equal to $-\tfrac{1}{2}S_{zz}$, indicating anisotropic rotation about the z-axis. There is experimental evidence for this phenomenon in the lyotropic liquid crystal formed from potassium 4,4-difluoromyristate in water.[142] Above 38°C this molecule gives L_α-type bilayers. The CF_2 fragment was used as a probe of segmental order by exploiting the F—F dipolar coupling and the chemical shift of ^{19}F, both of which are anisotropic. The existence of two "handles" allows a determination of all the diagonal components of the order-parameter tensor. Putting the x-axis along the F—F vector and the z-axis perpendicular to the CF_2 plane, the following results were obtained: $S_{xx} = -0.209$; $S_{yy} = -0.155$, and $S_{zz} = 0.364$. Obviously the segment is not reorienting isotropically around the director. Returning to the molecular dynamics study, we find that there is a correlation between θ, the angle of tilt of the chain as a whole, and the angle ϕ between the molecular yz plane and the plane containing the molecular z-axis and the bilayer normal (Figure 83). At high tilt angle, there is a clear preference for small ϕ. This means that the alkane chain tends to swing "sideways" in a motion that the authors liken to the opening of a penknife blade.

A particularly interesting result of the molecular dynamics treatment is the prediction of a time-dependent *collective* tilt of the chains. The motion of the chains is, on the average, symmetric about the bilayer normal, which is thus by definition the director. (Note that this statement is not inconsistent with the inequality of S_{xx} with S_{yy}.) However, the excursions of neighboring chains from the director are correlated. A quantitative measure of this correlation was constructed by using the vector S_i connecting the middle of the first C—C bond to the middle of the C_6—C_7 bond in chain i. The magnitude of the vector sum $T = \Sigma_i S_i$ gives a measure of the correlation of the directions of individual molecules. The angle θ_t between T and the director gives the average tilt of the group of molecules under consideration. Now both T and θ_t are time dependent and the computer simulation shows that they are strongly correlated — the larger the average tilt, the more the chains are aligned. This has a qualitative geometric explanation: If the heads of the chains are confined to constant area, then chain packing will be denser, the greater the tilt of a group of chains. This effect will be more pronounced at the "upper" part of the chains.

The molecular dynamics "experiment" tarnishes the image of the kink as a major contributor to chain disorder. The calculated occurrence of gauche bonds is about 25%. A kink contains two gauche bonds and a decane molecule has nine C—C bonds, so that if gauche bonds appear only in kinks there is a probability of roughly one kink per chain. The computer simulation gives only 0.20 kink per chain, close to the expected occurrence of kinks, if gauche bonds are distributed randomly along the

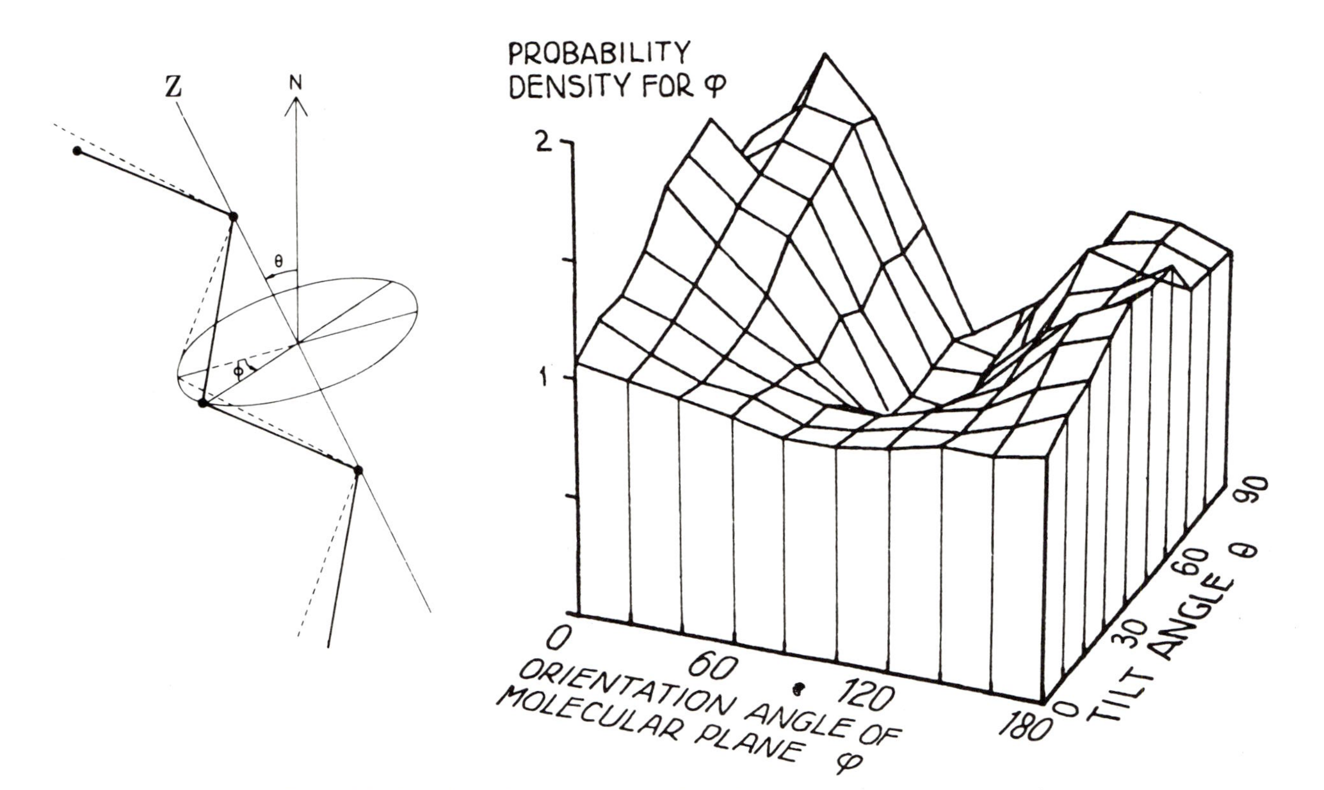

Figure 83. Correlation between θ, the tilt angle of the chain as a whole, and the angle ϕ, illustrated by the left-hand diagram and defined in the text. N is the bilayer normal and Z the molecular Z-axis. [From van der Ploeg, P., and Berendsen, H. J. C., *J. Chem. Phys.* 76:3271 (1982).]

length of the chains. Most gauche bonds are not associated with kinks, and neither are kinks of special significance as defects.

The molecular dynamics method starts with a more realistic physical model and gives a more detailed picture of the hydrophobic core than other methods. The number of molecules fed into the computer needs to be relatively small to obtain convergent results, at least in the case of an assembly of identical chains. The head groups are not considered in molecular detail, but for that matter they are not dealt with satisfactorily in any theory. It should be noted that none of the theories take into account the fact that most lipids of biological import possess two chains, and a given chain can thus interact both with its legal partner and with a neighbor. A good justification for not distinguishing between these inter and intra affairs is that X-ray diffraction reveals a commune — there is no evidence of pairing of chains in the packing patterns of liquid crystal bilayers.

The theories outlined in this chapter lay strong visual stress on energetic factors, in their equations the role of entropy tends to be overlooked, hidden as it is within the combinatorial problem that underlies the construction of the partition function. The molecular dynamics method gets by without even mentioning entropy in its formalism. To redress the balance, we emphasize that, hidden or not, entropy plays a central role in the chain-melting process. This can be seen by consideration of a related matter: the configuration of polymer chains in solution. A simple model for this problem is to assume that all configurations have the same energy — for hydrocarbons this is tantamount to ignoring the energy difference between a trans and a gauche bond. Probability theory allows the calculation of the probability of finding a chain with a given length. Since, on statistical grounds alone, most chains will have at least one bend, the chance of finding a fully extended straight chain will be very small and almost all chains in the system will have an end-to-end length well below the maximum possible length. This means that if we prepare an ensemble of fully extended chains in solution, we can expect to find an average contraction of the chains in the equilibrium state. This process, which is well known in polymer physics, requires and delivers no energy, but is driven by an increase in entropy, just as an ideal gas increases its translational entropy when expanding spontaneously into a vaccuum. The acyl chains of lipids present a slightly more complex problem. Theory predicts an entropy increase in the region of 14 R per mole consequent on the melting transition — a value hardly changed by taking into account the slight increase in translational entropy caused by the gentle disruption of the hexagonal-type head group lattice during melting.

9 Proteins, Cholesterol, and the Lipid Bilayer

There are few molecules that are not either adsorbed onto or dissolved in lipid bilayers, and the literature is replete with studies on contaminated phospholipids. In natural membranes the main "impurities" are proteins, glycoproteins, and cholesterol. In this chapter we look at the way in which such molecules perturb the lipid bilayer. We consider proteins first, avoiding biochemistry as far as possible. The influence of lipid composition on the activity of intrinsic enzymes has received attention, but the observed phenomena are not yet convincingly interpretable in physicochemical terms.

All membranes contain proteins, some of which have known biological functions. Many proteins can be readily separated from the membrane by mild chemical treatment that leaves the basic bilayer structure intact. Such proteins are probably linked to the surface of the membrane by a combination of chemical and physical bonding and are classified as *extrinsic proteins*. Some proteins of this kind may have purely structural roles, being responsible for maintaining the shape of the cell, or conceivably for fixing spatial relationships between *intrinsic proteins* — that is, those proteins assumed or known to transverse, partially or completely, the membrane from which they can generally be separated only by disrupting the lipid bilayer. Protein and lipid must exert mutually perturbing influences on one another. This interaction can be discussed within the framework of physical ideas used in earlier chapters but the field is poorly understood, and thus almost inevitably, controversial.

Electron microscopy and less direct spectroscopic methods provide ample evidence for the existence of a bilayer structure in all those cell membranes studied to date, even those containing very high proportions or proteins. We implicitly accept the phospholipid bilayer structure in the rest of this chapter. The simplest model of a bilayered membrane is the famous fluid mosaic model in which protein molecules "float" in a lipid sea.[143] The model had nothing specific to say about protein–lipid interactions but it has been known for many years that addition of protein to a lipid bilayer results in changes in the physical properties associated with lipids. For example, the chain-melting transition is usually broadened on the low-temperature side of the DSC or DTA curve and chain-order parameters are changed. There is also evidence of the influence of lipid on protein function, in particular, on the activity of membrane-bound enzymes that have, in many cases, been shown to be sensitive to changes in the lipid composition of the membrane. It is the hope of the physical chemist that such mutual disturbances will be largely accounted for in terms of a detailed theory of protein–lipid interaction. This hope is far from realized.

Lipids, being smaller molecules than proteins, are more susceptible to spectroscopic examination and the physical aspects of protein–lipid interactions have been examined largely from the lipid's viewpoint. The most provocative observations have been the outcome of magnetic resonance studies on membranes and bilayers containing either spin-labeled or deuterated lipids. A spin-labeled molecule dissolved in the hydrophobic core of a one-component bilayer above T_t shows an ESR spectrum that can be interpreted in terms of a single average orientation of the probe with respect to the local director. Spin-labeled phospholipids, for example, are oriented in much the same way as unlabeled lipids and the ESR spectra are consistent with rapid reorientation of the probe molecule around its long axis. The addition of other molecules to spin-labeled bilayers almost invariably results in spectral changes that often can be interpreted in terms of changes in the ordering of the spin label and, by inference, of the lipid chains.[48] A well-known example of direct relevance here is that of cholesterol, the addition of which to a spin-labeled bilayer results in spectral changes that can be attributed to an increase in chain order above T_t but a decrease below T_t (Chapter 5). Below T_t cholesterol acts as a defect preventing perfect crystallization of all-trans chains; above T_t the gyrating lipid chains are partially restrained in their movement by attractive dispersion forces pulling them toward the "flat" surface of the cholesterol molecule, which is aligned roughly parallel to the bilayer normal.

In the case of proteins, a new phenomenon is observed. The ESR spectra of lipid-spin-labeled bilayers frequently show two components, one being similar or identical to the spectrum in the absence of protein. The other component is often referred to as the immobilized spectrum since it has the broad spread and general line shape associated with spin labels in very viscous or frozen solution. This key observation was widely accepted as evidence for a layer of immobilized *boundary lipid* closely enough associated with the protein molecule to restrict severely the translational and rotational motion of the lipid chain.[144,145] The increase in the ESR "immobile" signal with increasing protein concentration lent credence to this simple interpretation. *It was assumed, either implicitly or explicitly, that boundary lipid was more ordered than bulk lipid.* The concept of boundary lipid was seized on by biochemists interested in the influence of lipids on membrane enzymes, but deuterium magnetic resonance was soon to undermine the hypothesis of a long-lived sheath of protein-bound lipid.

When deuterated lipids are incorporated into lipid bilayers containing proteins, the ^2H magnetic resonance spectra above T_t show only *one* component.[146,147] Furthermore, the quadrupole splittings, $\Delta\nu_Q$, from which order parameters may be deduced, are *less* in the presence of protein than in their absence. Perhaps the simplest interpretation of the spectra is that they result from diffusion of lipid molecules between bulk and "boundary" at a rate sufficient to average the NMR (but not the ESR) spectra associated with the two sites. If the orientational order parameters at the sites are S_{bulk} and S_{boundary}, then the observed order parameter is given by

$$S = p_{\text{bulk}}S_{\text{bulk}} + p_{\text{boundary}}S_{\text{boundary}}$$

where the p's are equilibrium occupation probabilities. If $S < S_{\text{bulk}}$, it follows that $S_{\text{boundary}} < S_{\text{bulk}}$, that is, that boundary lipid is *more* disordered than bulk, a conclusion contradicting that derived from ESR. In the next chapter we justify the appearance of two ESR signals and one NMR signal in terms of the characteristic time scales of the two spectroscopic techniques. With regard to the question of ordering or disordering at the protein surface, we earlier saw the danger of automatically associating low mobility with high order. An immobilized ensemble of lipids need not be more ordered than a relatively mobile ensemble, and in fact the immobilized ESR spectra are almost certainly due to a combination of restricted motion *and* disordering of chains at the surface of protein molecules. Since ESR probes carry a large organic group that has steric consequences not associated with deuterated lipids, we will accept the NMR evidence as being a more quantitatively reliable indication of the ordering of lipids. That the protein surface modifies the behavior of adjacent lipids is not in question; the controversy is centered on the cause and nature of this modification.

The introduction of a rough-surfaced protein molecule into a fluid hydrocarbon core is extremely unlikely to give a system containing large vacua around the pits and craters of the polypeptide surface. It takes about 25 erg to create a square centimeter of free surface in a typical fluid hydrocarbon. Thus lipid chains in bilayers can be supposed to accommodate themselves to the surface of intruding molecules. The forces holding lipid to protein certainly include dispersion forces, but could also involve electrostatic interactions involving the lipid head groups and polar or charged groups on protein. The boundary lipid hypothesis originally assumed that these forces were sufficient to create a *long-lived* lipid layer around membrane proteins.

A rather different explanation for immobilized lipid is based on purely geometrical arguments.[148,150] At low protein concentrations, above T_t, we can regard protein molecules in simple bilayers as being randomly distributed in two dimensions. Some experimental evidence is available on this point.[149] (In natural membranes there may well be a spatial organization of some intrinsic proteins, perhaps involving the aid of extrinsic proteins, but these matters are not for us.) We now assume that lipid molecules adjacent to protein molecules are *not* significantly translationally immobilized but have slightly more ordered chains, as for cholesterol. If the concentration of protein is raised, then chance alone, without the intervention of specific protein–protein forces, ensures that clusters of protein molecules will become statistically significant. The head groups of lipid molecules adjacent to two or three protein molecules are not expected to pack as closely as for bulk lipids. This is a purely geometric effect. The packing around a single protein may not be perfect either, but the disruption to the hexagonal head group lattice is generally far more serious in the vicinity of clustered proteins (Figure 84). The looser packing of the heads will be reflected in increased librational and rotational freedom of movement of the chains, that is, smaller order parameters. In addition, these lipids will usually have restricted translational freedom because they are "trapped" between protein molecules. If we are not to observe a separate signal for these molecules in the NMR spectra, then there must be a complete exchange of lipid between the different populations in a time less than 10^{-5} seconds (see next chapter).

Typical lateral diffusion constants, D, for lipids in bilayers at room temperature are in the vicinity of 10^{-8} cm^2 s^{-1} (again, see next chapter). Using the relationship $Dt = \overline{r}^2$ we find that in 10^{-5} seconds the root-mean-square displacement is about 30 Å, a distance probably sufficient to guarantee exchange between trapped and free lipid. (There are two implicit assumptions here — that diffusion coefficients are not affected

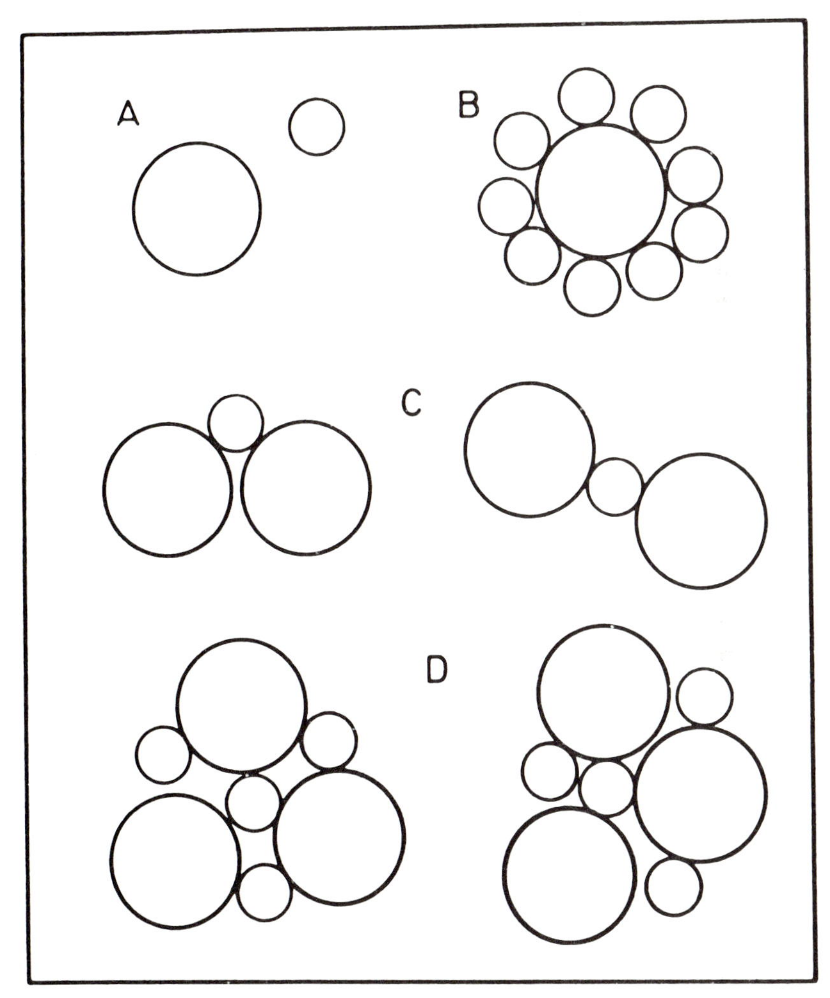

Figure 84. Schematic diagrams of the packing of lipid molecules (small circles) about protein molecules (large circles). **A.** Defines bulk lipid. **B.** Border lipid in dilute protein "solution". For clusters of two proteins, **C**, there will be lipids strongly hindered by two proteins. For three-protein clusters, **D**, there is strong disruption of the lipid hexagonal packing. [From Pink, D. A., et al., *Biochem.* 20:7152 (1981).]

by protein and that there is not a significant proportion of clusters containing far more than three protein molecules.) The characteristic time scale relevant to the ESR experiments is about 10^{-8} seconds. If exchange takes longer than this, it will be too slow to average the different signals from boundary and bulk lipid. Now for $t = 10^{-8}$ seconds the root mean square displacement is about 1 Å, too small a distance for exchange to occur. Thus we expect separate signals from different environments, in accordance with experiment.

The model outlined above provides an explanation for both immobilization and disordering without appealing to intermolecular forces, or significant translational restraints on a lipid adjacent to a single protein surface. The "trapped" lipid hypothesis has received support from a detailed analysis of the methyl group deuterium splittings in DMPC bilayers containing gramicidin A and cytochrome oxidase.[150] The model used assumed that the proteins are restricted to lie on a triangular lattice, with the interstitial space being filled with lipid. Three populations were defined: bulk lipid, lipid adjacent to one protein only, and lipid trapped between two or three proteins. The relative populations can be calculated as a function of the protein/lipid ratio, and the experimental plots for gramicidin of $\Delta\nu_Q$ against this ratio can be used to assign splittings to each type of lipid. The parameters are underdetermined but a reasonably convincing case can be made for the following values for $\Delta\nu_Q$: free lipid, ~3.6 kHz; lipid adjacent to one protein, ~6 kHz; and trapped lipid, ~0 kHz. Using these values and the calculated populations, and assuming complete exchange between all populations on a time scale of 10^{-5} seconds, it was possible to obtain good agreement with theory for the deuterium splittings in DMPC bilayers containing cytochrome oxidase (Figure 85). Indeed the same model gives an explanation of the changes in the ^{31}P NMR spectra of DMPC-cytochrome oxidase bilayers as a function of protein/lipid ratio. Here the assumption is that the head groups of trapped lipids only are affected by protein.

Further ammunition was supplied to the trapped-lipid school by the work of Davoust et al.[151] on bilayers with low protein/lipid ratios. An unsatisfactory aspect of the ESR detection of immobilized lipid was the need to use comparatively large concentrations of protein to obtain the immobilized component. Davoust et al. covalently affixed a spin-labeled lipid to the surface of rhodopsin and incorporated the protein into a bilayer. Since no bulk spin-labeled lipid interferes, it was possible to observe the signal of the spin label at low protein concentrations. The observed line shape was typical of a spin label possessing considerable motional freedom, and was in fact very similar to the spectrum of the spin-labeled lipid dissolved in a simple lipid bilayer. The label is not significantly immobilized by proximity to protein. However, an immobilized spectral component can be induced to appear by a number of procedures, including a raising of the protein concentration, cross-linking of the rhodopsin with disulfide bridges, and cooling the system. All these processes result in protein aggregation and, therefore, promote trapping. The weakness of the study is, of course, the fact that a covalently bound spin labeled lipid may be a poor steric stand-in for a normal lipid.

If the trapped-lipid hypothesis is correct, it implies that at low protein concentrations doubling the protein concentration will not significantly affect the amount of immobilized lipid. An entirely different

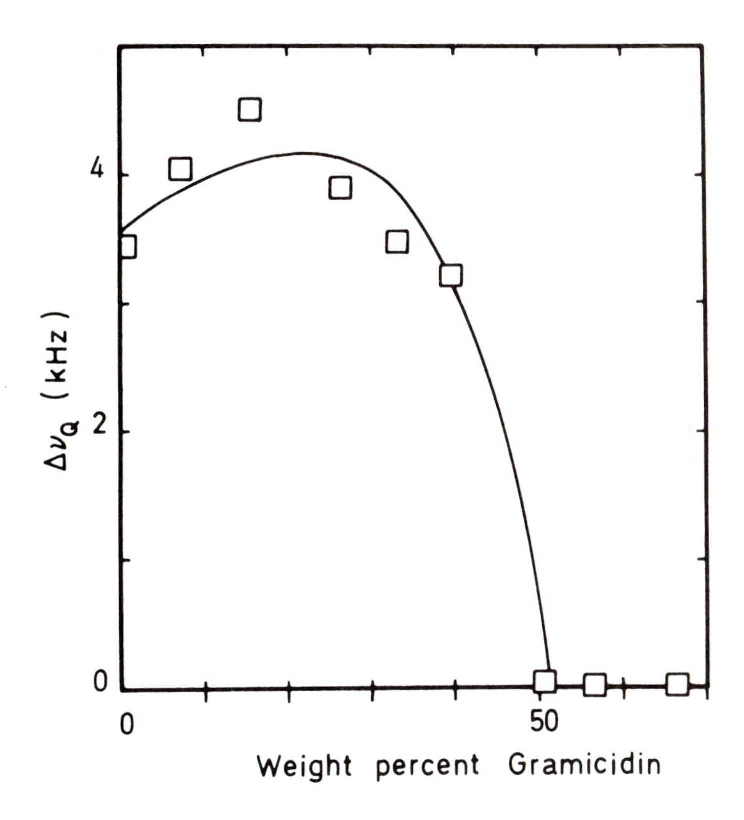

Figure 85. Calculated ^{2}H NMR quadrupole splittings for the methyl group of DMPC as a function of weight% of gramicidin A′ (solid line). □, data of Rice, D. M., and Oldfield, E., *Biochem.* 18:3272 (1979). [From Pink, D. A., et al., *Biochem.* 20:7152 (1981).]

prediction emerges from the original boundary lipid concept, namely, that the amount of immobilized lipid is proportional to the protein concentration, at least up to a concentration at which there is just sufficient lipid to cover each protein molecule. Thus a measurement of the effective number, n_l, of lipid molecules associated with each protein provides a means of differentiating between the theories. In a careful study[152] of the ESR spectrum of spin-labeled lipids in bilayers and membranes containing various proteins, the relative intensities of the motionally restricted and "fluid" spectra were recorded as a function of n_t, the protein/lipid ratio. Now n_t and n_l are related by

$$(n_f^*/n_b^*) = (n_t/n_l) - 1 \tag{9.1}$$

where (n_f^*/n_b^*) is the ratio of fluid to immobilized components in the spectra. (This easily derivable relationship holds only if the spin label is distributed in the same way as the unlabeled lipid.) Using (9.1) it was found that n_l remained fairly constant over a wide range of temperature and lipid/protein ratio in the DMPC-cytochrome oxidase system, with a value of 55 ± 5 mol/mol. A rough estimate of the number of lipids that can fit into the first shell around cytochrome oxidase can be made on the basis of molecular dimensions and the answer, 50, is within striking distance of that obtained experimentally. The conclusion drawn from these studies is that the immobilized signal comes from translationally restricted border lipid, the results being considered quantitatively inconsistent with trapping.

The trapped-lipid school claims that the above result can be explained in terms of mechanical hindrance of motion of *spin-labeled* lipid adjacent to protein due to the large size of the paramagnetic probe attached to the lipid.

The glycoprotein glycophorin found in the membrane of the red blood cell has received a great deal of spectrometric attention, the results of which will delight boundary lipid disciples. The lipids of the membrane can be extracted and reassembled into bilayers into which proteins can be incorporated. The ^{31}P NMR spectrum of this reconstituted system, which arises from the phosphorus in the head groups, is very similar to that of the bilayer without protein, no obvious effect of glycophorin being found.[153] However, intensity mesurements indicate that part of the phospholipid does not show up in the spectrum and that these hidden molecules increase in direct proportion to the glycophorin concentration, there being approximately 29 invisible phospholipids for each protein molecule over a protein/lipid ratio range of 1:50 to 1:200. The interpretation is in terms of boundary lipid, from which the absence of a spectrum is attributed either to a shortening of the nuclear spin-lattice relaxation time T_1 due to restricted motion of the lipid head groups or to an increase in the residual chemical shift anisotropy. The spectrum of the boundary lipid is presumably so broad as to be effectively unobservable.

Quantitative confirmation of the ^{31}P results comes from a study using DPPC labeled on the acyl chain with ^{13}C. In the ^{13}C NMR of a bilayer containing glycophorin two components were detected, one of which was severely broadened and had an intensity proportional to glycophorin content.[154] An estimated 30 lipids were associated with this peak for every glycophorin molecule. In both the ^{31}P and ^{13}C studies, the spectrum of the bulk lipid is effectively unchanged in shape by the presence of the protein. In general, if a molecule is exchanging between two environments, then for exchange rates above a certain minimum the observed spectrum is a weighted average of the spectra associated with each site. The fact that the ^{13}C and ^{31}P spectra are apparently not averaged with the broad "immobilized" peaks indicates that exchange between bulk and boundary lipid has a rate constant of less than 10^3 seconds. Bound phospholipid has been assumed to account for missing peak intensity in ^{31}P studies on several other protein–lipid systems.[153,155] It is significant that modification of the protein can sometimes cause the hidden phospholipid signal to come out of the closet. Thus removal of the exterior hydrophilic segment of glycoprotein by trypsin treatment of a virus membrane increases the intensity of the ^{31}P spectrum to a value similar to that in protein-free total lipid extracts. There seems little doubt that phospholipids interact with proteins in such a way as to strongly restrict the movement of the head groups. Cholesterol, which does not project into the head groups' living space, does not reduce the ^{31}P intensity in bilayers.

Raman spectra of glycophorin in chain-perdeuterated DPPC reveal that at a lipid/protein ratio of 125 : 1 there is a drastic change of conformation in part of the lipid, induced by the protein.[156] At temperatures well below the melting point, the lipid is in the trans state throughout the sample. On warming the bilayer, the $C-{}^2H$ stretching frequencies undergo shifts and intensity changes that reveal that the *perturbed lipid* undergoes a broad melting transition with a midpoint $\sim 15°C$ lower than T_t for the pure bilayer. This is presumably a noncooperative melting; the number of gauche bonds formed is the same as that in the phase transition of the bulk lipid that remains in the gel state during the premelting of the boundary lipid. At least 135 moles of perturbed lipid exist for each mole of protein, a far larger number than that found by NMR. The discrepancy may be partially attributed to the different time scales of the experiments.

Raman spectroscopy gives a subpicosecond exposure of the lipid chains, unaffected by motional averaging that can smooth out the difference between bulk lipid and the second or third shell of loosely bound conformationally perturbed lipid around the protein. ^{31}P and ^{13}C NMR detect only the strongly bound first shell. It is reasonable in general to believe that the perturbation of the lipid caused by the presence of the protein dies off slowly from the surface. Theory supports this view, as we will soon see. It should be noted that, in contrast to spin labels, the ^{13}C and ^{2}H isotopes hardly can be considered to perturb their environment significantly so that the experimental observations on glycophorin–lipid systems, particularly the constancy of the solvation number as a function of the protein/lipid ratio, strongly reinforce the straightforward boundary lipid concept, but are not easy to fit into a trapping scenario. Nevertheless, generalization is unadvisable. Glycophorin may behave similarly to other, but not all, proteins. It is interesting that glycophorin in DMPC or DPPC bilayers is randomly distributed at temperatures both above and below the transition temperature; it is not forced into fluid regions by the crystallization of the lipids as many other proteins are.[157] This suggests that the interaction between protein and lipid may have unusual features. (A similar situation occurs in some biological membranes in which cooling to low temperatures does not produce the protein aggregation — *patching* as it is known in the trade — common in most membranes. These exceptions involve bacteria with membranes containing unusually high proportions of branched chain lipids. The failure to patch may be due to the supposed looser packing of such lipids.)

The controversy over the nature of lipid–protein interactions continues, each side bringing favorable evidence for its own case. The basic difference between the two schools lies in their evaluation of the role of intermolecular forces, which the trapped-lipid school assume to be too weak seriously to restrict the translational motion of lipids. This may be true for some proteins, but not others; furthermore, it is conceivable that a rise in temperature could convert a boundary-lipid situation into a trapped-lipid system. It seems unlikely that there is a Rashomon situation, a single phenomenon with different interpretations. Nature is probably rich enough to afford both boundary and trapped lipid.

The complications attendant upon the addition of perturbing molecules to bilayers comprised of two or more lipids are many and fascinating. We might expect such a molecule to display a preference for one lipid over another in a binary lipid mixture, a preference dictated chiefly by steric and energetic factors. In mixtures of DPPC and DPPE (1:1) showing phase separation, cholesterol partially abolishes the melting transition in DPPC.[158] In pure DPPC this phenomenon also occurs, as we saw above, and the interpretation of the observations on mixed lipids is that cholesterol is preferentially dissolved in the DPPC. A steric explanation has been proffered by Israelachvili and Mitchell[31b] who suggest that the tapered form of DPPC complements the inverted truncated cone shape of cholesterol (cf Chapter 4) but is not compatible with the shape of DPPE.

There is growing evidence of the preferential interaction of proteins with lipids in mixed systems. It has been known for many years that changing the lipid composition of membranes can affect the activity of enzymes, although in some cases this could simply be due to the role of head groups in binding coenzymes or substrates. More convincing evidence comes from studies of the type carried out by Boggs et al.[159] on mixtures of phosphatidylserine and DPPC, which show a broad transition at a temperature intermediate between that of the individual lipids. Addition of lipophilin shifts the transition temperature upward and at 34% wt protein the system melts at T_t for pure DPPC. This suggests selective binding of phosphatidylserine. Preferences of this kind are probably largely determined by head group–protein interactions, but the role of the chains has not received much attention and the influence of unsaturated chains seems worth investigation. Natural membranes always contain several kinds of lipids and the total spatial organization of the membrane will depend upon the mutual adjustment of all the molecules, which will be decided partially by steric factors and partially by specific and nonspecific forces between lipids and between lipids and proteins. Little is known about these matters. A start has been made in attempting to order the preferences of a given protein for different lipids by extracting binding constants from spin-label data by the use of a multiple equilibria approach.[160] The drawbacks inherent in the use of bulky probes should be borne in mind in weighing the results of such studies, but the very fact that there are selective associations suggests very strongly that selective attractions exist.

It may well be that interactions between lipid head groups and polar areas in proteins are a major factor in determining the average lifetime of a lipid molecule on the surface of a protein; there is no reason to suppose that a phosphatidylethanolamine group will interact with a given protein in the same way as a phosphatidylcholine group, or that different proteins will display the same behavior with a given lipid. The tendency has been to interpret the results of experiments on a single protein as if they could be generalized to all proteins. Simple geometric analysis forces us to accept the phenomenon of trapping at high protein concentrations. Nevertheless, until more is done to estimate protein–lipid forces, it would be rash to dismiss the possibility that a protein can create a strongly modified layer of lipid around itself solely by the use of intermolecular forces, and that such a layer is the source of the immobilized spectrum seen in spin-label studies and the broadened or undetectable spectra in ^{13}C and ^{31}P NMR studies.

Whatever the status of the trapped-lipid theory, the existence of clusters and the associated phenomenon of protein aggregation are expected to have an effect on bilayer properties. On cooling any protein–lipid system slowly from the liquid state, we can expect protein aggregation since crystallization of lipid will tend to force protein out of the crystalline (gel state) areas. Furthermore, there may be specific forces tending to bring protein molecules together (see below). (This "explanation" of aggregation is a hand-waving version of the thermodynamic treatment of binary lipid mixtures given in Chapter 4. It is the best that we can do at present.) Evidence for the existence of protein aggregation is forthcoming from electron microscope photographs of freeze-fractured reconstituted bilayers quenched from temperatures below T_t.[161] Quenching from higher temperatures freezes the bilayer too fast for protein aggregation to occur. (Phase separation implies the possibility of lateral diffusion in the bilayer, this phenomenon is discussed in Chapter 10.) The lipid trapped in the protein patches will freeze at a lower temperature than that in the areas of pure lipid due to imperfect packing. This is one proposed explanation of the broadening of the gel–liquid crystal transition found in protein–lipid mixtures and the lowering of the transition enthalpy. The same qualitative ideas explain why fluorescent probes trapped in protein patches become mobile before those in the bulk lipid. It is tempting to give a similar explanation for the fact that Arrehenius plots of membranal enzyme activity often show a break well below the melting point of the bilayer. However, it is still not certain to what extent such breaks depend on the lipids rather than on intrinsic properties of the proteins.

Cholesterol — a comparatively simple impurity — dissolves easily in bilayers, its polar group laying near the lipid head groups. The effect of cholesterol on bilayer thickness and area has been discussed in terms of simple steric concepts and its influence on chain-order parameters is likewise amenable to interpretation in terms of the shape and rigidity of the molecule. Nevertheless, the nature of lipid–cholesterol mixtures is still not clear. A great deal of effort has been expended in attempting to determine phase diagrams, but no final conclusions have been drawn. Considering the medical importance of cholesterol and its strong influence on membranes, the phases present in cholesterol–lipid mixtures are of more than academic interest. Addition of cholesterol to a bilayer results in a progressive lowering of the transition temperature and a decrease in the transition enthalpy. At a cholesterol/lipid molar ratio of $\sim 1:2$ ΔH vanishes, a phenomena similar to that found for many protein–lipid systems. The differential scanning calorimetric plots on which these conclusions are mainly based show a strong broadening with increasing cholesterol concentration (Figure 86). In a very careful

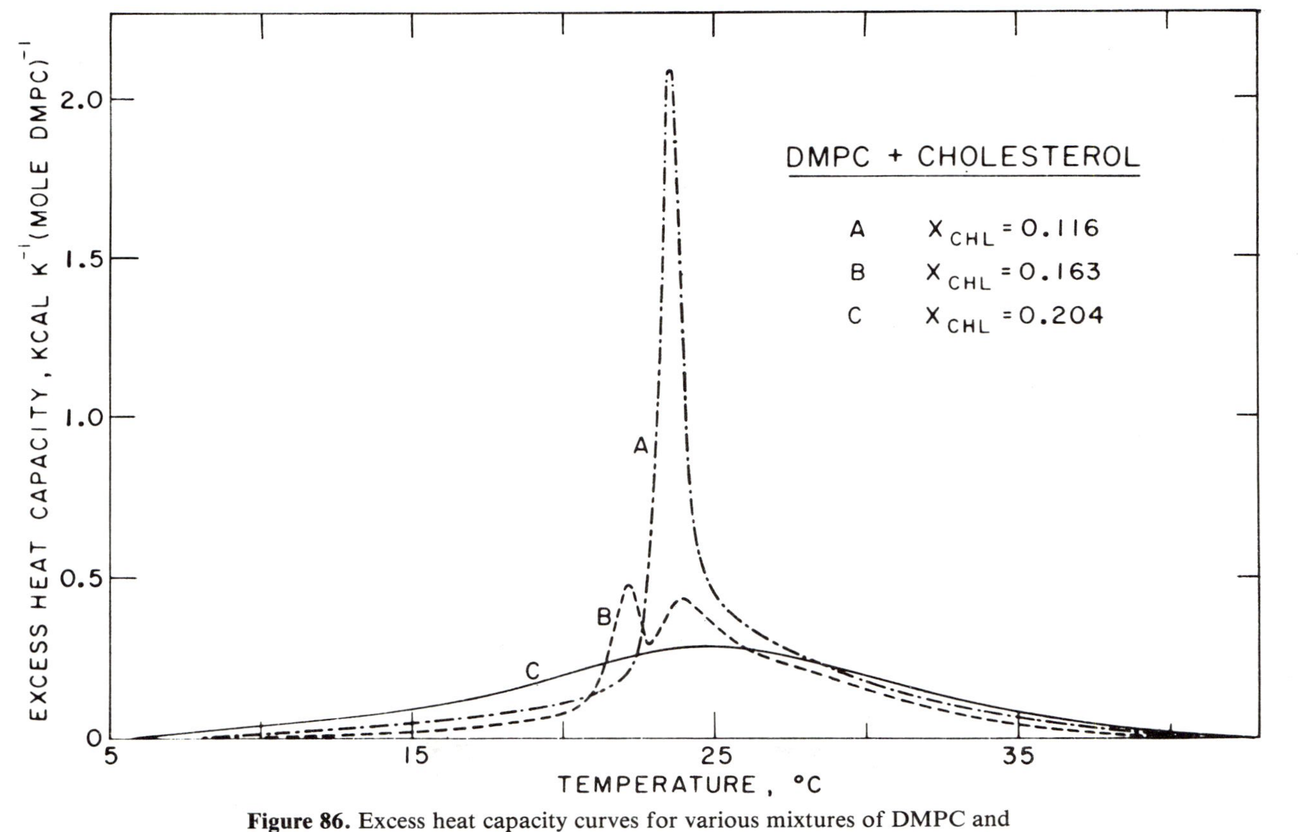

Figure 86. Excess heat capacity curves for various mixtures of DMPC and cholesterol. [From Mabrey, S., et al., *Biochem.* 17:2464 (1978).]

study[162] it was found that for DPPC–cholesterol mixtures the peak could be resolved into two components, one narrow and one broad, with the former being undetectable at more than 20 mol% cholesterol. The broad peak moves to higher temperatures and sinks beneath the waves at ~45 mol% cholesterol. The corresponding plots for DMPC show three peaks over the cholesterol concentration range 13 to 20 mol%. No convincing explanation of this result has appeared. It is easy to check that the Gibbs phase rule does not permit three solid phases in a binary system. There have been two general interpretations of the broadening and eventual loss of the transition. The first historically was in terms of lateral phase separation into pure lipid areas and cholesterol/lipid areas. The lipids in the latter phase exist as small areas partially separated by rigid walls of cholesterol. Melting of this lipid is presumed to involve small cooperative units, that is, the melting of one island has little or no effect on the next. The thermodynamics of melting, which was glanced at in Chapter 6, predicts a broad phase transition. Support for the slow onset of disorder in at least part of the mixture comes from ^2H magnetic resonance on DPPC[163] and on DMPC labeled at three different chain positions.[164] At least one study on the effects of protein on a bilayer casts doubt on the need to assume separation into two phases. In a Raman and Fourier transform infrared study of DMPC layers containing glycophorin, the lipid transition was found to broaden and shift to lower temperatures and the mobility of the acyl chains was increased by the protein.[165] Unfortunately for the two-phase theory, no significant change was found in conformational order, either in the gel or liquid crystal phase. Neither was there any evidence of immobilized lipid or of phase separation. For that matter, no direct evidence is available for two phases in cholesterol–lipid mixtures in the composition range above 20 mol% over which the transition broadens and vanishes.

The necessity for a two-phase system has been challenged on the strength of statistical considerations of the arrangement of cholesterol and lipid molecules.[166] The argument is similar in spirit to that applied to protein–lipid mixtures. Random arrays of lipid and cholesterol were constructed and the proportion of lipid adjacent to cholesterol molecules was estimated. Assuming that in the melting process only nonadjacent lipid contributes to the transition enthalpy, it was found that the decrease in enthalpy with increasing mol% cholesterol closely followed experimental results. An entirely different explanation of the DSC plots arises from theoretical work to be described later.

We now look at what theory has to say about protein–lipid systems. Efforts have been concentrated on what the protein does to the lipid rather than on the at present almost impossible task of accounting for lipid perturbation of protein function. The simplest model for a protein molecule is of an object large by comparison with a lipid molecule and having a surface that can be divided into hydrophobic and polar regions. (It has been shown that some intrinsic proteins have a preponderance of hydrophobic groups on the surface in contact with the bilayer core. A number of proteins are so strongly bound to their lipid neighbors that protein and associated lipid pass hand in hand through the protein purification procedure.) Since the early ESR studies spoke in terms of ordering at the protein surface, theoreticians incorporated into their models the requirement that the walls of the protein be rigid and parallel to the bilayer normal, at least within the membrane core. The chains in the boundary lipid were supposed to be constrained in the manner suggested for cholesterol, so that above T_t reorientation of chain segments is strongly discouraged, as is chain tilting. Thus boundary lipid was supposed to be partially immobilized and ordered, and in turn to exert a restraining and progressively diminishing effect on the mobility and order of successive shells of lipid. To derive quantitative results, a variety of theoretical approaches have been employed.

Marcelja[167] uses a modification of his mean field theory to estimate the field acting at each lipid segment. Only adjoining molecules are allowed to interact. In focusing on a specific lipid chain next to a protein, the interaction with neighboring lipids is given by (8.5) and the protein is taken to exert a constant potential on the lipid. Feeding the potentials into a self-consistent set of equations for the molecular field results in a solution that predicts a falling-off in the value of the order parameters with increasing distance from the protein. The order-parameter correlation length increases with decreasing temperature, which is a fancy way of saying that the influence of the protein at a given point in the bilayer can be reduced by increasing thermal motion. An interesting consequence of the theory arises in the case of proximate protein molecules for which the separate regions of modified lipid overlap. Assuming that lipid chains adjacent to proteins are more extended than those in the bulk, there will be a gain in configurational entropy if two protein molecules come together and eject part of their lipid coats into the bulk. Thus the change in entropy is a driving force bringing protein molecules together and can be described as a lipid-mediated protein–protein attraction reminiscent of the meniscus-mediated attraction of floating bodies. Of course, in any overall treatment of aggregation other factors must enter, such as the deterrant effect of translational entropy (entropy of mixing), electrostatic and dispersion forces between proteins, and, in biological membranes, the influence of extrinsic proteins.

Another type of theory[168] treats the lipid core as a continuum. The free energy of the system in a given infinitesimal volume placed outside the protein is expressed as an expansion in one order parameter:

$$G = Tu^2/2 - u^3 + u^4/2 + |\nabla u|^2/2 \qquad (9.2)$$

This is that part of the free energy that depends on the order parameter u, which is defined as

$$u = (A_f - A)/(A_f - A_s) \tag{9.3}$$

where A_f, A_s are the molecular areas at the bilayer surface in the "fluid" and "solid" phases. The order parameter so defined is not directly related to those defined in Chapter 5, but is rather a kind of averaged parameter that goes from 1 to 0 in heating the bilayer through T_t. The fact that volume changes are rather small (Chapter 3) means that area and thickness changes are approximately reciprocally proportionate and thickness is certainly related to chain order. (Expansions in order parameters are associated with theories of phase transitions due to Landau[169] and de Gennes[43].)

The gradient term in (9.2) expresses the fact that work must be done to produce spatial inhomogeneities in the order parameter. This is obvious for simple nematics where intermolecular dispersion forces must be overcome by the creation of a misalignment. For chains one can imagine that differences in order, and therefore average conformation, will destroy the "fit" between neighboring chains. In (9.2) T is a reduced temperature chosen to be equal to 1 at T_t. At thermodynamic equilibrium G is at a minimum. In the absence of impurities (e.g., proteins) ∇u vanishes and two minima are found for G corresponding to two values of u. The relative depths of these minima depend on the temperature, a phase transition occurring for that temperature at which the minima lie at the same level. Since

$$\partial G/\partial u = Tu - 3u^2 + 2u^3 \tag{9.4}$$

and

$$\partial^2 G/\partial u^2 = T - 6u + 6u^2 \tag{9.5}$$

there are two minima at $T_t(T = 1)$, both having $G = 0$, namely, $u = 0$ and $u = 1$, which physically just means that at T_t the fluid and solid (liquid crystal and gel) are in equilibrium. Below T_t the solution with smaller u (less order) is higher in energy, and vice versa above T_t.

The protein enters the problem as a rigid cylinder at the surface of which the order parameter has a fixed value u_0 chosen arbitrarily to fall between 0 and 1. Thus the so-called protein in this theory is in fact merely a boundary condition on the solutions of equation (9.2). Again two stable solutions are found but now the difference in free energies is a function not only of temperature, but also of protein concentrations. The relevance of concentration is clear when it is realized that for dilute protein the order parameter changes montonically from u_0 to the bulk value of u while in concentrated protein the order parameter reaches a turning point between two protein molecules, at which point u is different from its bulk value in dilute solution.

A general but not very surprising result of the theory is that in the gel state the order parameter rises monotonically from the protein surface to its value of unity in the bulk while in the liquid crystal state u falls from u_0 to a bulk value of zero. The gel state is not too interesting from the biological point of view but Jahnig[170] has pointed out that the diffusion rate of proteins in gel-state bilayers is near the rate in the liquid crystal state and that a possible explanation is that the immediate environment of the protein in the gel is disordered and fluidlike so that the protein melts its way throught the gel in a manner recalling the passage of a wire through a block of ice.

The main qualitative results of the theory are that proteins can change the lipid-phase transition temperature and reduce the transition enthalpy. A point of considerable interest is the prediction of a critical point in the plot of transition temperature against protein concentration. At this point the transition enthalpy goes to zero and beyond the point no sharp transition occurs. Such behavior has been found for at least one protein and also, as we saw above, for cholesterol. However, the explanation, whether in terms of two phases or not, was not in terms of a critical point and the question arises as to whether critical phenomena are observed in simple phospholipids. The answer is yes, for *monolayers* spread on water. The compression of such monolayers at a fixed temperature gives pressure–area isotherms reminiscent of the classic van der Waals' plots for gas–liquid equilibria. While the phospholipid isotherms do not have horizontal sections, they do have cusps and inflections, which allows a critical point to be defined at the temperature at which the inflections disappear.

Whether a critical point occurs for bilayers is not known — experiments involving controlled lateral pressure on bilayers have not been performed as a function of temperature.[133] The continuum theory provides an explanation of the broad and sharp peaks in the experimental DSC plots in terms of the variation of the mean order parameter with temperature; lateral phase separation is not required. The behavior of u_{mean} as a function of T, the reduced temperature, is shown in Figure 87. The clifflike drop in u_{mean} is responsible for the sharp peak in the heat absorbtion in the DSC curves. For a system in which one tenth of the surface area is protein, the plot of u_{mean} is characterized by a sharp drop at $T = 1.13$ and an overall decline over the whole temperature range shown. It is this gradual loss of order over a wide temperature range that is, in this theory, the cause of the broad component in the DSC curves. (The authors applied their results to the case of cholesterol and came to the conclusion that it was unnecessary to invoke the presence of two phases to explain the experimental results. Subsequently it was found that cholesterol–DPPC mixtures containing <0.20 mol fraction of cholesterol *do* exhibit lateral phase separation into alternating bands of pure lipid and cholesterol–lipid solid solutions.[171] Thus in the case of cholesterol the broad peak in the DSC curves may be identified with the melting of the mixed bands, but there is still no universally accepted explanation of the behavior of this system.)

No reference is made in any of the theories to the structure of the impurity. (Indeed the continuum model ignores the structure of the lipids.) Few structural or mechanistic facts are available with which even to start guessing at the effect of lipid environment on protein function and conformation. A handful of proteins have received a great deal of attention, rhodopsin being perhaps the outstanding example, but the

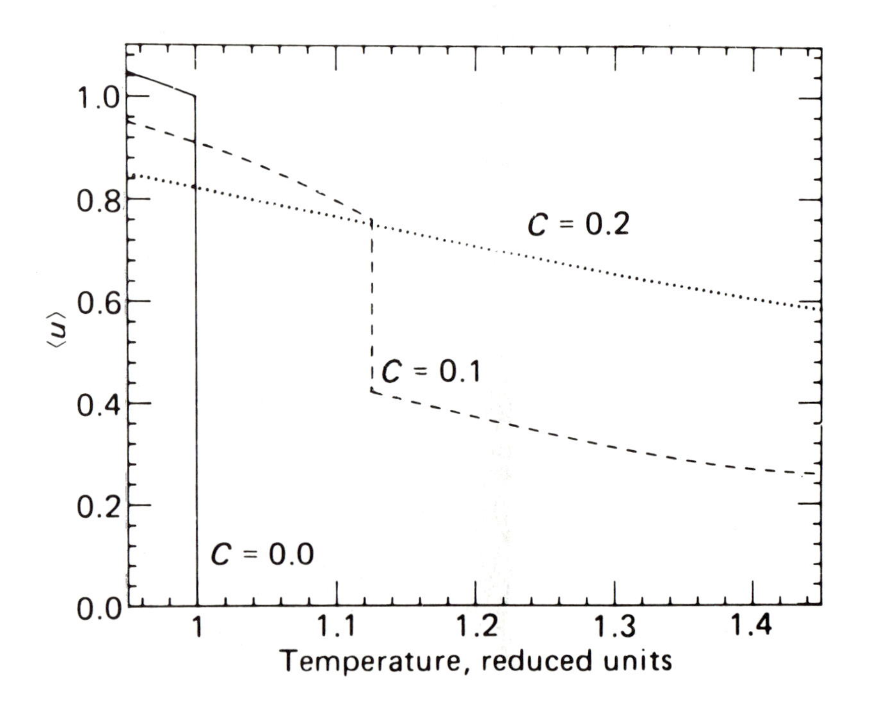

Figure 87. Dependence of the mean value of the order parameter in a membrane on temperature for three protein concentrations for the model described in the text. C is the fraction of the surface area covered by protein. The value of the order parameter at the protein–lipid interface was taken as 0.75. On the reduced temperature scale used in the calculations, the phase transition of the pure lipid is at $T = 1$. [From Owicki, J. C., and McConnell, H. M., *Proc. Nat. Acad. Sci.* 76:4750 (1979).]

mode of action of membrane enzymes is certainly not ripe for a physicochemical summary. Nevertheless, as we have seen, there are aspects of protein–lipid systems that may be at least partially explicable in terms of extremely simple models. The lattice model treatment of quadrupolar deuterium splittings is a particular example of the reduction of a complex system to a binary mixture of near-structureless bodies. A further example of the provocative conclusions that may be drawn from considerations of packing is the study of clustering in protein–lipid mixtures described by Freire and Snyder.[172] In fact the results apply to any planar binary mixture, and although allowance was made for attractive or repulsive forces between proteins, the results for molecules interacting only via hard-core repulsions are sufficient to illustrate the general conclusions arrived at.

Figure 88 illustrates three binary mixtures of different percentage composition. The stippled areas, in the case of protein–lipid mixtures, represent border lipid. At moderate concentrations of protein, the border (or annular) lipid forms a connected path of the dimensions of the system. At very high concentrations, the classification into border and free lipid breaks down. The authors define a pair connectedness function $C_{BB}(d_{\max})$ that measures the probability that the physical extension of a given compositional domain is of the order of the dimensions of the system. The dependence of $C_{BB}(d_{\max})$ on the protein/lipid ratio depends on whether intermolecular forces are taken into account, but the general form of the function is always the same as the curves shown in

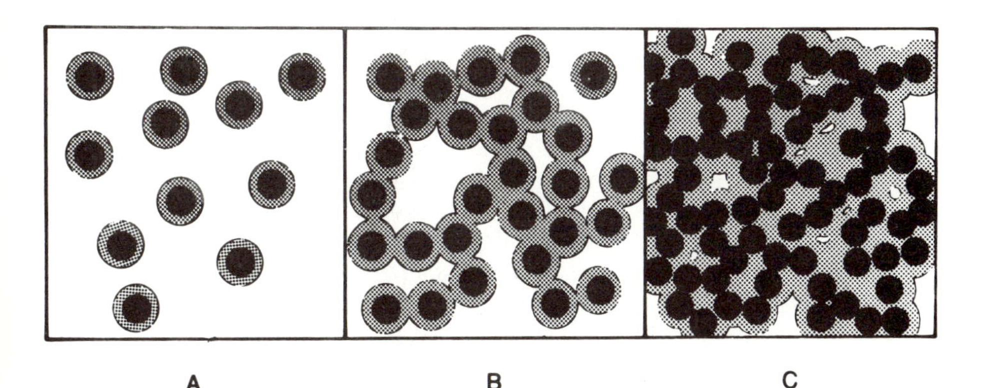

Figure 88. Model for protein–lipid organization in a bilayer. **A.** Protein molecules (black circles) plus annular lipid (stippled areas), which is isolated and dispersed. **B.** The annular lipid is now connected and forms a continuum. Free lipid (white areas) is disrupted into isolated compartments. **C.** At very high protein density both the annular and free lipid are disrupted into isolated patches. [From Freire, E., and Snyder, B., *Biophys. J.* 37:617 (1982).]

Figure 89. The outstanding feature is the sudden change in connectivity that occurs over a narrow range of protein/lipid ratio. The significance of this jump in connectivity may be judged by the fact that in both gramicidin–DMPC and cytochrome oxidase–DMPC the deuterium quadrupole splittings, $\Delta\nu_Q$, reach a maximum at a protein/lipid ratio that coincides with that at which the border lipid becomes connected and the free lipid is broken into isolated domains. As we saw, the trapped-lipid model, which is also a "geometric" model likewise predicts a maximum in the plot of order versus protein/lipid ratio. (The formation of a single network of a given component is known as percolation, a phenomenon to which we return in the next chapter.)

One of the lessons of this study and others that have emphasized the geometric consequences of high "impurity" concentration is that the distinction between border (or trapped) lipid and bulk lipid loses meaning above certain critical concentrations. All the lipid becomes strongly modified by the protein. This may well be the case in most biological membranes. For this reason it is sometimes convenient to have a measure of the average state of the lipid and a concept often used for this purpose is *microviscosity*, values for which are derived from experimental determinations of the rotational or translational diffusion of probes. Detailed discussion of these matters is postponed to the next chapter where the dangers of simplistic interpretations will become apparent. The most common methods of measuring microviscosity are by the use of spin labels or fluorescent probes. The results, in general, fit in with the conclusions drawn from other experiments, but the extraction of viscosities from the diffusional data is of doubtful value. As an example, we consider the effect of the addition of cholesterol on the behavior of the fluorescent probe 1,6-diphenyl-1,3,5-hexatriene (DPH), an elongated molecule that dissolves in the hydrophobic core of bilayers.[173] The motion of the molecule within the core is sometimes described as "wobbling in a cone" and analysis of the nanosecond fluorescence measurements provides estimates of the angle of the cone, that is, the rough range of angles through which the probe diffuses about its local director. The rate of diffusion also emerges from the analysis and is frequently converted into a microviscosity by use of a Stokes-type equation. Cholesterol strongly reduces the angular freedom of the probe above T_t, consistent with the familiar ordering of the lipid chains induced by cholesterol. However, little effect was found on the value of D_w, the diffusion constant describing the rate of motion of the probe within its confining cone. This is in strong contrast to the marked lowering of the *translational* diffusion constants of lipids produced by the addition of cholesterol. On the other hand, rotational diffusion of probes is often hampered by the addition of protein,[174] a result interpreted as indicating

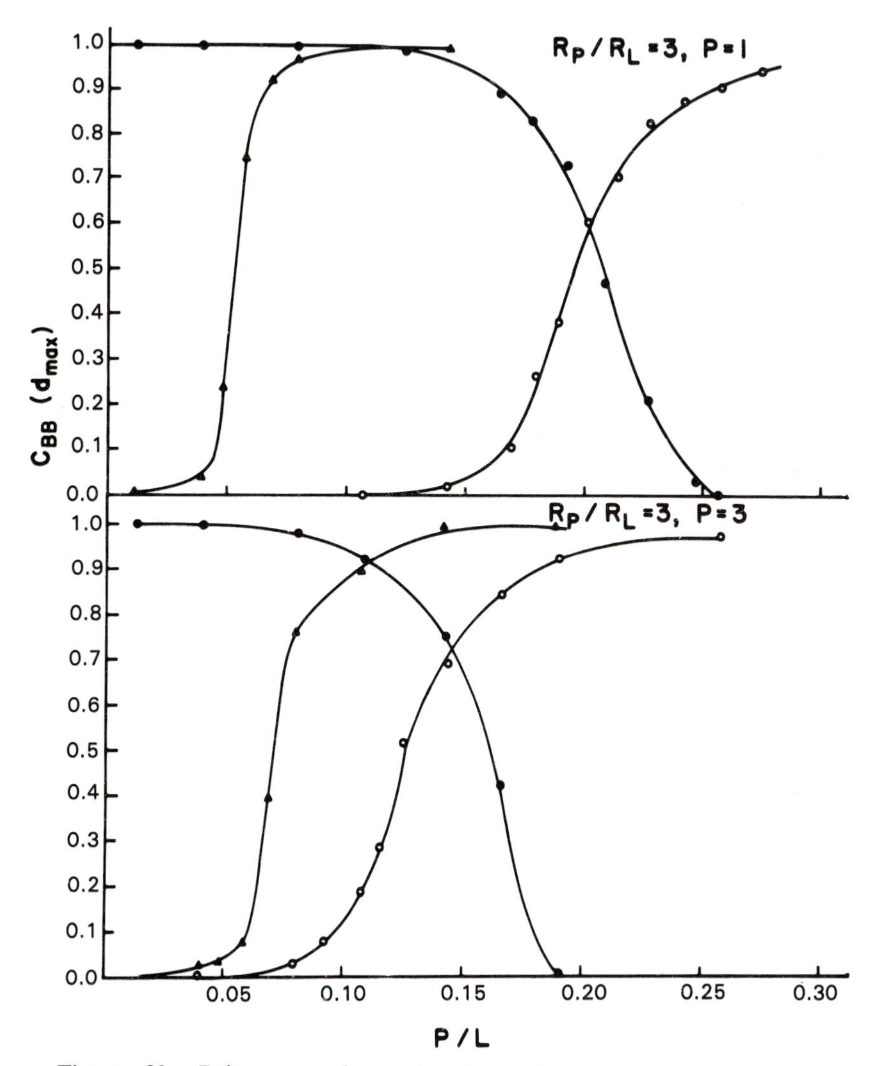

Figure 89. Pair-connectedness function at maximum separation, $C_{BB}(d_{max})$, for the model membrane of Figure 88 for P/L ratios of 1 and 3. △, annular lipid; • annular plus free lipid; ○, protein. [From Freire, E., and Snyder, B., *Biophys. J.* 37:617 (1982).]

a rise in the microviscosity. It should be clear that extraction of a property called microviscosity from rotational diffusion data is chancy.

Our discussion of protein–lipid systems has at times implicitly accepted the possibility of lateral diffusion of membrane components. In the next chapter we take an overall view of what is known about the motion of molecules in membranes.

10 Motion

It has been known for many years that the membranes of living cells can flow. Hydrodynamic flow of a biological membrane in the liquid crystal state can be expected, and its quantitative analysis is the province of fluid dynamics. In this chapter we will be mainly concerned with rather different kinds of motion that have been revealed more recently by spectroscopic and optical experiments on bilayers and natural membranes. We refer to lateral diffusion of lipids across bilayers, rotation and translation of proteins and other microscopic forms of motion that can be broadly classified as diffusional and involve the displacement or rotation of complete molecules with respect to a set of coordinates passing through the local center of mass. Motions of these kinds certainly occur in natural membranes and are implicated in several important biological phenomena such as capping, a process familiar in immunology involving the aggregation of membrane proteins. Here we concentrate on the basic molecular mechanisms responsible for motion rather than its biological consequences. Experiments in this field are usually summarized in the form of diffusion coefficients (see Chapter 12).

For isotropic diffusion in two dimensions

$$D_{\text{trans}} = \overline{r^2}/4t \qquad (10.1)$$

where $\overline{r^2}$ is the mean-square displacement of a diffusing particle in time t. For rotation about a single axis

$$D_{\text{rot}} = \overline{\theta^2}/2t \qquad (10.2)$$

where $\overline{\theta^2}$ is the mean-square angular rotation in time t. Operationally these definitions refer to an experiment on an ensemble of particles and are based on a microscopic view of motion. In practice two-dimensional diffusion is usually discussed in terms of a Fick-type equation:

$$\frac{\partial c}{\partial t} = D_{\text{trans}}\left(\frac{\partial^2 c}{\partial x^2} + \frac{\partial^2 c}{\partial y^2}\right) \tag{10.3}$$

where diffusion is assumed to be isotropic in the plane ($D_{\text{trans}} = D_x = D_y$) and c is the concentration at a given point. This equation rarely holds over a wide concentration range but is sufficient for most purposes. Equation (10.3), as will become clear in Chapter 12, expresses the belief that the diffusion rate is proportional to the concentration gradient.

A similar expression can be constructed for angular diffusion, with the anisotropic distribution of molecules being the hypothetic driving force,

$$\frac{\partial n}{\partial t} = D_{\text{rot}} \cdot \frac{\partial \theta^2}{\partial n^2} = \frac{1}{\tau_r} \cdot \frac{\partial^2 n}{\partial \theta^2} \tag{10.4}$$

where n is the number of molecules having a given axis within the molecule falling in the range θ to $\theta + d\theta$ and τ_r is a relaxation time. The equations are for rotation about one axis. Measurements of rotational diffusion are often made by optical measurements and it is traditional among workers in this field to use relaxation times rather than diffusion constants which accounts for the alternative form given in (10.4). The relaxation time for a physical or chemical process is a measure of the rate at which a system returns to equilibrium from a nonequilibrium state. For a sphere of radius r in a continuum having viscosity η,

$$\tau_r = \frac{4\pi}{3} \eta r^3 \tag{10.5}$$

this equation has been used, often without much justification, to deduce values of η from diffusion experiments. Equation (10.4) presupposes a single relaxation time and leads inexorably to an exponential time dependance of n, which is not always observed. One simple cause for macroscopic deviation from exponential decay is the existence of more than one microscopic environment for the diffusing molecule. In such a case, the decay curve could be the superposition of two or more exponentials.

The experimentalist presents the values of D_{trans} and D_{rot} (or τ_r), which the theorist attempts to explain. Thus we first look at the qualitative facts, then at what theory proclaims, and finally at some experimental methods. We also take the opportunity to expand our understanding of some of the motional phenomena discussed in previous chapters.

Protein molecules undergo both *lateral* and *rotational* diffusion although many proteins certainly do not diffuse laterally since they are integral components of specialized areas on the cell surface such as synaptic junctions. Since all natural membranes are probably at least partly in the liquid crystal state at physiological temperatures, the absence of lateral protein diffusion is usually attributed to bonding to structures that could be on the surface of the cell or elsewhere, for example, in the cytoplasm. Evidence for such clamping will be given below.

Protein rotation in membranes has been firmly established by both optical and electron spin resonance spectroscopy. Where the data are accurate enough to provide a check on theory, they seem to support the thesis that rotation occurs about the bilayer normal. For a protein with a predominantly hydrophobic surface within the membrane and a hydrophilic head, or heads, projecting from the membrane, rotation about axes perpendicular to the normal might well be very unfavorable energetically. Rotational rates are reduced by lowering the temperature or raising the protein-to-lipid ratio. The nature of the lipid is also relevant; at 40°C rhodopsin in DPPC has $\tau_r \simeq 10^{-4}$ seconds, while in disk membranes, its natural habitat, $\tau_r \simeq 10^{-5}$ seconds.[175]

Lipids diffuse laterally at readily detectable rates above T_t and at considerably lower rates below T_t. Lipid molecules can also leave one side of a bilayer and appear at the other side; this process is dubbed *flip-flop* and involves an as-yet-unknown mechanism. We already know that lipids rotate in bilayers (Chapter 3).

The rates of all the microscopic motions listed here are sensitive to the composition of the membrane, and for this reason many studies have concentrated on simple artificial systems such as one-component lipid bilayers or single proteins incorporated into bilayers. These systems are just about accessible to theory.

The lipids of hydrated bilayers in the liquid crystal state diffuse laterally; typical diffusion constants are in the range $D_{\text{trans}} = 1$ to 10^{-8} cm^2 s^{-1} at 30°C. This means that a lipid molecule can traverse a distance of 10^5 Å, a typical cell diameter, in about 20 seconds. D_{trans}, not surprisingly, depends on the lipid head group, the chain length, the presence of proteins or other molecules in the bilayer, the ions in the ambient solution, and, of course, the temperature. The curvature of the bilayers may also be relevant in that it affects the packing of the lipids. The influence of impurities can be complex. Thus addition of cholesterol in small amounts, up to ~10 mol%, increases the rate of lateral diffusion, but at higher concentrations the rate decreases below the value observed in the pure lipid.[176] Conceivably, small amounts of cholesterol produce defects in the lipid lattice aiding movement. Translational diffusion is slowed down by the presence of intrinsic proteins. As an example, 15% gramicidin reduces D_{trans} in DMPC from 8×10^{-8} to 2×10^{-8} $cm^2 s^{-1}$ at 30°C. Lowering the temperature of the pure lipid below T_t reduces D_{trans} drastically: to 10^{-10} $cm^2 s^{-1}$ at 20°C in DMPC. Diffusion in mixed lipid systems and in protein–lipid systems needs considerably more experimental study before enough data are available to supply an adequately broad testing ground for theory. It is becoming clear that the spatial organization of membrane constituents has a strong effect on lateral diffusion, as we will see at the end of this chapter.

A typical value quoted for the rotational diffusion constant of a lipid is $\sim 10^3$ s^{-1} at 30°C. This means that a lipid molecule in a bilayer undergoes, on the average, one complete rotation in the time that it diffuses through one molecular diameter, ~4.5 Å.

Flip-flop motion of phospholipids was originally detected by showing that spin-labeled lipids could cross from one monolayer to the other in a lipid bilayer. The rates measured may not reflect the flip-flop rate of unlabeled lipid since the spin label perturbs its environment. It is generally found that the half-time at physiological temperatures is of the order of hours to days. It is interesting that the implied life-time of a lipid in a monolayer is in the range 10^4 to 10^5 seconds, which is that deduced for the residence time of a lipid molecule in a bilayer in aqueous solution.[178] One might regard the molecule as having a roughly equal chance of jumping out of the bilayer into water or jumping inward. Energetically, both processes have their problems. In one case the so-called hydrophobic force has to be overcome, and in the other the polar head group has to bury its way through the hydrophobic core. If (and there has been scepticism on this point) the head group really has to penetrate an otherwise almost unperturbed hydrophobic core, it may be assisted by the head group's water of hydration, the hydrated polar group being the analog of an ion plus an ionophore (Chapter 14).

It may be rash to assume from the lethargic pace of flip-flop observed in simple bilayers that this process has no biological significance. A hint to the contrary is contained in the fact that biosynthesis of microsomal phosphatidylcholine and phosphatidylethanolamine is reported to occur only on the cytoplasmic side of the membrane,[179] and yet these lipids appear outside the microsome in the living cell. Tracer studies suggest that the time scale for translocation of phospholipids across certain cell membranes occurs with a half-life of the order of hours at physiological temperatures.[180] It is interesting in this connection that a considerable portion of the phosphorus in the membranal phospholipids of microsomal membranes displays a narrow ^{31}P NMR spectrum typical of *isotropic* motion. The extracted lipids, on the other hand, give spectra showing the anisotropic motion associated with bilayers. The authors conjecture that in the intact membrane there may be transitory organizational forms for lipids other than the bilayer structure, one possibility being that the hexagonal H_{II} form exists momentarily within the bilayer.[179] As explained in Chapter 5, the H_{II} phase gives a narrower ^{31}P signal than the bilayer. It was suggested that these transitory forms provide a mechanism for flip-flop motion. However, the rates do not match; movement would have to be on a time scale of less than microseconds to give isotropic phosphorus NMR signals.

On turning to proteins we find a far wider range of translational diffusion constants than in the case of lipids. Some proteins in natural membranes appear to be laterally immobile. The highest values for D_{trans} are for small proteins. For gramicidin C in DMPC layers at 30°C, $D_{trans} \simeq 1.4 \times 10^{-8}$ cm^2 s^{-1}.[177] This represents a rate not too far below that found for lipids. Many intrinsic proteins diffuse far more slowly with $D_{trans} \simeq 10^{-10}$ cm^2 s^{-1}. Such slow rates are inexplicable in terms of the size of the proteins. Webb and his co-workers[177] have shown that when the actin cytoskeleton is unlinked from a variety of slow-moving membrane proteins, their diffusion rates increase to values above 10^{-9} cm^2 s^{-1}. Thus in striated muscle cell the acetylcholine receptor protein is apparently immobile, but after decoupling from actin, $D_{trans} = 2 \times 10^{-9}$ cm^2 s^{-1}. The organization of membrane proteins is outside our scope, but we do have something to say about the magnitude of diffusion constants.

The theory of diffusion in bilayers, and in particular of protein diffusion, is tarnished by compromise. The model commonly used has a strong resemblance to the continuum model of protein–lipid interaction (Chapter 9). The bilayer appears as an infinite plane of viscous liquid separating semi-infinite regions of liquids that sometimes have been taken to be of comparatively low viscosity. The protein molecule is represented by a cylinder with its axis normal to the bilayer and its motion limited to lateral displacement and rotational diffusion about the axis (Figure 90). The dislike end surfaces of the cylinder are coplanar with the surface of the "membrane". The system is now treated as a problem in hydrodynamics.[181,182] A useful dimensionless parameter with which to characterize the system is $\epsilon = (\mu_1 + \mu_2)\, a/\eta h$. The solution of the dynamic equations give expressions for D_{trans} and D_{rot} for all ϵ. Since in practice ϵ is expected to be less than unity, we can use approximate forms:

$$D_{trans} = \frac{kT}{4\pi(\mu_1 + \mu_2)a} \cdot \epsilon[\ln(2/\epsilon) - \gamma + 4\epsilon/\pi - (\epsilon^2/2)\ln(2/\epsilon)] \tag{10.6}$$

where γ is Euler's constant (0.5772) and $\epsilon \lesssim 1$.

$$D_{rot} = \frac{kT}{4\pi(\mu_1 + \mu_2)a^3} \cdot \frac{\epsilon}{1 + 8\epsilon/3\pi} \tag{10.7}$$

and $\epsilon \lesssim 2$.

The translation and rotation of the cylinder depend on the external viscosities to different extents. An increase in μ_1 or μ_2 results in an increase in ϵ, a decrease in D_{trans}, and a smaller decrease in D_{rot}. Apparently rotation is largely determined by the viscosity of the membrane. Upon dividing (10.6) by (10.7), we see that $D_{trans}/D_{rot}\ a^2$ is a function of ϵ only, and thus ϵ can be directly determined from measurements of diffusion coefficients and the dimensions of the cylinder. $(\mu_1 + \mu_2)$ can then be found from (10.6) or (10.7), and finally η can be estimated. The results are not encouraging. Two difficulties arise: For the molecules for which data are available (rhodopsin and erythrocyte band 3 protein), the membrane has a calculated viscosity of around 0.2 P and the environment an average viscosity of about 1 P. Most biological membranes, including the visual receptor and erythrocyte membranes, have viscosities ranging from 0.3 to 1.3 P as estimated by rotational diffusion measurements on fluorescent probes. The viscosities of dilute aqueous solutions are nearer 0.01 P than 1 P. Furthermore, it is impossible to find a set of values for a, h, $(\mu_1 + \mu_2)$ that give a simultaneous fit to the experimental values of D_{trans} and D_{rot}. The theory's one apparent success is the value of D_{trans} ($10^{-8}\ cm^2 s^{-1}$) obtained by using $\bar{\mu} = (\mu_1 + \mu_2)/2 = 0.01$ P and $\eta = 2$ P; these viscosities at least are in the range expected for a bilayer in an aqueous milieu. Using the same viscosities and putting $a = 5$ Å and $h = 50$ Å, the

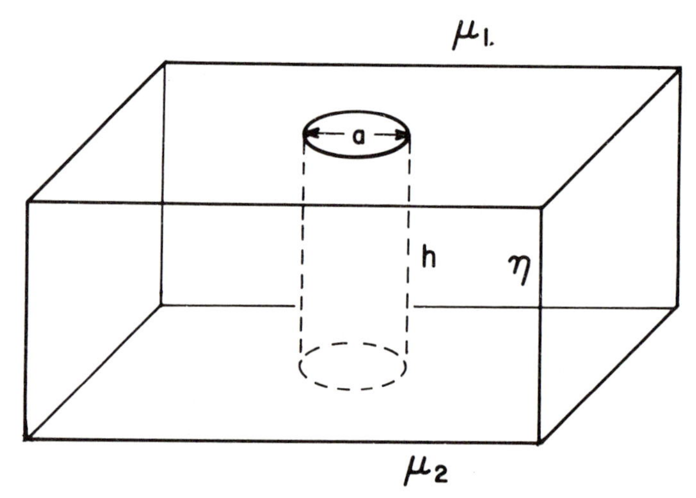

Figure 90. Model used in the calculation of diffusion of protein (represented by a cylinder) in a bilayer represented by a continuum of coefficient of viscosity η bordered by media with coefficients of viscosity μ_1 and μ_2.

calculated value of D_{rot} is $0.13 \times 10^8 \text{ s}^{-1}$, which is a factor of about sevenfold below experiment.

It is usually easier to criticize a theory than to construct one, but it is clear that there are several gross oversimplifications in this theory. The use of a rigid-body-in-a-continuum model for lipid diffusion is unsatisfactory. For molecules that are much larger than those of the solvent, simple models often give good answers; The Stokes–Einstein equations usually work quite well for small molecules in water but can hardly be expected to apply to flexible chains dissolved in identical flexible chains. For a large protein dissolved in lipid, one might hope that the continuum model would be more applicable, but now the choice and the relevance of the solvent viscosity are problematic. Even for much simpler systems, the macroscopic viscosity is sometimes almost irrelevant since it reflects the way in which solvent molecules interact with each other rather than with the solute. The estimated microviscosity for a given system may vary by nearly two orders of magnitude, if determined by rotational diffusion of different probe molecules. The fact, noted in the previous chapter, that proteins sometimes diffuse almost as fast in the gel as in the liquid crystal is another warning that viscosity is not always what is seems to be. The rotation of a hydrophobically surfaced protein in a fatty solvent may well be practically uninhibited by tangential stress — the solvent cannot get a grip. In this case the calculated D_{rot}, based on a macroscopic viscosity, will be too small. The abolishment of tangential stress has a rather small effect on D_{trans}. As an example, putting $\mu = 1$ P, $\mu' = \mu_1 = \mu_2 = 0.01$, $a = 50$ Å, and $h = 75$ Å, we find $D_{trans} \simeq 0.68 \times 10^{-8} \text{ cm}^2 \text{s}^{-1}$ and D_{trans} (no tangential stress) $\sim 0.89 \times 10^{-8} \text{ cm}^2 \text{s}^{-1}$. These values were calculated using convenient approximate expressions given by Saffman and Delbruck.[181]

$$D_{trans} = \frac{kT}{4\pi\mu h}\left[\log\left(\frac{\mu h}{\mu' a}\right) - \gamma\right] \tag{10.8}$$

$$D_{trans}\text{ (no tangential stress)} = \frac{kT}{4\pi\mu h}\left[\log\left(\frac{\mu h}{\mu' a}\right) - \gamma + \frac{1}{2}\right] \tag{10.9}$$

The continuum model can give simple qualitative explanations of some experimental facts that seem strange at first sight. For example, the addition of cholesterol has a strong effect on D_{trans} for proteins but little, if any, effect on D_{rot}. This is so because on addition of a rather large molecule the effective viscosity of a continuum increases for translational but not rotational motion, as the reader can appreciate by experimenting in a swimming pool before and during the school vacations.

The model described is basically macroscopic; it could be applied to beer barrels. The assumption of a continuum is in line with the philosophy behind Fick's equation and its spiritual descendants such as the Nernst–Planck equation used in Chapter 12. Such equations can be given a microscopic explanation in terms of random walks, but the walks themselves describe a continuum of resting places. An alternative approach to diffusion is based on the activated rate theory that supposes, in this particular application, that molecular motion can be envisaged as a series of jumps between discrete sites separated by energetic barriers. Chemists will be familiar with the use of this concept in chemical kinetics. Lateral diffusion in a bilayer is conceived of in terms of the exchange of occupied and vacant sites in a two-dimensional crystal by a simple hopping mechanism. The for a square lattice

$$D_{\text{trans}} = \frac{l^2}{2\tau} \qquad (10.10)$$

where l is the lattice side and $1/\tau$ the average jumping rate. The value of the diffusion constant will be determined by the free energy of activation, the geometry of the lattice, and the temperature. The model seems more appropriate to the gel than to the liquid crystal state, but in the gel state there could be defects such as the dislocations found in many crystals and along which diffusion is much faster than in perfectly packed areas. Even in a defect-free sample, any attempt to calculate an activation energy for jumping between sites in the liquid crystal state would have to take into account the by now familiar complex dynamic state of the lipid chains in the hydrophobic core. This is not a feasible project at present; the accurate theoretical estimation of the energy barrier in such a system is beyond us. Thus, as far as diffusion in bilayers is concerned, the activated rate theory is to be regarded as an inducement to fit parameters rather than as a means of theoretically estimating rate constants or activation parameters.

In summing up the status of theory in this field, we can discern a major failure in the inability to provide a method for simultaneously accounting for the rotational and translational diffusion constants of a given molecule. Upon looking back at the complex picture of bilayer structure that emerges from previous chapters, it would be surprising if a simple geometric model would suffice to explain the diffusion of amphiphilic molecules in a medium that consists partly of a relatively fluid core and partly of a polar or charged layer that may interact with foreign molecules via electrostatic forces. The identification of μ_1 and μ_2 with the viscosity of an aqueous medium could well lead to paradoxes if η is given the value expected for the lipid core. It conceivably could be more difficult for a protein molecule to rotate or translate within the polar head groups than within the core, in which case it might be advisable to take μ_1 and μ_2 to be larger than η.

We now turn to experiment, this being the only chapter in which we refer to spectroscopic techniques by more than name. We concentrate almost exclusively on those aspects of experiment that determine the *range* of motional frequencies observable by a given technique. As we saw in the discussion of border lipid (Chapter 9), the instrument can prejudice our thoughts on the system.

Motion has been studied by both transient and steady-state experiments. In the former methods we prepare a system in a state of macroscopic nonequilibrium and observe its relaxation toward equilibrium. Examples include fluorescence photobleaching recovery (FPR), fluorescence depolarization subsequent to pulse photoselection, and spin-pulse determinations of electron and nuclear spin-relaxation times. Steady-state methods usually involve the line-shape analysis of the spectra of systems at thermodynamic equilibrium. A whole range of physical techniques has been applied to membranes in recent years, including time-dependent neutron scattering, picosecond resonance Raman spectroscopy, and electron-spin resonance saturation transfer spectroscopy. It would be out of character for this book to contain a discussion of the details, limitations, and advantages of all these methods. We will attempt only to explain the importance of time scales in spectroscopy, and also have a look at one type of nonspectroscopic technique.

To illustrate the limitations imposed by a specific type of experiment, we consider the measurement of rotational diffusion using fluorescence depolarization.[183]

Molecules almost invariably have directional preferences for absorption and emission of light; the transition moments have fixed orientations with respect to a molecular coordinate system. The probability that a photon of light will be absorbed is proportional to $\cos^2 \theta$, where θ is the angle between the absorption transition moment and the electric vector of the radiation, which we label by z. Experiments on membranes are often carried out on randomly oriented absorbers (chromophores), and the irradiation of such a system by polarized light at an appropriate frequency will result in the preferential excitation of those molecules that have small values of θ (Figure 91). In general, each excited molecule has components of its transition moment both perpendicular and parallel to z, and in a random sample the ratio of the x, y, and z excited moments is 1 : 1 : 3 at the time of excitation. The process of preparing an ensemble of this kind is termed *photoselection*. Two time-dependent phenomena now occur: rotational diffusion of the molecules and emission of light. The time scales characterizing these processes depend on the molecule and its environment.

We consider two extreme cases. First we suppose that appreciable rotation occurs in a much shorter time than the average lifetime of the

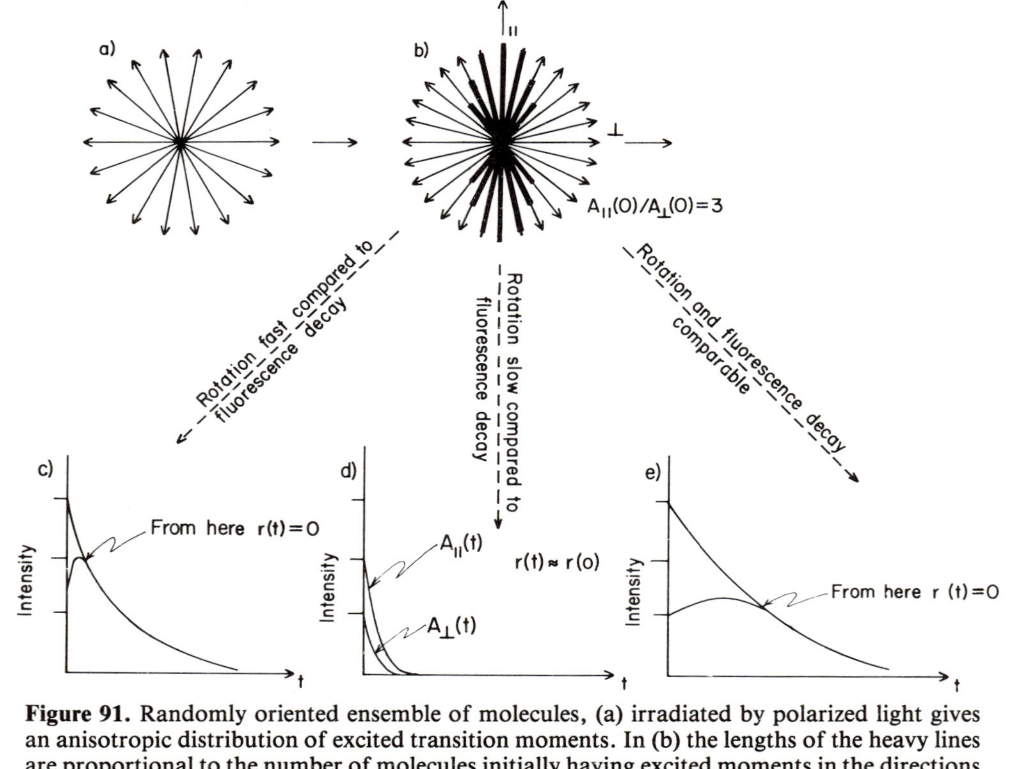

Figure 91. Randomly oriented ensemble of molecules, (a) irradiated by polarized light gives an anisotropic distribution of excited transition moments. In (b) the lengths of the heavy lines are proportional to the number of molecules initially having excited moments in the directions indicated. The relative rates of molecular rotation and fluorescence decay determine the characteristics of the emitted light. $r(t)$ is defined in equation (10.11). No account is taken of anisotropic diffusion.

excited state. In this situation the excited transition moments will randomize, that is, relax to an isotropic distribution before any significant emission of light occurs. When light is emitted, it will be unpolarized and we can deduce only that rotation is fast, but we cannot say how fast. The second case is that of a molecule for which rotation occurs at a much slower rate than the return to the electronic ground state. The molecules will all emit light before they can move from the initial radiation-induced angular distribution of transition moments. The emitted light will have the same polarization as the absorbed light and the experiment only tells us that the excited lifetime τ_e is much shorter than the rotational correlation time τ_r. The intermediate case is the most informative; if $\tau_e \simeq \tau_r$, rotation will carry the excited transition moments into changing angular distribution patterns while emission of light occurs. If we write $A_\parallel$ and $A_\perp$ for the intensities of the emitted light polarized along z and x, respectively, then the *fluorescence anisotropy* is defined by

$$r(t) = \frac{A_\parallel(t) - A_\perp(t)}{A_\parallel(t) + 2A_\perp(t)} \tag{10.11}$$

If the emission and transition moments are parallel, then for the light emitted immediately after excitation the ratio $A_\parallel(0)/A_\perp(0)$ will be given by the absorbtion ratio 3 : 1, and $r(0)$ by $(3-1)/(3+2) = 0.4$, a value found experimentally. For a molecule undergoing isotropic rotational diffusion, $r(t)$ decays to zero:

$$r(t) = r(0)\exp(-t/\tau_r) \tag{10.12}$$

For anisotropic systems, however, such as liquid crystals or bilayers, the chromophores need not necessarily relax to an isotropic distribution and $r(t)$ is then given by,

$$r(t) = [r(0) - r(\infty)]\exp(-t/\tau_r) + r(\infty) \tag{10.13}$$

where $r(\infty)$ is the fluorescence anisotropy for very long times. Now at times long enough to restore orientational equilibrium, the value of the the polarization is determined by the fluctuations of the probes about their average equilibrium orientation and it has been shown that [184]

$$r(\infty) = 2/5\langle P_2(\cos\theta)\rangle^2 = 2/5\langle(\tfrac{3}{2}\cos^2\theta - \tfrac{1}{2})\rangle^2 \tag{10.14}$$

which evokes memories of the definition of order parameters [equation (5.1)] and suggests that $r(\infty)$ contains information on *probe* ordering. To evaluate the average of $P_2(\cos\theta)$, we need to know the angular distribution function $\omega(\cos\theta)$ giving the probability of finding the probe with a certain value of $\cos\theta$ with respect to its director. Often $\omega(\theta)$ is taken to be a Gaussian.

The competition between radiative and directional decay is more clearly evident in the equation for the fluorescence anisotropy under conditions of *continuous* irradiation:

$$r_s = [r(0) - r(\infty)]/[1 + \tau_e/\tau_r] + r(\infty) \qquad (10.15)$$

r_s is the steady-state fluorescence anisotropy. If $\tau_r \gg \tau_e$, $r_s = r(0)$. If $\tau_r \ll \tau_e$, $r_s \to r(\infty)$. In either case, it is only possible to put an upper or lower limit on τ_r. If, however, τ_e and τ_r are similar in magnitude, then measurements of r_s, $r(0)$, and $r(\infty)$, together with the value of τ_e, allow τ_r to be determined.

The general lesson is that, in order to measure the rates of physical processes, the time scale of a spectroscopic technique must be compatible with that of the motion under study. For example, most fluorescent chromophores have an average lifetime in the excited state of 10^{-7} to 10^{-9} seconds and thus are not suitable for measurements on the rotation of proteins for which τ_r in membranes is likely to fall in the microsecond to millisecond range. Among the probes that are suitable for studying these slow motions are molecules that can be excited to comparatively long-lived triplet states. The exploitation of fluorescence decay kinetics to follow motion in the subnanosecond range is made possible by the use of very short excitation pulses, for example, from a synchrotron.[185] The rotation of tryptophan residues on a protein have been detected in this way.

A related aspect of the curbs imposed by spectroscopic time scales is contained in the apparent paradox concerning the nature of boundary lipid. To understand the different findings of NMR and ESR, we consider the wider problem of a molecular moiety having a choice of two conformations or sites associated with different spectral characteristics. An example is a spin label exchanging positions between bulk and boundary lipid. We simplify by supposing that there is an equal probability of a molecule occupying either site and that there is only one spectral absorption line at each site. For extremely slow exchange rates, the observed spectrum (Figure 92) consists of two lines of equal intensity at frequencies we label ν_1 and ν_2 while at very high exchange rates a single averaged line is observed at a frequency $(\nu_1 + \nu_2)/2$. If the average lifetime at either site is τ, then it can be shown that the condition for observing a single line is that $\tau\Delta\nu \ll 1$, where $\Delta\nu = \nu_1 - \nu_2$. For very long average residence times for which $\tau\Delta\nu \gg 1$, two separate lines are observed at the frequencies for which they would appear if there were no exchange. A deuterium NMR spectrum of a chain-labeled phospholipid preparation might typically contain lines separated by up to 100 kHz. These lines will be averaged if $\tau \times 10^5 \ll 1$, that is, if $\tau \ll 10^{-5}$ seconds. On the other hand, the frequency difference between the ESR spectral lines of spin-labeled lipids in bulk and boundary can reach nearly

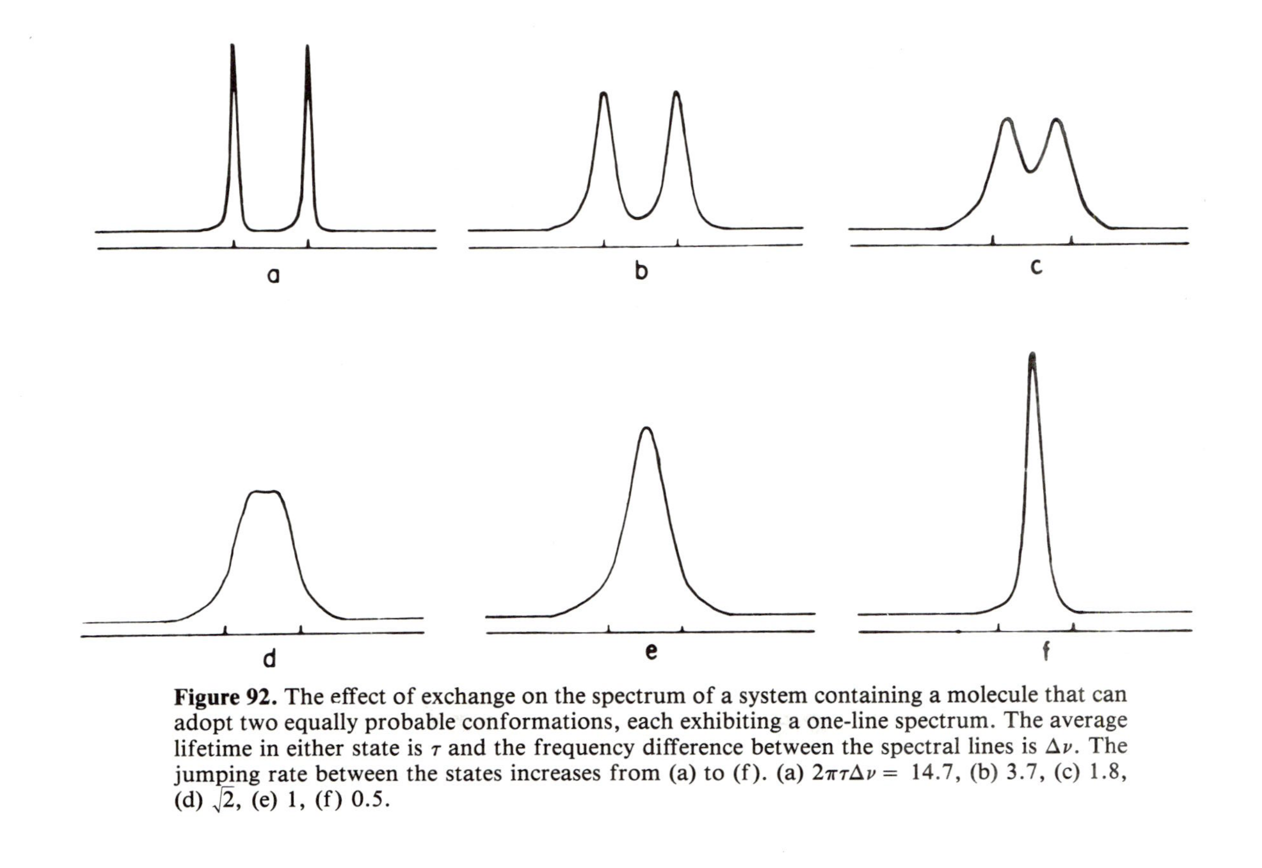

Figure 92. The effect of exchange on the spectrum of a system containing a molecule that can adopt two equally probable conformations, each exhibiting a one-line spectrum. The average lifetime in either state is τ and the frequency difference between the spectral lines is $\Delta\nu$. The jumping rate between the states increases from (a) to (f). (a) $2\pi\tau\Delta\nu = 14.7$, (b) 3.7, (c) 1.8, (d) $\sqrt{2}$, (e) 1, (f) 0.5.

10^5 kHz, which means that τ must be less than $\sim 10^{-8}$ seconds to average the two lines. If 10^{-8} seconds $< \tau < 10^{-5}$ seconds, the ESR spectra will show separate signals from bulk and boundary but the deuterium NMR signal will be averaged.

This seems a plausible explanation of the experimental facts and independent support for this interpretation is provided by at least two types of experiment. Deuterium NMR spin-relaxation measurements on phospholipids in bilayers containing proteins suggest that there is a motion with a rate constant of at least 10^6–10^7 s^{-1}. A related type of observation is contained in the line-width measurements made by Marsh et al. on spin-labeled lipids in bovine rod outer-segment membranes. For exchange rates for which $\tau\Delta\nu$ is neither much larger nor much smaller than unity, the line widths or line shapes of the spectra (Figure 92) can be used to obtain values for τ. This is a very common procedure in nuclear magnetic resonance, and to a lesser extent in electron spin resonance, and has been used in the above system to estimate an exchange rate of 10^6–10^7 s^{-1}, consistent with other experimental evidence.

Another, but far less obvious, case of compatibility arises in the case of spin relaxation. For example, the lifetime of a magnetic nucleus in a given energy level depends strongly on the time-varying fields in which it is bathed by its surroundings. These fields may arise from neighboring magnetic nuclei or from electric charge distributions around the nucleus. As a result of vibrational, rotational, and translational motion, these fields fluctuate randomly. Those parts of the field varying at transition frequencies of the nucleus will induce the nucleus to jump between its available energy levels and thus shorten its residence lifetime. This is why measurements of nuclear spin-lattice relaxation times, T_1, can provide motional information. If the random field variations induced by motion do not contain a significant contribution from the right frequencies, T_1 will be insensitive to motion. Thus vibrations of bonds are far too fast to affect T_1. The incompatibility here involves two frequencies characteristic only of the system. T_1 and other relaxation times have been measured for carbon (^{13}C), hydrogen, and phosphorus in bilayer phospholipids.[186] The interpretation of the results is not always straightforward; a single relaxation time is insufficient to determine a molecular motion. Nevertheless, relaxation studies have been, and will continue to be, useful supplements to more direct approaches.

It is fortunate, in view of the great range of lateral diffusion rates, that there are measuring techniques not significantly limited by any time scale, except perhaps the patience of the observer. A particularly useful method involves the use of selective photobleaching to study lateral diffusion.[187] The range of rates at present measurable by this technique is very large, stretching down to diffusion constants as small as 10^{-11} $cm^2 s^{-1}$. In essence, the method consists of irradiating with a short, intense light pulse a small area of a membrane containing fluorescent molecules that can be irreversibly photochemically bleached. Diffusion, or other processes, will lead to a recovery in the concentration of fluorescent probes in the bleached region at a rate that can be monitored by the use of an exciting beam so strongly attenuated as to cause, at the most, insignificant bleaching. The kinetics of recovery show very different time dependence for diffusion as against hydrodynamic flow, thus allowing the processes to be differentiated. In addition, the observation of incomplete recovery can be used, with careful checks, to establish the existence of immobile membrane components. A variety of clever variations have been dreamed up and the reader is referred to references.[187-189]

An interesting technique for the measurement of translational diffusion is fluorescence correlation spectroscopy.[190] The principle, which is more fully explained in Chapter 17, is based on the fact that the number of probes in a given small area of bilayer will fluctuate about a mean value due to random exits and entrances of individual probes into and from the area. The fluctuations in number density occur solely because the probes can move and the average rate of fluctuation clearly depends on the average rate of lateral diffusion of the probes. Experimentally, fluorescence is excited by a laser of constant intensity and the variations in the intensity of emitted light are monitored. The method is best suited to fast diffusional rates such as those typical of lipids in the liquid crystal state of bilayers. Experience shows that over the longer lengths of time needed to measure the slower diffusion rates typical of proteins there are systematic fluctuations in fluorescence intensity, the nature of which are not clear. This limitation on the applicability of fluorescence correlation spectroscopy is a consequence of the instability (mechanical, chemical, or physiological) of the observed system and differs completely from the limitations inherent in the magnetic resonance techniques discussed earlier where the spectroscopic time scale was the determining factor.

Many of the techniques mentioned above are suitable, and have been used, for the detection and measurement of intramolecular motion. In particular, attention is now being focused on the conformational fluctuations of proteins, and eventually we may learn through these methods how the bilayer modifies the form and action of membrane proteins in situ.

The problem of describing diffusion in biological membranes is rendered almost impossible by our ignorance of membrane structure. Nevertheless, progress has been made in understanding diffusion in heterogeneous bilayers. The presence of foreign molecules in a bilayer will affect diffusion in a way that depends partly on the topology of the system. If the distribution of impurities is random, it is convenient to use Monte Carlo methods to simulate diffusion.[191] The random distribution of, say cholesterol can be modeled and lipid diffusion simulated by a biased random walk in which the direction of a translational step is determined by the occupancies of the sites around the diffusing molecule and the type of molecule occupying each site. On increasing the impurity concentration, percolation (the formation of a network of comparable size to the system) sets in suddenly (Figure 89) and the effect of diffusion will clearly be to hinder severely movement of lipid over long distances. Even for relatively small amounts of foreign molecules, diffusion could be significantly affected in certain circumstances. For example, a diffusing membrane component might bind to an isotropically distributed but widely spaced collection of immobile structures in the membrane. If the kinetics of binding and dissociation is fast compared with the time taken to diffuse between sites, then over a long period of time the observed diffusion constant will be given by $D = XD_0$, where X is the fraction of time that the molecule is unbound and D_0 is the free diffusion constant.[192]

An entirely different type of heterogeneous system is that described by Owicki and McConnell.[192] There is evidence that below T_t DPPC and cholesterol (<20 mol%) adopt a structure consisting of parallel alternating zones in the form of ridges of pure gel-state DPPC and fluid plains consisting of ~20 mol% cholesterol in DPPC. Clearly the diffusion of a molecule perpendicular to these strips will be slow compared with diffusion along the liquid strips. The authors set up diffusion equations and show that the solutions are compatible with the results of photobleaching experiments.

It is necessary to take diffusion into account in the description of many biological membrane processes. As a simple example, consider the membrane-bound enzyme Ca^{++}- activated adenosinetriphosphatase, which, from biochemical evidence, appears to have a turnover time of about one tenth of a second. This is a measure of the time that a substrate molecule spends in the vicinity of the active site. Since spectroscopy shows that the average lifetime of a lipid molecule on the surface of the protein is certainly less than 10^{-5} seconds, we cannot subscribe to the once fashionable model based on the essential role in enzyme action assigned to a supposedly long-lived sheath of lipid around the protein. By contrast, the high lateral mobility of cytochrome *c* oxidase in mitochondrial membranes (probably due to the high percentage of unsaturated chains and the virtual absence of cholesterol) is said to be well within the time structure of known rates of electron transfer and energy transduction at physiological temperatures.[193] We have concentrated on lateral diffusion in bilayers and membranes because less rotational data exist and the role of protein rotation is yet to be elucidated.

11 The Donnan and Nernst Equilibria

In this short chapter, we record some elementary aspects of ionic equilibria that will be useful in the following chapter, which deals with the nonequilibrium process of transport. We concern ourselves with the partitioning of ions and molecules between two phases.

An *uncharged* molecule distributes itself between two phases in accordance with the thermodynamic requirement that its chemical potential be the same in both phases:

$$\mu_i^{(1)} = \mu_i^{(2)} \tag{11.1}$$

For ideal solutions under isothermal and isobaric conditions:

$$(\mu_i^{(1)})^{\circ}_{T,P} + RT \ln x_i^{(1)} = (\mu_i^{(2)})^{\circ}_{T,P} + RT \ln x_i^{(2)} \tag{11.2}$$

where the first term on each side is a standard chemical potential and x_i is a mole fraction. Rearranging (11.2) we obtain

$$\frac{x_i^{(1)}}{x_i^{(2)}} = \exp(\Delta\mu_i^{\circ}/RT) \equiv \beta(x_i) \tag{11.3}$$

which means that at fixed temperature and pressure the partition coefficient, $\beta(x_i)$, is completely defined by the standard chemical potentials.

The effects of charges in the solvents do not appear explicitly. Considering aqueous systems, if one phase contains, say, fixed charges, such as those on the head groups in bilayers, there will certainly be an effect on, among other things, (1) the average orientation and the polarization of the neighboring polar water molecules, and (2) the distribution of ions in the bordering aqueous solution (see the discussion of the double layer in Chapter 7). Both of these effects will alter the values of $(\mu_i^{(1)})^\circ$ and $(\mu_i^{(2)})^\circ$ from those found in the absence of charge. Nevertheless, it is generally supposed that the variation of the partition function for uncharged molecules is not significant over the range of variation of surface charge expected during the life of a membrane. The partition coefficient also can be affected by changes in the concentrations of other uncharged molecules, including its own concentration. The use of activities formally takes care of these phenomena, but such refinements are rarely justified in the light of the experimental and theoretical uncertainties that accompany most work on membranes.

For charged molecules the situation is complicated by the presence in the electrochemical potential of a term giving the potential energy of a charge in an electric field. This field need not have an external origin but may arise, for example, as a consequence of space charge densities generated within the system. The condition for equilibrium is now

$$\tilde{\mu}_i^{(1)} = \tilde{\mu}_i^{(2)} \tag{11.4}$$

which, again for ideal solutions, we write as

$$\begin{aligned}(\mu_i^{(1)})_T^\circ + P^{(1)}\bar{V}_i + RT \ln x_i^{(1)} + z_i F\phi^{(1)} \\ = (\mu_i^{(2)})_T^\circ + P^{(2)}\bar{V}_i + RT \ln x_i^{(2)} + z_i F\phi^{(2)}\end{aligned} \tag{11.5}$$

where we have discarded the isobaric constraint of (11.2). Two assumptions have been made here, namely, that partial molar volumes are independent of solvent and pressure and that ϕ is independent of position within one phase. The first assumption is "nearly valid," and the second is true if we keep away from the border between the phases and thus avoid the double layer. Equation (11.5) does not completely determine the state of the system but the condition of electroneutrality leaves no room for ambiguity:

$$\sum_i z_i x_i^{(1)} = \sum_i z_i x_i^{(2)} = 0 \tag{11.6}$$

This is a *macroscopic* condition. It should be realized that deviations from electroneutrality, although vanishingly small, can occur, as in the case of double layers. We now take a series of progressively more complex cases, starting with a system in which (a) pressure differences can be ignored or are absent, and (b) $(\mu_i^{(1)})_T^\circ = (\mu_i^{(2)})_T^\circ$, which is true if (1) and (2) are the same solvent.

With these restrictions,

$$\frac{x_i^{(1)}}{x_i^{(2)}} = \exp(z_i \Delta\phi / RT) \tag{11.7}$$

where $\Delta\phi = \phi^{(2)} - \phi^{(1)}$. This is the celebrated *Nernst equilibrium equation.*[194] For a membrane permeable to all the ions in the system, the equilibrium state in the absence of an external field is given by $x_i^{(1)} = x_i^{(2)}$ so that $\Delta\phi = 0$. We can therefore regard $\Delta\phi$ in such a system as the externally imposed potential difference needed to maintain a difference in concentration across the membrane. (Electrochemists will see a concentration cell lurking in the shadows, or maybe an overpotential.)

In many systems, such as colloid solutions, ion exchange resins, membranes, and cells, there are at equilibrium unequal concentrations of a given ion in two adjoining phases, giving rise to a potential gradient. In the absence of active transport — biochemical pumping — the normal origin of the concentration differences is the presence of ionizable molecules incapable of passing through the phase interface. A biological example is a large peptide unable to penetrate a membrane. A second example, immediate to our interests, is the array of ionized groups that exist in all membranes — ionized head groups or membrane proteins. Although membranes are the center of interest for us, the following treatment is general.

The system considered consists of two solutions separated by a partition permeable to all ions except one, Y^{n-}, an anion with average negative charge ne situated in phase 1. The other ions are univalent; the solvent is denoted by w, suggesting, but not necessarily denoting, water. Equation (11.6) gives

$$\begin{aligned} x_+^{(1)} - x_-^{(1)} - n x_Y^{(1)} &= 0 \\ x_+^{(2)} = x_-^{(2)} &= x \end{aligned} \tag{11.8}$$

Successively applying (11.5) to cations, anions, and solvent:

$$(P^{(1)} - P^{(2)})\bar{V}_+ + F(\phi^{(1)} - \phi^{(2)}) = RT \ln(x_+^{(2)}/x_+^{(1)}) \tag{11.9}$$

$$(P^{(1)} - P^{(2)})\bar{V}_- - F(\phi^{(1)} - \phi^{(2)}) = RT \ln(x_-^{(2)}/x_-^{(1)}) \tag{11.10}$$

$$(P^{(1)} - P^{(2)})\bar{V}_w = RT \ln(x_w^{(2)}/x_w^{(1)}) \tag{11.11}$$

Notice that since the polyanion Y^{n-} appears on only one side of the partition, we cannot write for it an equality corresponding to those for permeating species.

If we add (11.9) to (11.10), we find:

$$\ln \frac{x_+^{(2)} x_-^{(2)}}{x_+^{(1)} x_-^{(1)}} = \frac{(P^{(1)} - P^{(2)})(\bar{V}_+ + \bar{V}_-)}{RT} \tag{11.12}$$

For negligible pressure differences, such as usually obtained in biological or biochemical systems, the right side of the equality vanishes and we are left with

$$x_+^{(1)}x_-^{(1)} = x_+^{(2)}x_-^{(2)} = x^2 \tag{11.13}$$

the second equality following from (11.8). If we again neglect ΔP, equations (11.9) and (11.10) give

$$\Delta\phi = -\frac{RT}{F}\ln\frac{x_+^{(2)}}{x_+^{(1)}} = -\frac{RT}{F}\ln\frac{x_-^{(1)}}{x_-^{(2)}} = -\frac{RT}{F}\ln r \tag{11.14}$$

Using (11.8) and r as defined by (11.14), the reader can obtain

$$r^2x_+^{(2)} - x_+^{(2)} - rnx_Y^{(1)} = 0 \tag{11.15}$$

The quadratic equation in r has a positive solution:

$$r = \frac{nx_Y^{(1)}}{2x_+^{(2)}} + \left\{\left(\frac{nx_Y^{(1)}}{2x_+^{(2)}}\right)^2 - 1\right\}^{1/2} \tag{11.16}$$

The value of r can be substituted into (11.14) to find $\Delta\phi$. The equilibrium defined by (11.14) and (11.16) is termed a *Donnan equilibrium.* Those searching for a more thorough account of the Donnan equilibrium are recommended to consult reference 194a.

The relevance of the above manipulations to the transport processes to be analyzed later lies in the differing natures of β as defined in (11.3) and r, defined by (11.16). Both are partition coefficients, but while r is a function of ionic concentration, β is concentration independent. Consider the model membrane of Figure 93 in which a monovalent cation in solutions I and II has concentrations c^{I} and c^{II}, and a *concentration-independent* partition coefficient β. Then the value of $\Delta\phi$ at the two borders is given by

$$\Delta\phi_{\mathrm{I}} = \frac{RT}{F}\ln(c^{\mathrm{I}}/\beta c^{\mathrm{I}}) = -\frac{RT}{F}\ln\beta$$

$$\tag{11.17}$$

$$\Delta\phi_{\mathrm{II}} = \frac{RT}{F}\ln(\beta c^{\mathrm{II}}/c^{\mathrm{II}}) = \frac{RT}{F}\ln\beta$$

The illustration shows the changes in potential in crossing the membrane. It is seen that since $\Delta\phi$ depends only on β — that is, a *ratio* of concentrations — it has the same absolute value on both sides of the membrane. We now impose an internal potential gradient on the system. One source of such a gradient, as we shall see in the following chapter, is diffusion of ions across the membrane. Again referring to the illustrations, we see that the internal potential difference across the membrane, ϕ_D, is equal

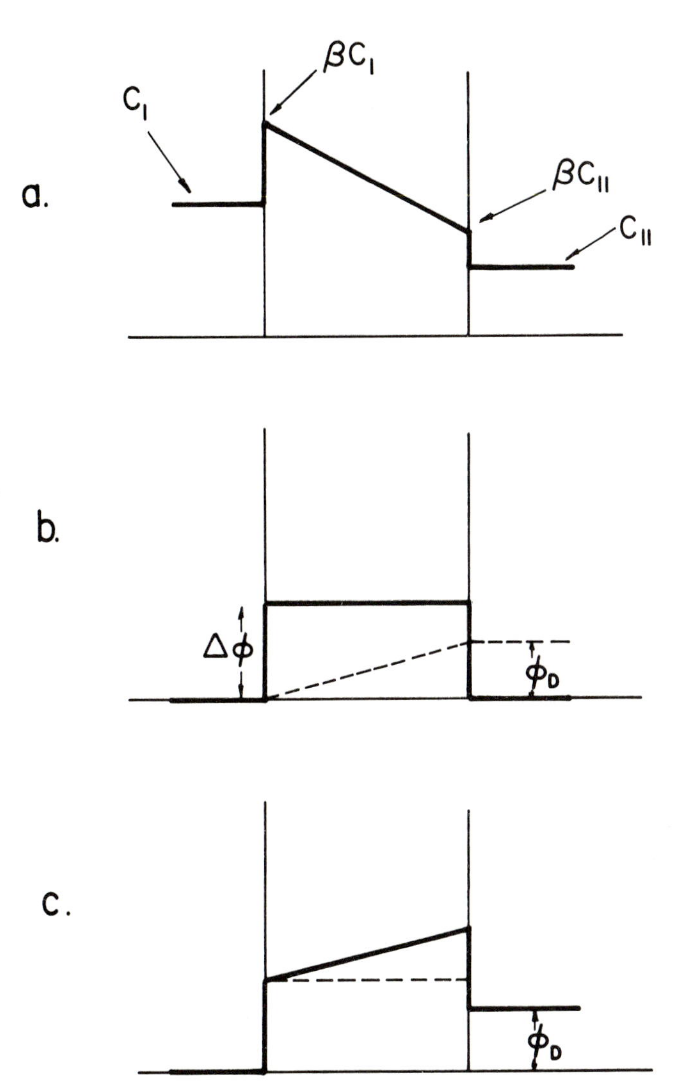

Figure 93A. Concentration profile across a membrane with identical partition functions for the two surfaces. **B.** Potential profile (solid line) due to the Nernst potential. The dashed line represents an additional potential, for example, a diffusion potential. **C.** Sum of the potentials in **B** is indicated by the solid line, showing that for this case the membrane potential ϕ_D is independent of $\Delta\phi$, the Nernst potential.

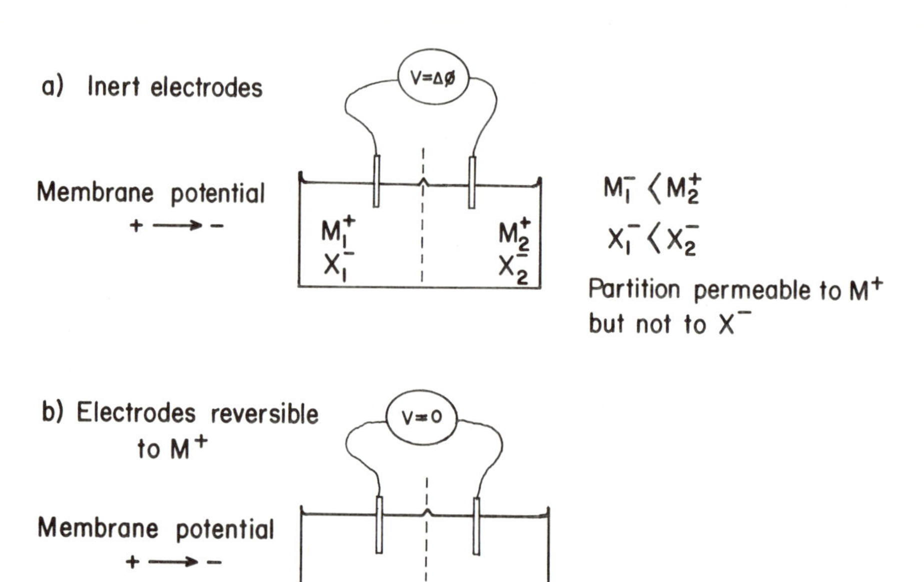

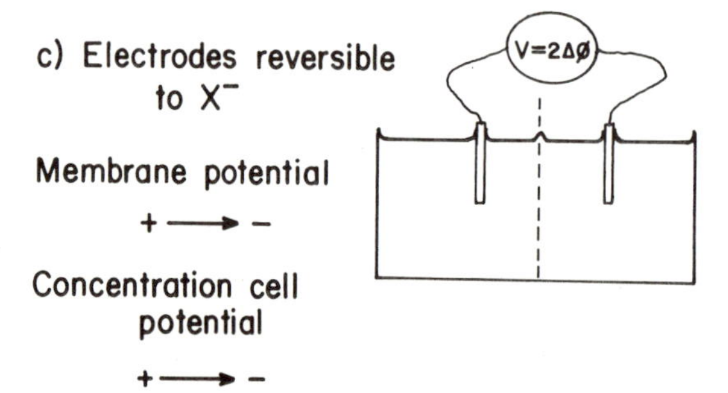

c) Electrodes reversible to X^-

V=2Δø

Membrane potential

+ ⟶ −

Concentration cell potential

+ ⟶ −

Figure 94. Dependence of the measured potential across a semipermeable membrane depends on the nature of the electrodes.

to the *membrane potential*, the total change in crossing the membrane,

$$\phi_D = \phi_M \tag{11.18}$$

an equation that will be of use later. The argument breaks down if the border potential is due to a Donnan equilibrium, in which case, unless $c^{\mathrm{I}} = c^{\mathrm{II}}$ for the cation on either side of the membrane, $\Delta\phi_{\mathrm{I}} \neq \Delta\phi_{\mathrm{II}}$. This is of relevance to the derivation of the Goldman–Hodgkin–Katz equation dealt with in the next chapter.

A note on the *observed* potential difference across a semipermeable membrane may be of help to some readers. Referring to Figure 94, we consider a membrane permeable to M^+ and impermeable to X^-. Let the concentration of M^+ be greater on the right than on the left so that r, defined by (11.14), is greater than unity. A Donnan equilibrium will be set up with a potential difference across the membrane given by $\Delta\phi = -(RT/F)\ln([M^+](2)/[M^+](1)) = -(RT/F)\ln r$. Since $\Delta\phi$ is defined by $\Delta\phi = \phi_2 - \phi_1$, it follows that $\phi_2 < \phi_1$, which simply means that the gradient of the potential is in the direction required to maintain the illustrated charge imbalance. It is $\Delta\phi$ that will be registered on a voltmeter connected to *inert* electrodes placed in the two compartments. If we now substitute electrodes that are *reversible to* M^+, the system will attempt to equalize the concentrations of M^+ by passage of electrons from left to right in the external circuit. In fact, we have a concentration cell with a potential difference across the electrodes given by $(RT/F)\ln r$, which exactly compensates the Nernst potential. The observed potential difference across the electrodes therefore is zero. If we choose to use electrodes that are reversible to X^-, we will have a concentration cell driven by the anions. The resulting potential, $(RT/(-1)F)\ln r$, acts in the same direction as the Nernst potential and the observed potential drop across the electrodes is $2\Delta\phi$.

In measuring a Donnan potential, the electrodes can be placed anywhere in the bulk solutions on opposite sides of the membrane. The potential in the bulk is not, except very close to the surface, a function of the distance from the membrane surface. This is true because the Donnan potential arises as a consequence of the differing compositions of the bulk solutions. In contrast, the potential produced in the bathing medium by surface charge effectively dies out after a few Debye lengths. If the two surfaces of a membrane differ in surface charge density, a transmembrane potential gradient will result in addition to that due to Donnan potentials. There is as yet no completely satisfactory method for separating these two contributions to the membrane potential in cells.[194b]

12 Transport – The Nernst-Planck Approach

The living cell requires for its proper functioning materials that originate outside the cell, and as a result of its functioning produces materials that leave the cell. The passage of molecules across the cell membrane is known in many cases to be a purely diffusive process, describable in principle by the same physical laws that control the diffusion of molecules in solution or across artificial membranes, or the migration of atoms in solids. There are, however, numerous examples of so-called *active transport,* which involves the use of energy obtained from chemical processes to drive molecules across membranes, frequently against their concentration gradients. In this chapter we will not be concerned with these biologically more interesting processes, but rather with the (at first sight), simpler processes that can be classified under the heading of diffusion.

In looking back at the chapters in which the molecular structure of membranes was outlined, it is evident that it will not be easy to give a microscopic account of the passage of even a simple spherical ion, such as Na^+, through the gyrating polar head groups and flaying hydrocarbon chains of a one-lipid bilayer. The fact that a cell membrane contains many kinds of phospholipid together with assorted proteins and other molecules further complicates the theoretician's task and has resulted in simplified theoretical and experimental approaches to transport. Experimentalists have devoted a great deal of attention to artificial bilayers composed of one phospholipid, while theoreticians have tended to construct models that are susceptible to mathematical treatment. We shall describe the main models currently fashionable and we will see that while differing physical models sometimes lead to the same transport equations, there are sometimes independent justifications for preferring one model over another. We will avoid enumerating the almost endless variations on basic models, attempting rather to confine ourselves to general principles. We try to maintain a critical stance throughout.

Diffusion, the process that will occupy much of our attention, can be approached in a number of ways.

1. Historically the first approach was to derive phenomenologic equations from simple diffusion processes. August Fick[194] formulated his first and second laws in terms of concentration gradients in the diffusing substance. For one-dimensional diffusion his first law states that

$$J_i = -D_i \frac{\partial c_i}{\partial x} \tag{12.1}$$

where J_i is the *flow* of substance i in units of the *mass* of i passing through unit *area* in unit *time*. The law states that this flow is proportional to the concentration gradient at the same point. The *diffusion constant* D_i is a function of the chemical nature of the diffusing molecule i and the other components in the system. Often one component is in great excess and is termed the solvent. Later we develop the consequences of Fick's law but now we point out a major shortcoming. Equation (12.1) implies that flow ceases when the concentration gradient is zero. This condition is not sufficient to cover situations where an electric or gravitational field is imposed on the system, or when there is a second solute that itself has a concentration gradient. Thermodynamics provides a condition for equilibrium that is the basis of a more general approach to diffusion.

2. The condition for thermodynamic equilibrium in the absence of mechanical forces is that the electrochemical potential for a component i is the same throughout the system. If we write, again for one dimension,

$$J_i \,\alpha\, \frac{\partial \tilde{\mu}_i}{\partial x} \tag{12 2}$$

then flow will cease when the condition for equilibrium is obeyed. Equation (12.2), which is associated with Nernst, has been the basis of much of the published work on diffusion theory.

3. Since diffusion is an example of the *approach* to equilibrium, a system in which macroscopic diffusion is observed cannot be at equilibrium. It follows that the use of classical thermodynamics to discuss flows is esthetically unpleasing at the least, and possibly theoretically suspect at the worst. *Nonequilibrium thermodynamics,* on the other hand, is a theory that addresses itself specifically to the relationships between flows and thermodynamic parameters and has enabled sweeping generalizations to be made concerning the connection between different physical phenomena. As a thermodynamic theory it has nothing to say about molecular mechanisms, but it often provides conditions with which other theories must comply.

All of these theories avoid microscopic considerations; they refer neither to the structure of the solution nor to the process by which molecular translation occurs. There have been a number of microscopic approaches; we list three, again simplifying to one dimension.

4. It is assumed that a molecule in an isotropic solution can undergo in unit time a number of translational steps. The process is random in that the direction of the individual steps is unpredictable, being subject only to the constraint that there is an equal probability of jumping to the left or to the right. This is the famous random-walk problem of probability theory. The easiest way to visualize the physical consequences is to imagine a sheet of molecules initially lined up in a solvent and then allowed to execute their own random walks. It is intuitively obvious that after some time the original distribution will be destroyed, and molecules will be found distributed at various distances from the plane. The probability of finding a molecule at a given distance from the plane can be shown to obey the normal or Gaussian distribution. Fick's first law leads to the same distribution. The statistical approach to diffusion was developed mainly by Einstein[195] and Smulochowski.[196] Although the theory refers to the translational movement of individual molecules, it does not attempt to delineate any specific physical picture for translation beyond the general idea of molecular buffeting. The theory is essentially an excercise in the theory of probability, but it does stress that there is no need to consider a "diffusional force" as the source of molecular migration. In an inhomogeneous mixture of two ideal gases where there are no intermolecular forces, diffusion will occur solely as a result of molecular motion. Thermodynamics would attribute diffusion in this system to the tendency of the system to lower its entropy, but this is merely a reflection of the connection between probability and entropy.

5. Langevin, in his research on Brownian motion[197], postulated an *equation of motion* for a molecule in solution,

$$m \frac{dv}{dt} = \zeta v + f(t) \tag{12.3}$$

which states that the force on a molecule is a combination of a time-averaged frictional force that is proportional to the molecular velocity and a randomly fluctuating force $f(t)$ that averages to zero over periods of time that are long compared with the rate of fluctuations. Some of the theoretical apparatus used in the development of the theory is explained in Chapter 17 in another context. The theory shows that the friction coefficient may be obtained from a knowledge of the random force (and vice versa), and that the diffusion coefficient, not surprisingly, is related to the randomly varying microscopic velocity of the molecule. The microscopic nature of the theory allows the derivation of other properties, such as viscosity.

6. A model that superficially differs entirely from those listed above is that originated by Eyring[198] on the basis of the activated rate theory of chemical reactions. As applied to diffusion, this model uses the idea of discrete sites separated by potential barriers and initially found its main application in the treatment of diffusion in solids. Recently this approach has been extensively and successfully applied to membrane transport processes.

In this book we will concentrate almost exclusively on theories 2 and 6, the general thermodynamic approach and activated rate theory. We also introduce the basic ideas of irreversible thermodynamics and give a specific illustration hinting at its power. We bypass the mechanistic model of Langevin since its application to as complex a "solvent" as a lipid bilayer in the end would boil down to parameter fitting. We are still a long way from understanding how molecules break through the layer of head groups or pass through the liquid crystal hydrocarbon core.

We define our objective at the outset. We will attempt to find relationships between the flow of molecules through membranes and the differences in concentration and electric potential across the membrane. The experimental fact that under many conditions there is a potential difference $\Delta\phi$ across natural membranes is of the greatest importance in the working of cells. Potential gradients can also occur in artificial membranes where they can be caused only by physical processes and we will study the nature of these processes, which include diffusion. We start, however, with a less complex problem, the diffusion of an uncharged molecule across a homogeneous slab of material. Although this is hardly a good model for a membrane, it will allow us to define some much used concepts and to appreciate some of the problems common to the solution of diffusion problems.

We assume Fick's first law (12.2). It should be realized that this is a *local* law referring to the flow *at a given point* in terms of the concentration gradient at the same point. In solutions containing a single solute, we hope that the ratio of flow to gradient $J_i/(\partial c_i/\partial x)$ is a constant at a given temperature and pressure and *independent of solute concentration.* If so, D_i, the diffusion coefficient, is a constant, but the strong deviation from ideality of concentrated solutions warns us to be careful. We will take D_i to be constant throughout the homogeneous slab. Referring to Figure 95, the change in *mass* of solute i in time δt in the infinitesimal slice bounded by the planes is

$$\delta m_i = J_i(x)A\delta t - J_i(x+\delta x)A\delta t \tag{12.4}$$

where the first and second terms on the right give the rate at which material enters the slab and the rate at which it leaves. The change in *concentration* of i is given by

$$\delta c_i = \frac{\delta m_i}{V} = \frac{\delta m_i}{A\delta x} = [J_i(x) - J_i\ (x+\delta x)]\ \frac{\delta t}{\delta x}$$

so that

$$\frac{\delta c_i}{\delta t} = \frac{J_i(x) - J_i(x+\delta x)}{\delta x} \tag{12.5}$$

Taking the limit as $\delta x \to 0$,

$$\frac{\partial c_i}{\partial t} = -\frac{\partial J_i}{\partial x} \tag{12.6}$$

This is a *continuity equation.* In words it says that the rate of change in concentration at a given point must be connected to the difference between inward flow and outward flow. If the two flows are the same, then at that point $\partial J_i/\partial x$ is zero and the concentration is constant.

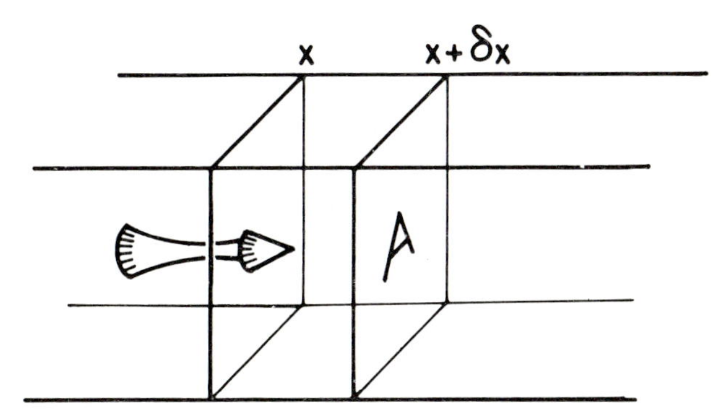

Figure 95. Diagram to aid in the derivation of the continuity equation (12.6).

Substituting (12.1) into (12.6):

$$\frac{\partial c_i}{\partial t} = \frac{\partial}{\partial x}\left\{\frac{D_i \partial c_i}{\partial x}\right\} = \frac{D_i \partial^2 c_i}{\partial x^2} + \frac{\partial D_i}{\partial x}\frac{\partial c_i}{\partial x} \tag{12.7}$$

Assuming that D_i is concentration independent means that $\partial D_i/\partial x = 0$, giving

$$\frac{\partial c_i}{\partial t} = D_i \frac{\partial^2 c_i}{\partial x^2} \tag{12.8}$$

which is *Fick's second law.*

Now suppose that i has fixed concentrations c_i^{I} and c_i^{II} on the two sides of the slab in Figure 96A and that we have left the system long enough for it to reach a steady state, that is, the concentration profile of i within the slab is constant, as is the flow through the slab. This means that $\partial c_i/\partial t = 0$ at all points, which, from (12.8), gives

$$\frac{\partial^2 c_i}{\partial x^2} = 0 \tag{12.9}$$

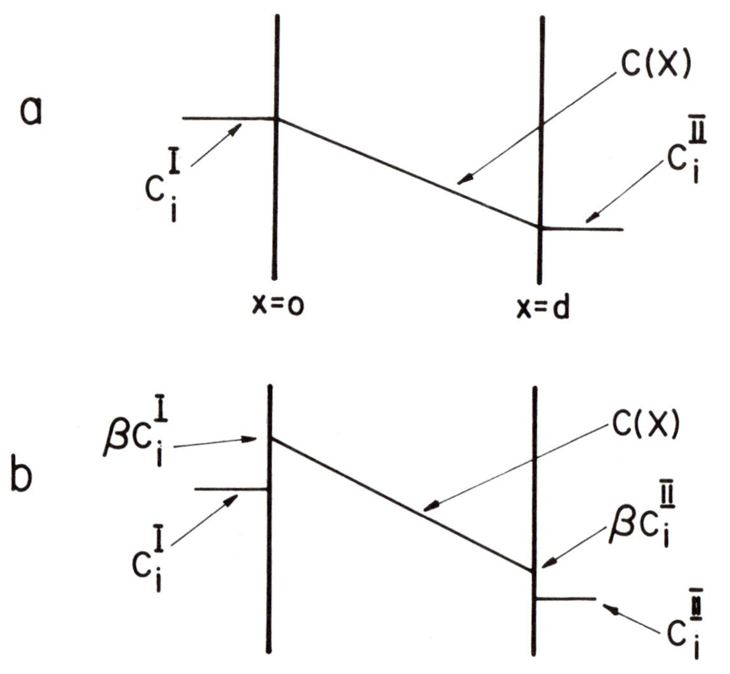

Figure 96. Diagram to illustrate the derivation of the flow equations (12.11) and (12.14).

The boundary conditions for this equation are

$$c(x) = c_i^{\text{I}} \qquad \text{for } x = 0$$

$$c(x) = c_i^{\text{II}} \qquad \text{for } x = d$$

and the solution gives a linear dependence of $c(x)$ on x,

$$c(x) = \frac{c_i^{\text{II}} - c_i^{\text{I}}}{d} \cdot x + c_i^{\text{I}} \tag{12.10}$$

The flow is given by Fick's first law:

$$J = -D_i \frac{\partial c_i}{\partial x} = -\frac{D_i}{d}(c_i^{\text{II}} - c_i^{\text{I}}) = \frac{D}{d}\Delta c \tag{12.11}$$

where $\Delta c = c_i^{\text{I}} - c_i^{\text{II}}$. We consider an instructive special case, supposing that *at equilibrium i* does not dissolve equally well in the solvent and the slab. Then it is possible to define a partition coefficient β, giving the ratio of concentrations in the two media. A diffusing system is not at equilibrium, but if the net rate of diffusion is much less than the rate of migration of individual molecules across the borders between the media, it is usual to assume that concentrations are still controlled by the partition coefficient determined at equilibrium, so that just inside the borders

$$\begin{aligned} c_i(x) &= \beta c_i^{\text{I}} \qquad \text{for } x = 0 \\ c_i(x) &= \beta c_i^{\text{II}} \qquad \text{for } x = d \end{aligned} \tag{12.12}$$

Using these concentrations we find that

$$c(x) = \frac{\beta}{\text{d}}(c_i^{\text{II}} - c_i^{\text{I}})x + \beta c_i^{\text{I}} \tag{12.13}$$

which is plotted in Figure 96B. The flow equation is

$$J_i = -D_i \frac{\partial c_i}{\partial x} = -\frac{D_i\beta}{d}(c_i^{\text{II}} - c_i^{\text{I}}) = \frac{D_i\beta}{d}\Delta c = P\Delta c \tag{12.14}$$

where P is the permeability coefficient in units of meter sec^{-1}. The determination of the flow thus leads only to a knowledge of the product of D_i and β if d is known.

The passage of *charged molecules* through the simple system of solvent–slab–solvent has concerned a large number of theoreticians from Nernst and Planck onward. One complication is the electric potential gradient produced by the process of diffusion. In fact, this complication has been the main object of theoretical and experimental attention because of the importance of membrane potentials in biology and junction potentials in electrochemistry. Junction potentials are entirely due to diffusion; membrane potentials, as will later become apparent, contain other contributions. Although we will obtain an expression for the flow, this will not be our final object, which is *an expression for the diffusion potential in terms of the concentrations and mobilities of the diffusing ions.*

In considering the problem of the passage of charged species across membranes, from the start we take into account the possibility of the existence of an electric field across the membrane.

We will find it convenient to regard diffusion as arising from the action of a force on the diffusing molecules, which is an acceptable idea for the case of an electric field acting on an ion, but is less obvious for the diffusion of a neutral molecule. Since diffusion ends at equilibrium, when $\tilde{\mu}_i$, the electrochemical potential of the diffusing species, is the same everywhere, it appears that a gradient in $\tilde{\mu}_i$ can be identified as an effective force. It should be clearly recognized that there need be no actual net physical force to produce diffusion.

Accepting the idea of a generalized force, we can write: Flow is proportional to force times concentration. This is good solid common sense; it says that the flow of matter through unit area in unit time depends on how much matter there is around and how hard you push each molecule. It should be appreciated that the force referred to does not include the opposing frictional force of the solution. Under steady flow conditions, Newton tells us, the net average force acting on each molecule is zero. If we write

$$ma = X - \zeta v \tag{12.15}$$

then for steady flow $a = 0$ and v is a constant equal to X/ζ. X is the "force" that we are talking about. By using a proportionality constant u_i, we have the local equation

$$J_i = u_i f_i c_i \tag{12.16}$$

where u_i is called the *mobility* and can easily be seen to have dimensions of velocity per unit force. Substituting for f_i, and limiting ourselves to one-dimensional diffusion,

$$J_i = -u_i c_i \frac{\partial \tilde{\mu}_i}{\partial x} \tag{12.17}$$

To proceed we need an expression for $\tilde{\mu}_i$. Under conditions of constant temperature and pressure

$$\tilde{\mu}_i = (\mu_i)^\circ_{T,p} + RT \ln a_i + z_i F\phi \tag{12.18}$$

The last term is the potential energy of a mole of ions of charge z_i at a point where the elecric potential equals ϕ. Specializing to the case of an ideal solution of an *uncharged* molecule, we have

$$\tilde{\mu}_i = (\mu_i)^{\circ}_{T,p} + RT \ln c_i \qquad (12.19)$$

which, when placed in (12.17), gives

$$J_i = u_i RT \frac{\partial c_i}{\partial x} \qquad (12.20)$$

On comparing with (12.1), we find

$$D_i = RT\, u_i \qquad (12.21)$$

an equality due to Einstein. [It is a point of general interest that this relationship connects a property of the system in the absence of an external force to a parameter u_i specifying the response to an external force. The theoretically inclined will perhaps remember that the diffusion coefficient is related to the autocorrelation function of the molecule's velocity and the mobility to the autocorrelation function for the random force beating on the molecule. Both velocity and random force appear in Langevin's equation (12.3).]

We return to (12.17) and substitute the ideal solution form of (12.18)

$$J_i(x) = -u_i c_i(x) \left\{ \frac{RT}{c_i(x)} \frac{dc_i(x)}{dx} + z_i \frac{Fd\phi(x)}{dx} \right\} \qquad (12.22)$$

where the variable x has been written explicitly to stress once again that this is an equation for the flow at one specific point. Equation (12.22) is the most common form of the *Nernst–Planck equation,* the starting point for many theoretical treatments of ion transport. The equation has a simple structure; the first term is a Fick-type contribution to the flow, which, in the ideal solution form (12.22), depends only on concentration gradients (i.e., on entropy). The diffusion constant of Fick's equation has been replaced by the mobility, using Einstein's equation. The second term is the contribution to the flow from the electric field $\hat{E} = -d\phi/dx$.

[To give all their due, Smulochowski[196] derived an equation now bearing his name for the motion of a particle undergoing Brownian motion in the presence of an external force

$$\frac{\partial c}{\partial t} = D \frac{\partial^2 c}{\partial x^2} - UX \frac{\partial c}{\partial x} \qquad (12.23)$$

where the mobility U and the force X refer to one particle. Using (12.6) and assuming D and B to be independent of position, we find

$$J = -D \frac{\partial c}{\partial x} + UXc \qquad (12.24)$$

which has the same physical structure as (12.22).]

We now tackle the Nernst–Planck equation. Planck[199] integrated the equation but his derivation is labyrinthine and the results are given in terms of a parameter that has to be determined by graphically or numerically solving a transcendental equation. The problem, and the result, were simplified by Henderson,[200] who assumed that the concentration of all ions in the slab obey the relationship (12.10), that is, he postulated linear concentration profiles (Figure 96). The Henderson equation, has been much used because, in contrast to Planck's solution, it gives an explicit, although approximate, expression for the diffusion potential:

$$\Delta\phi_D = \frac{\sum_i |z_i^+|\, u_i^+ \Delta c_i^+ - \sum_i |z_i^-|\, u_i^- \Delta c_i^-}{\sum_i (z_i^+)^2 u_i^+ \,\Delta c_i^+ + \sum_i (z_i^-)^2 u_i^- \Delta c_i^-} \times \frac{RT}{F} \ln \frac{\sum_i (z_i^+)^2 u_i^+ c_i^+(0) + \sum_i (z_i^-)^2 u_i^- c_i^-(0)}{\sum_i (z_i^+)^2 u_i^+ c_i^+(d) + \sum_i (z_i^-)^2 u_i^- c_i^-(d)} \tag{12.25}$$

where z^+ and z^- are the number of charges on diffusing cations and anions: $c(0)$ and $c(d)$ are the the concentrations of the relevant species just inside the borders of the slab; Δc is the difference in concentration of a given species across the slab, and u is a mobility. Before we leave this equation, we note that it can be applied to liquid–liquid junction potentials if we picture the junction as being an idealized zone of thickness d outside of which the concentrations are constant and inside of which linear concentration profiles obtain.

The Nernst–Planck equation has been solved by Morf[201] to give an expression for the diffusion potential that is exact but as easy to use as the Henderson equation. We present the derivation of Morf's results here, going into some detail in the proofs of his equations. Those who wish to find out who done it by turning to the last page may jump to equations (12.38) and (12.41).

First multiply (12.22) by $\exp(z_i F\phi(x)/RT)$:

$$J_i(x)\exp(z_iF\phi(x)/RT) = -[\exp(z_iF\phi(x)/RT)]\,u_iRT\,\frac{dc_i(x)}{dx} - [\exp(z_iF\phi(x)/RT)]\,z_ic_i(x)u_iF\,\frac{d\phi(x)}{dx} \tag{12.26}$$

But

$$\frac{d}{dx}[c_i(x)\exp(z_iF\phi(x)/RT)] = \exp(z_iF\phi(x)/RT)\,\frac{dc_i(x)}{dx} + c_i(x)\exp(z_iF\phi(x)/RT)\,\frac{z_iF}{RT}\,\frac{d\phi(x)}{dx} \tag{12.27}$$

Comparing the last two equations:

$$J_i \exp(z_i F\phi(x)/RT) = -RTu_i \frac{d}{dx} [c_i \exp(z_i F\phi(x)/RT] \quad (12.28)$$

This is still a local equation. It is now assumed that u_i is independent of x, which is true for the homogeneous slab of our model but probably untrue for a bilayer. In the steady state, the flow across the slab is constant, and it follows that the flow at all points in the slab is constant and equal to the flow at all other points. Under these conditions we can integrate across the slab taking J_i outside the integration sign:

$$J_i = \frac{-RTu_i[c_i(d) \exp(z_i F\phi(d)/RT) - c_i(0) \exp(z_i F\phi(0)/RT]}{\int_0^d \exp(z_i F\phi(x)/RT)dx} \quad (12.29)$$

This is a form of Kramer's equation[202]. It cannot give a value for J_i unless we know the function $\phi(x)$. Later we will assume $\phi(x)$ to be linear in x and obtain another approximation for the diffusion potential.

The following physical assumptions are made by Planck and accepted by Morf.

(a) The electric current due to diffusion is zero, a condition that can be expressed by

$$j = \left\{ \sum_i |z_i^+| J_i^+ - \sum_i |z_i^-| J_i^- \right\} F = 0 \quad (12.30)$$

where the units of j, the current density, are Coulomb area^{-1} time^{-1}.

(b) Electroneutrality holds at all points in the system:

$$\sum_i |z_i^+| c_i^+(x) = \sum_i |z_i^-| c_i^-(x) \qquad \text{for all } x \quad (12.31)$$

Morf limits his treatment to one class of mobile cations and one class of mobile anions, that is, the cations all have the same charge, the anions all have the same charge, but $|z^+|$ is not necessarily equal to $|z^-|$. (This restraint is less restrictive than that of Planck, whose solution applies only to monovalent ions.) The limitation allows (12.30) to be rewritten as,

$$\frac{j}{F} = |z_i^+| \sum_i J_i^+ - |z_i^-| \sum_i J_i^- = 0 \quad (12.32)$$

The next step is to define *mean mobilities:*

$$\bar{u}^+ = \frac{\sum_i J_i^+}{\sum_i J_i^+/u_i^+} = \text{const}(x)$$

$$\bar{u}^- = \frac{\sum_i J_i^-}{\sum_i J_i^-/u_i^-} = \text{const}(x) \quad (12.33)$$

For example, if the cations are Na^+ and K^+, then

$$\overline{u^+} = \frac{J_{Na^+} + J_{K^+}}{J_{Na^+}/u_{Na^+} + J_{K^+}/u_{K^+}}$$

Substituting (12.33) into (12.32),

$$\frac{j}{F} = |z_i^+| \sum_i (J_i^+/u_i^+)\overline{u_i^+} - |z_i^+| \sum_i (J_i^-/u_i^-)\overline{u_i^-} \tag{12.34}$$

Equations (12.34) and (12.22) give

$$\begin{aligned} \frac{j}{F} = \overline{u_i^+}\,|z_i^+|&\left\{-\sum_i \frac{RT}{F}\frac{dc_i^+(x)}{dx} - \sum_i |z_i^+|\,c_i^+(x)\,\frac{d\phi(x)}{dx}\right\} \\ -\overline{u_i^-}\,|z_i^-|&\left\{-\sum_i \frac{RT}{F}\frac{dc_i^-(x)}{dx} - \sum_i |z_i^-|\,c_i^-(x)\,\frac{d\phi(x)}{dx}\right\} = 0 \end{aligned} \tag{12.35}$$

Rearrangement results in

$$\frac{d\phi(x)}{dx} = -\frac{RT}{F}\,\frac{\overline{u_i^+}\sum_i |z_i^+|\,(dc_i^+(x)/dx) - \overline{u_i^-}\sum_i |z_i^-|\,(dc_i^-(x)/dx)}{\overline{u_i^+}\sum_i (z_i^+)^2 c_i^+(x) + \overline{u_i^-}\sum_i (z_i^-)^2 c_i^-(x)} \tag{12.36}$$

This expression can be integrated after using (12.31) to factorize the denominator:

$$\begin{aligned} \overline{u_i^+}\sum_i (z_i^+)^2 c_i^+(x) + \overline{u_i^+}\sum_i (z_i^-)^2 c_i^-(x) &= z_i^+\,\overline{u_i^+}\sum_i z_i^+ c_i^+(x) + z_i^-\,\overline{u_i^-}\sum_i z_i^- c_i^-(x) \\ &= (z_i^+\,\overline{u_i^+} + z_i^-\,\overline{u_i^-})\sum_i z_i^+ c_i^+(x) \end{aligned} \tag{12.37}$$

Finally, the integration gives, for the diffusion potential,

$$\Delta\phi_D = \frac{\overline{u^+} - \overline{u^-}}{|z^+|\,\overline{u^+} + |z^-|\,\overline{u^-}} \cdot \frac{RT}{F} \ln \frac{\sum_i c_i(0)}{\sum_i c_i(d)} \tag{12.38}$$

[For those who have not followed the derivation, $\overline{u^+}$, $\overline{u^-}$ are mean mobilities defined in (12.33)] Since the concentrations are measurable, we need only the mean mobilities to determine $\Delta\phi_D$. We now derive an expression connecting the mean mobilities and diffusion potential. To reduce the typographic verbiage, we define the symbol $A_i(x) = \exp(z_i F\phi(x)/RT)$, which converts (12.29) to

$$J_i = u_i RT \left[\frac{c_i(d)A_i(d) - c_i(0)A_i(0)}{\int_0^d A_i(x)dx}\right] \tag{12.39}$$

Substituting in (12.33), we obtain

$$u^+ = \frac{\sum_i J_i^+}{\sum_i (J_i^+/u_i^+)} = \frac{-RT\sum_i \left\{u_i^+ [c_i^+(d)A_i(d) - c_i^+(0)A_i(0)] \Big/ \int_0^d A_i(x)dx\right\}}{-RT\sum_i \left\{[c_i^+(d)A_i(d) - c_i^+(0)A_i(0)] \Big/ \int_0^d A_i(x)dx\right\}} \tag{12.40}$$

Since, in $A(x)$, z_i is the same for all i, the integral can be taken out of the summation and cancelled, which leads immediately to the required result:

$$\overline{u^+} = \frac{\sum_i u_i^+ c_i^+(d) \exp(z_i F\Delta\phi_D/RT) - \sum_i u_i^+ c_i^+(0)}{\sum_i c_i^+(d) \exp(z_i F\Delta\phi_D/RT) - \sum_i c_i^+(0)} \tag{12.41}$$

where we have used $\Delta\phi_D = \phi(d) - \phi(0)$. A similar expression holds for $\overline{u^-}$.

Equations (12.38) and (12.41) can be solved iteratively to give exact values for the diffusion potential. In practice, little effort is needed to obtain self-consistent solutions of the two equations.

Having got over the mathematics, it is time to have a look at the physics to uncover the source of the diffusion potential. The equations derived hold for two types of model, in both of which there are zones of constant but differing composition separated by a zone in which diffusion occurs. The liquid–liquid junction model will not concern us despite its importance in other fields. The slab model is, at this stage, our primitive substitute for a membrane. The *diffusion* potential is only one source of the *membrane* potential, which can include contributions for the charges on the head groups and a boundary potential, both of which we discuss later. The zones of constant composition included in our model are to be visualized as lying just inside the edges of the membrane. The concentrations of ions in these zones are, under steady-state conditions, constant but not necessarily the same as the concentrations outside the membrane. The diffusion potential arises from diffusion between these zones.

It is instructive to write equation (12.38) for the case of a single 1 : 1 electrolyte, say NaCl,

$$\Delta\phi_D = \frac{u_{Na^+} + u_{Cl^-}}{u_{Na^+} + u_{Cl^-}} \frac{RT}{F} \ln \frac{[Na^+]_0}{[Na^+]_d}$$

where we have used the fact that $[Na^+]_0 = [Cl^-]_0$ etc. $\Delta\phi_D$ is zero under two circumstances: if $[Na^+]_0 = [Na^+]_d$ or if $u_{Na^+} = u_{Cl^-}$. The first condition merely reflects the absence of a diffusion potential when there is no net diffusion. The second condition is more interesting since it shows that the potential depends on differences in mobility between positive and negative charge. This is the physical basis of $\Delta\Phi_D$. Consider a solution containing NaCl placed in contact with pure water. Diffusion will carry both types of ion into the water but due to differences in mobility, the distribution of the ions will not be identical. At 25°C in water the mobilities of Na^+ and Cl^- are 5.19 and 7.91×10^{-8} m^2s^{-1} V^{-1}, respectively, so that the sodium ions will tend to be left behind by the chloride ions. This means that the chloride ions will be in excess to the right of the border in Figure 97 and soidum ions will be in excess to the left. A sketch of the distribution of the ions some time after the commencement of diffusion is shown in Figure 97A. Because of the different concentration profiles of Na^+ and Cl^-, there will be a net space charge density $\varrho(x)$ throughout the diffusion zone. Common sense dictates that this density will be zero at some distance from the border, and so the general shape of $\varrho(x)$ will be as sketched. (An approximate form can be obtained from the solution of the diffusion equation and the assumption that the ions undergo independent diffusion.) If we have a system for which $\varrho(x)$ is known, we can use Poisson's equation (Chapter 7) to derive the electric

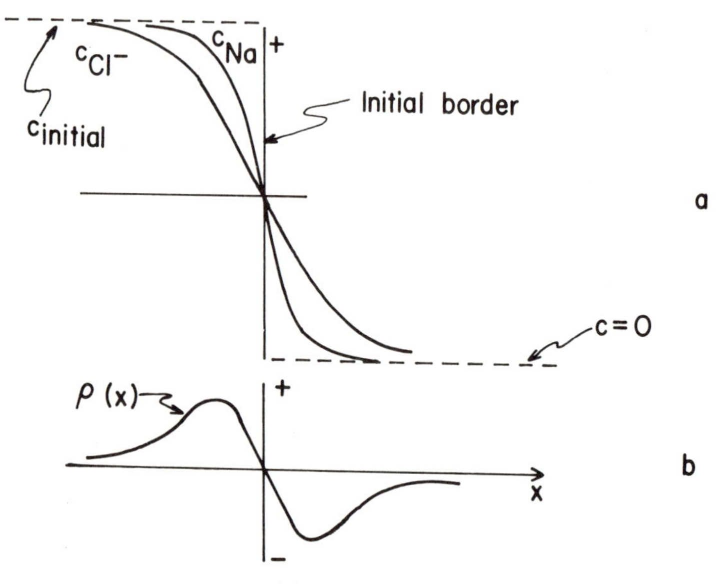

Figure 97. A. Schematic illustration of the concentrations of Na^+ and Cl^- ions in the neighborhood of a boundary initially defined by the interface of pure water and a salt solution. **B.** General form of the distribution of spatial charge density in the vicinity of the boundary.

potential. For one dimension,

$$\frac{\partial^2\phi(x)}{\partial x^2} = -\frac{4\pi\varrho(x)}{\epsilon} \tag{12.42}$$

Without requiring the analytical form for $\varrho(x)$, we can see that the general forms for $\partial^2\phi/\partial x^2$, $\partial\phi/\partial x$, and ϕ are those shown in Figure 98. Now, $-\partial\phi(x)/\partial x$ is a measure of the electric field at point x. We see that the electric field reaches a maximum in going through the border and drops to zero on either side. This makes sense since at the border there is a net positive charge immediately to the left and a net negative charge to the right. Integrating $\partial\phi(x)/\partial x$, we find that $\phi(x)$ is constant at large distances from the border and falls continuously in going from left to right through the border. The difference between the values of $\phi(-\infty)$ and $\phi(\infty)$ is $\Delta\phi_D$, the diffusion potential. The electric field resulting from diffusion acts so as to slow down the more rapidly diffusing chloride ions and to accelerate the slower sodium ions, thus ensuring that sodium chloride diffuses as a married couple should.

The reader may be concerned by the discrepancy between our assumption of electroneutrality in solving the Nernst–Planck equation

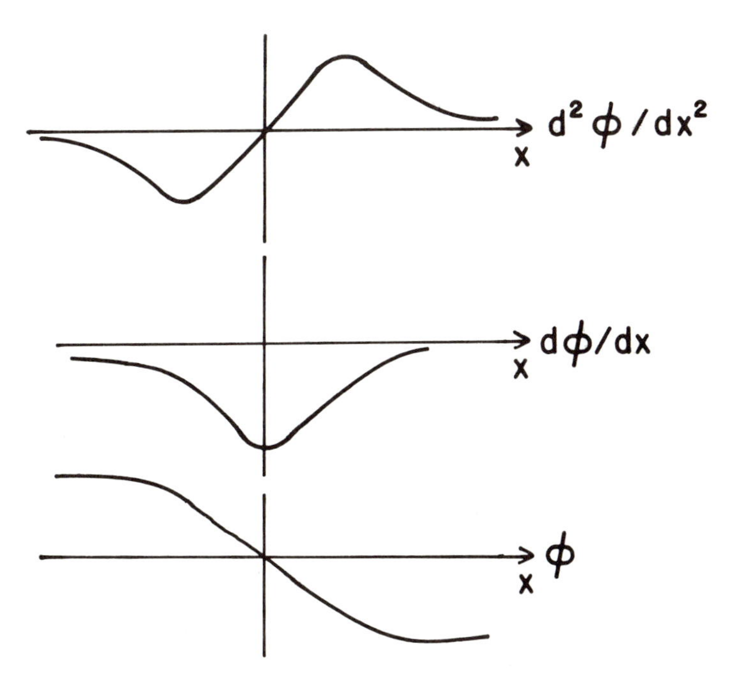

Figure 98. Potential and its first and second derivatives in the neighborhood of a border across which ionic diffusion is occurring.

and our demonstration of a net space charge in the diffusion zone. The answer is that the difference in cation and anion concentrations necessary to result in a significant space charge is too small to be measured by present analytic techniques, and even if measurable, is so tiny that we are easily justified in employing the mathematical condition of electroneutrality (12.31).

It is important to realize that in the above treatment of the Nernst–Planck equation, and in nearly all other theories based on the equation, it is assumed explicitly or implicitly that $\phi(x)$ is independent of the concentration of ions in the membrane. This is physically improbable at high concentrations and we come back to this problem later.

Keeping our earlier promise, we follow up the consequences of putting $d\phi(x)/dx = $ constant. This is the *constant field approximation* of Goldman[203]. (The assumption of linearity is least dangerous for small values of $\Delta\phi_D$. Planck[204] showed that if the total concentration of ions is the same on both sides of the membrane, the field is always constant.) In what follows it is convenient to choose $\phi(0)$ to be the zero of potential, so that for the constant field approximation $\phi(x) = (\Delta\phi_D/d)x$. Substituting this expression into the integral in (12.29), we have

$$\int_0^d \exp(zF\phi(x)/RT)dx = \int_0^d \exp(zF\Delta\phi_D/RT\mathrm{d})x\, dx$$

$$= \left(\frac{RTd}{zF\Delta\phi_D}\right) [\exp(z\mathrm{F}\Delta\phi_D/RTd)x]_0^d \quad (12.43)$$

$$= \left(\frac{RTd}{zF\Delta\phi_D}\right) [\exp(zF\Delta\phi_D/RT) - 1]$$

Thus the solution of (12.29) is

$$J_i = -\frac{u_i[c_i(d)\exp(z_iF\Delta\phi_D)/RT - c_i(0)]}{[d/z_iF\Delta\phi_D]\ [\exp(z_iF\Delta\phi_D/RT) - 1]} \quad (12.44)$$

This is the *Goldman equation* for a constant field. The concentrations $c_i(0)$, $c_i(d)$ are those at two points just *within* the slab. In dealing with diffusion across a membrane, we are more interested in the concentrations *outside* the membrane since these are more accessible to measurement. The relationship between the concentrations inside and outside one surface of a membrane was discussed in Chapter 11. In 1949 Hodgkin and Katz chose to use concentration-independent partition coefficients for the diffusing ions,[205]

$$c_i(d) = c_i^{\mathrm{II}}\,\beta_i \quad \text{and} \quad c_i(0) = c_i^{\mathrm{I}}\,\beta_i \quad (12.45)$$

This choice, as explained in Chapter 11, leads to the equality,

$$\Delta\phi_D = \Delta\phi_M \quad (12.46)$$

that is, the diffusion potential equals the membrane potential. Substituting (12.46) into (12.44) and writing $P_i = D_i\beta_i/d = u_iRT\beta_i/d$:

$$J_i = -\frac{P_i z_i F\Delta\phi_M}{RT}\left[\frac{c_i^{\mathrm{II}}\exp(z_iF\Delta\phi_M/RT) - c_i^{\mathrm{I}}}{\exp(z_iF\Delta\phi_M/RT) - 1}\right] \tag{12.47}$$

For diffusion in the absence of an externally applied force, the zero-current condition (12.30) applies. Substituting (12.47) into (12.30), we obtain an expression that is replete with indices and is easier to grasp for a specific and famous case, the membrane potential due to the diffusion of a mixture of NaCl and KCl:

$$j = 0 = \frac{\{P_{\mathrm{K}}c_{\mathrm{K}}^{\mathrm{I}} + P_{\mathrm{Na}}c_{\mathrm{Na}}^{\mathrm{I}} + P_{\mathrm{Cl}}c_{\mathrm{Cl}}^{\mathrm{II}} - [P_{\mathrm{K}}c_{\mathrm{K}}^{\mathrm{I}} + P_{\mathrm{Na}}c_{\mathrm{Na}}^{\mathrm{I}} + P_{\mathrm{Cl}}c_{\mathrm{Cl}}^{\mathrm{II}}]\exp(\Delta\phi_M F/RT)\}\Delta\phi_M F/RT}{1 - \exp(F\Delta\phi_M/RT)} \tag{12.48}$$

Since $(\Delta\phi_M F/RT)/(1 - \exp(F\Delta\phi_M/RT)) \neq 0$, it follows that

$$\Delta\phi_M = \frac{RT}{F}\ln\frac{P_{\mathrm{K}}c_{\mathrm{K}}^{\mathrm{I}} + P_{\mathrm{Na}}c_{\mathrm{Na}}^{\mathrm{I}} + P_{\mathrm{Cl}}c_{\mathrm{Cl}}^{\mathrm{II}}}{P_{\mathrm{K}}c_{\mathrm{K}}^{\mathrm{II}} + P_{\mathrm{Na}}c_{\mathrm{Na}}^{\mathrm{II}} + P_{\mathrm{Cl}}c_{\mathrm{Cl}}^{\mathrm{I}}} \tag{12.49}$$

This classic result, which Hodgkin and Katz obtained on the basis of Goldman's constant field approximation, is usually known as the Goldman–Hodgkin–Katz (GHK) equation. It is one of the most used equations in membrane physiology, and is frequently employed to estimate values for permeability coefficients, or their ratios. It is beyond the scope of this book to enter into detailed criticism of the GHK equation, but the readers' suspicions have probably been aroused by the constant field approximation on which the equation is based. The symbol $\phi(x)$ that appears in the Nernst–Planck equation is there to represent the *total* potential at a given point. We have assumed $\phi(x)$ to be of linear function, even though we know that at least one contribution to $\phi(x)$ is definitely not linear, namely, the image force discussed in Chapter 7. Worse is to come. It can be shown[206] that the GHK equation, as derived from the Nernst–Planck equation, gives *voltage-dependent* permeabilities unless the potential energy surfaces for different ions are related to each other by highly restrictive conditions. For example, for sodium and potassium the potential energy profiles, exclusive of contributions due to externally imposed fields, must obey

$$\phi_{\mathrm{Na}}(x) = \phi_{\mathrm{K}}(x) + (RT/F)\ln[D_{\mathrm{Na}}(x)/D_{\mathrm{K}}(x)] + \text{constant} \tag{12.50}$$

The chance that this relationship holds is fairly small.

For the above and other reasons we do not undertake an analysis of the GHK equation. It is perhaps best to regard the equation as a means of defining permeability ratios under the pertaining experimental conditions. The equation is not consistent with a great deal of experimental data. Some discrepancies will be touched on subsequently. One condition under which the equation might be expected to be particularly unreliable is for a high density of fixed charges on the surface of the membrane. From Chapter 11 we know that fixed charges result in concentration-dependent partition functions, in contrast to the assumption of Hodgkin and Huxley contained in (12.45). As we shall see, there are many aspects of ion transport that cannot be explained by the classic Nernst–Planck approach, even without approximations, unless considerable modifications are made to the original model.

One unsatisfactory feature of the exact solution of the Nernst–Planck equation is that we end up with $\Delta\phi_D$ but not with an explicit form for $\phi(x)$, the potential energy profile of the membrane. The problem is not trivial, as we will later see. There is, however, an important equation, widely used by experimentalists, which can be derived from the Nernst–Planck equation but which is independent of the shape of $\phi(x)$. To arrive at this relationship, we need the concept of a *unidirectional flow*, which is the flow of a component i through the membrane when the concentration of i on one side of the membrane is zero. From (12.29), putting $\phi(0)=0$ as usual and $c_i(d)=0$, we have for the unidirectional flow J_i^{od}:

$$J_i^{0d} = \frac{RT\, u_i c_i(0)}{\int_0^d \exp(z_i F\phi(x)/RT)dx} \tag{12.51}$$

Correspondingly, for $c_i(0) = 0$,

$$J_i^{d0} = \frac{-RT\, u_i c_i(d)\exp(z_i F\Delta\phi/RT)}{\int_0^d \exp(z_i F\phi(x)/RT)dx} \tag{12.52}$$

In these two expressions $\phi(x)$ is the potential due to all causes, internal or external to the membrane, and $\Delta\phi$ is the difference in potential in going from 0 to d within the membrane. The ratio of the magnitudes of the unidirectional flows is

$$\frac{J_i^{0d}}{J_i^{d0}} = \frac{c_i(0)}{c_i(d)}\exp(-z_i F\Delta\phi/RT) \tag{12.53}$$

The hidden assumption in canceling the integrals is that the form of $\phi(x)$ is the same in both experiments. This can be assured by having exactly the same concentrations of ions in the measurements of the two unidirectional flows but using tracer amounts of isotopes to follow the flow. If two isotopes are available—for example, ^{22}Na and ^{24}Na—it is possible to follow both J_{Na}^{0d} and J_{Na}^{d0} simultaneously by initially adding each isotope to a different side of the membrane. Equation (12.53) is known as the flux-ratio equation and is attributable to Ussing.[207] Its great value lies in the fact that it is independent of the form of $\phi(x)$. Being based on the Nernst–Planck equation, the Ussing equation can be expected to hold when transport is due solely to passive diffusion and measurements of flux ratio have been used extensively to prove the presence of electrodiffusion. If active transport operates, that is, there is "pumping" of specific ions based on metabolic processes, the Nernst–Planck equation, and hence the Ussing equation, cannot be expected to hold. *The converse unfortunately is not true.* Deviations from the predictions of the flux-ratio equation have been frequently observed in the absence of active transport, and have led to much building of models to replace the simple electrodiffusion model of Nernst. In the next chapter we look at one major class of alternative model.

The rather negative light thrown on the Nernst–Planck approach in the previous paragraphs is not intended to obscure the importance of the GHK equation in explaining, often quantitatively, the dependence of membrane potential on ionic concentrations. A great amount of experimental data fits the GHK equation—see reference 208, for example. However, the fact that the GHK formalism was a historic advance in the understanding of membrane potentials does not imply that its physical basis is all embracing, a fact not lost on the theory's architects.

Irrespective of their degree of success in accounting for experimental observations, all theories of ion transport predict that the passage of ions across membranes will result in a potential difference across the membrane. Inspection of the form of $\phi(x)$ given in Figure 98 reveals that the diffusion potential ϕ_M can be detected by placing electrodes in the bulk solution on either side of the membrane. However, care must be exercised in calculating diffusion potentials, by whatever method, not to assume too readily that the concentrations appearing in the relevant equations—for example, (12.49)—are the bulk concentrations of the ions. The relationships (12.45) relate ion concentrations *just* inside and outside the surfaces of the membrane. The ion concentrations near a charged surface are, as we saw in Chapter 7, not the same as the bulk concentrations, with the ratio between the two quantities depending on both the surface charge density and the composition of the aqueous medium. It should be realized that because of the interdependence of surface potential and solution composition, a difference in the ionic concentrations between the two sides of a membrane can result in differing surface potentials and, therefore, a potential drop across the membrane even in the absence of diffusion. However, in this case we are concerned with two double layers and the potentials in the solutions on both sides fall to zero after a few Debye lengths so that the potential difference will not be detectable by electrodes placed in the bulk solutions. We see that the definition and measurement of membrane potential require some care.

13 The Rate Theory of Transport

The Nernst–Planck electrodiffusion equation was formulated and solved well before experimental studies on membrane transport had provided accurate enough data to be tested against theory. It was natural, therefore, that the equation, or one of its approximate solutions such as the GHK equation, was adopted by the experimental pioneers. One who did not follow the herd was Danielli[209] who proposed that the passage of ions through membranes proceeded by a succession of jumps over energy barriers (Figure 99). His sharp physical insight provided an alternative to the Nernst–Planck conception, but the general idea of discontinuous diffusion was not formalized until Eyring and his school developed the activated rate theory in terms of which they discussed rate processes ranging from diffusion in solids to membrane transport.[210,211] They realized that membrane potential could affect energy barriers, and they obtained an equation that converges with the GHK equation as the number of (identical in their case) barriers increase. In 1955 Hodgkin and Keynes found that the concentration of potassium outside a squid axon had an effect on the escape of radioactive potassium through the membrane. In looking at (12.51) one will see that for a unidirectional flux only the concentration of the tracer should be relevant. Hodgkin and Keynes[212] proposed that K^+ ions move in single file through a channel

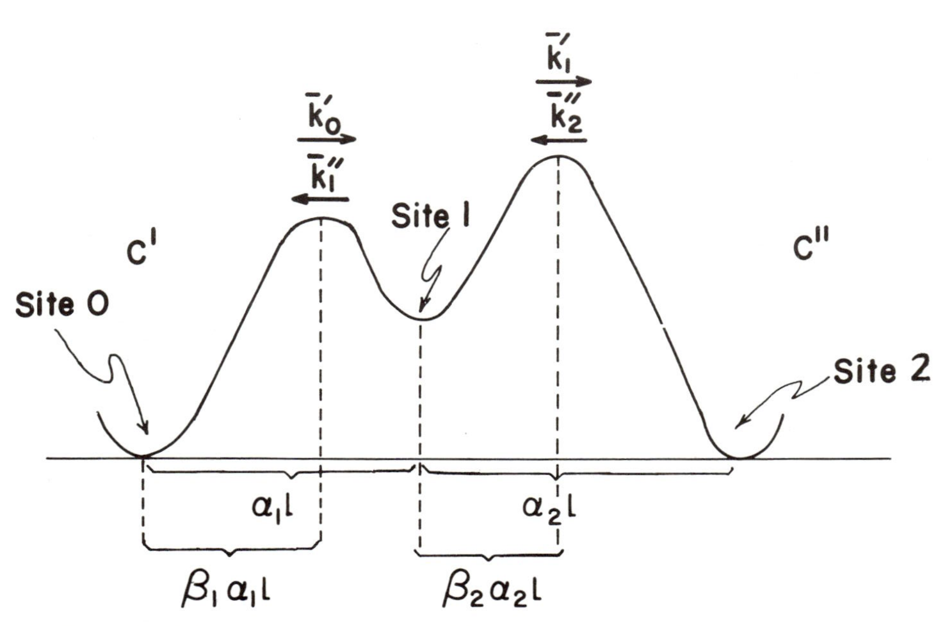

Figure 99. Schematic representation of a simple potential profile for an ion crossing a membrane via a channel. Sites 0 and 2 are at the channel openings and are separated by a distance *l*. The *k*'s are first-order rate constants and the *c*'s are ion concentrations outside the membrane.

containing a series of saturable binding sites. The model of Danielli and Eyring had found a use. Many other experiments have given results incompatible with the Nernst–Planck equation, at least in its original form. In this chapter we survey the basic ideas of the rate theory of ion transport and show that, even in its simplest form, the theory is capable of providing qualitative explanations of phenomena for which simple diffusion theory cannot account.

The simplest nontrivial model for rate theory is that of a "channel" with two energy barriers between which lies a valley deep enough for an ion to reside for longer than an average vibration period (Figure 99). (The molecular structure of the channel is irrelevant at this stage but there is enough chemical work for us to rest assured that well-defined molecules with the properties of channels do exist.[213]) Ions are assumed to be adsorbed at two sites, one on each side of the membrane at the channel openings. Transport then occurs by a "hopping" of ions over the barriers. The probability of a given channel-opening being occupied by an ion is taken to be proportional to the concentration of that ion in the adjoining bulk solution:

$$p_0 = \nu c^{\mathrm{I}}$$

$$p_2 = \nu c^{\mathrm{II}} \tag{13.1}$$

where the subscripts 0 and 2 refer to the two openings. The reader should now consult Figure 99 to appreciate the meaning of the fractions α_1, α_2, β_1, β_2, and the assignment of the first-order rate constants $\bar{k}'_0$, $\bar{k}'_1$, $\bar{k}''_1$, and $\bar{k}''_2$.

Initially the potential drop across the membrane is put at zero. We now derive a useful relationship between the flow rate constants. At equilibrium, that is, under conditions of no net flow across the membrane, the flow of an ion over a barrier in a given direction must equal the flow of the same ion in the opposite direction. Thus, if the occupancies of the valleys are $\bar{p}_0$, $\bar{p}_1$, and $\bar{p}_2$ respectively,

$$\bar{p}_0\bar{k}'_0 = \bar{p}_1\bar{k}''_1$$

and

$$\bar{p}_1\bar{k}'_1 = \bar{p}_2\bar{k}''_2 \tag{13.2}$$

Eliminating $\bar{p}_1$ between these equations, we find

$$\frac{\bar{p}_0\bar{k}'_0}{\bar{k}''_1} = \frac{\bar{p}_2\bar{k}_2}{\bar{k}'_1} \tag{13.3}$$

If the openings of the channel are now supposed to be identical as far as adsorption of ions is concerned, then at equilibrium (remember

$\Delta\phi = 0$), $\bar{p}_0 = \bar{p}_2$, and from (13.3)

$$\frac{\bar{k}_0' \bar{k}_1'}{\bar{k}_1'' \bar{k}_2''} = 1 \tag{13.4}$$

a result we shortly require and which could have been obtained by appealing to the principle of microreversibility.

A linear potential gradient is now imposed on the channel (Figure 100). The height of the energy profile is altered by the addition of the potential energy of an ion, $ze\phi$, in the electric field. To find the effect of this on the rate constants, we appeal to absolute rate theory. The rate constant for the surmounting of a barrier of height ΔE is given by

$$k = Ae^{-\Delta E/kT} \tag{13.5}$$

where ΔE is the energy barrier for one ion. If the height of the barrier is now increased by δE, and the factors that enter into A, usually entropic, are left unchanged, then the new rate constant is

$$k' = Ae^{-(\Delta E + \delta E)/kT}$$

and the ratio of the rate constants is

$$\frac{k'}{k} = Ae^{-\delta E/kT} \tag{13.6}$$

It is essential in calculating δE to take into account any change in the

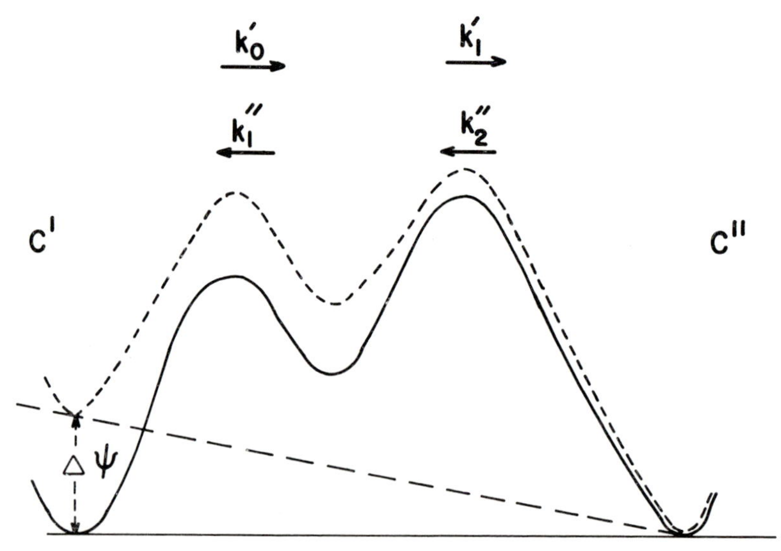

Figure 100. The potential profile of Figure 99 with the addition of a linear potential gradient giving a skewed resultant potential.

energy of the initial state. If this warning is heeded, the reader should be able to arrive at the following relationships:

$$\begin{aligned} k_0' &= \bar{k}_0' \exp[\beta_1\alpha_1\Delta\phi ze/kT] \\ k_1' &= \bar{k}_1' \exp[\beta_2\alpha_2\Delta\phi ze/kT] \\ k_1'' &= \bar{k}_1'' \exp[(\beta_1 - 1)\alpha_1\Delta\phi ze/kT] \\ k_2'' &= \bar{k}_2'' \exp[(\beta_2 - 1)\alpha_2\Delta\phi ze/kT] \end{aligned} \tag{13.7}$$

where the bare-headed k's are the rate constants in the presence of the potential gradient. We now deduce another simple relationship:

$$\frac{k_0' k_1'}{k_1'' k_2''} = \frac{\bar{k}_0' \bar{k}_1'}{\bar{k}_1'' \bar{k}_2''} \cdot \exp[\beta_1\alpha_2 + \beta_2\alpha_2 - (\beta_1 - 1)\alpha_1 - (\beta_2 - 1)\alpha_2] \cdot (ze\Delta\phi/kT)$$

$$= \frac{\bar{k}_0' \bar{k}_1'}{\bar{k}_1'' \bar{k}_2''} \cdot [\exp ze\Delta\phi/kT]$$

Therefore

$$\frac{k_0' k_1'}{k_1'' k_2''} = \exp zu \tag{13.8}$$

where (13.7) has been used and u is defined as

$$u = e\Delta\phi/kT \tag{13.9}$$

An expression is now derived for the flux of an ion through the channel as a function of its concentration on either side and $\Delta\phi$.

Under stationary-state conditions, that is, when the flow is constant, the net flux J_n over any barrier n must be constant and independent of n. For barriers 1 and 2, indicated in Figure 100,

$$J_1 = k_0' p_0 - k_1'' p_1 = J \tag{13.10}$$

$$J_2 = J_1 = k_1' p_1 - k_2'' p_2 = J \tag{13.11}$$

where J is the flow through the channel.

From (13.1) and (13.10):

$$p_1 = (k_0' \nu c^{\mathrm{I}} - J)/k_1''$$

and when this is substituted into (13.11), the result, after again using (13.1), is

$$J = k_2'' \nu \left[c^{\mathrm{I}} \cdot \frac{k_1' k_0'}{k_1'' k_2''} - c^{\mathrm{II}} \right] / (1 + k_1'/k_1'')$$

or from (13.8):

$$J = k_2'' \nu [c^{\mathrm{I}} \exp zu - c^{\mathrm{II}}]/(1 + k_1'/k_1'') \tag{13.12}$$

This expression for the flux should be compared with Kramer's equation (12.29) and (12.47), which is essentially Goldman's equation. The parallel is obvious. The rate theory expression can be suggestively rearranged as follows:

$$J = 1 - \exp(-zF\Delta\phi/RT) \cdot \frac{\nu k_0'}{1 + k_1''/k_1'} \left[\frac{c^{\mathrm{I}} \exp(zF\Delta\phi/RT) - c^{\mathrm{II}}}{\exp(zF\Delta\phi/RT) - 1}\right]$$

$$= (PzF\Delta\phi/RT)\left[\frac{c^{\mathrm{I}} \exp(zF\Delta\phi/RT) - c^{\mathrm{II}}}{\exp\ zF\Delta\phi/RT) - 1}\right] \tag{13.13}$$

where we have changed e and k to F and R to facilitate comparisons. Equation (13.13) is identical with Goldman's equation. The expression for P is

$$P = \frac{1 - \exp(-zF\Delta\phi/RT)}{\exp(zF\Delta\phi/RT} \cdot \frac{\nu k_0'}{1 + (k_1''/k_1')} \tag{13.14}$$

The Nernst–Planck equation, based on continuous diffusion, leads to an expression for flux of the same form as the activated rate theory, based on discrete jumps. The permeability coefficient in rate theory is voltage dependent through the rate constants and $\Delta\phi$.

Alternatively placing $c^{\mathrm{I}} = 0$ and $c^{\mathrm{II}} = 0$, we obtain the unidirectional fluxes

$$J^{\mathrm{I,II}}(c^{\mathrm{I}} = 0) = \frac{(PzF\Delta\phi/RT)c^{\mathrm{II}}}{\exp(zF\Delta\phi/RT) - 1} \tag{13.15}$$

$$J^{\mathrm{II,I}}(c^{\mathrm{II}} = 0) = \frac{(PzF\Delta\phi/RT)c^{\mathrm{I}}}{1 - \exp(-zF\Delta\phi/RT)} \tag{13.16}$$

The flux-ratio equation is given by

$$\frac{J^{\mathrm{II,I}}(c_{\mathrm{II}} = 0)}{J^{\mathrm{I,II}}(c_{\mathrm{I}} = 0)} = \frac{c_{\mathrm{I}}}{c_{\mathrm{II}}} \cdot \exp(-zF\Delta\phi/RT) \tag{13.17}$$

This is identical with (12.53) derived from the Nernst–Planck equation, but the agreement is not to be wondered at. Ussing, in his original derivation of the flux-ratio equation, showed that it held independent of the structure of the membrane *if the ions move independently* and diffusion is passive, that is, active transport is not involved. The most general form of the equation is

$$\frac{J_i^{0d}}{J_i^{d0}} = \exp(\Delta\tilde{\mu}_i/RT) \tag{13.18}$$

(For the superpedantic, the equation is strictly valid only if there is no solvent flow associated with ionic flow, but this is a complication beyond our scope in this text.)

It is interesting to apply the flux expressions to the transport of 1 : 1 electrolyte in the absence of an external field. Under these conditions the total electric current is zero:

$$j = z_+ F J_+ + z_- F J_- = 0$$

therefore;

$$J_+ - J_- = 0 \tag{13.19}$$

Substituting from (13.13), being careful to put in the right values (±1) for z, and using $(e^{-u} - 1)/(1 - e^u) = e^{-u}$, we find for the specific case of NaCl:

$$\Delta\phi = \frac{RT}{F} \ln \frac{P_{\mathrm{Na}} c_{\mathrm{Na}}^{\mathrm{II}} + P_{\mathrm{Cl}} c_{\mathrm{Cl}}^{\mathrm{I}}}{P_{\mathrm{Na}} c_{\mathrm{Na}}^{\mathrm{I}} + P_{\mathrm{Cl}} c_{\mathrm{Cl}}^{\mathrm{II}}} \tag{13.20}$$

which looks like the GHK equation. In fact, the equations are only superficially identical; the permeability coefficients in (13.20) are voltage dependent, which they are not in the GHK equation — compare (12.14) with (13.14).

The flux-ratio equation and a GHK-type equation have been derived by assuming either continuous diffusion in a constant field or discrete jumps over barriers. We must appeal to experiment if we are to weigh the merits of these two models.

As mentioned previously, the GHK equation fails to accommodate much experimental data. But the rate theory in the form that we have presented is also incapable of rationalizing all the evidence. Neither approach, for example, explains the fact that there are deviations from the flux-ratio equation in the *absence* of active transport. We will now dig a little deeper into the two theories and show that both can be modified in ways that bring them closer to the laboratory and that also make physical common sense. Before we start, it is worth reminding ourselves briefly of the physical background. The subject is ion transport and for energetic reasons (see Chapter 7 on image forces and Born energy) it is not to be expected that ions, either bare or hydrated, will easily cross the hydrophobic core of a bilayer. Experiment has shown that two major classes of compounds facilitate ion transport. We can classify them crudely as tunnels and ferryboats or channels and carriers. Our concern here is with channels; carriers will be discussed in Chapter 14. There are molecules which, when incorporated into lipid bilayers, result in enormous increases in permeability to ions, and which, as a result of intense investigation, can be said almost certainly to span bilayers from side to side. The classic example is gramicidin A, a linear peptide with known primary structure (Figure 101). It has been suggested that the pore produced in a membrane by gramidicin A is approximately 30 Å in length and about 4 Å in internal diameter.[214] An alkali metal passing through this channel would certainly interact electrostatically with the carbonyl group dipoles of the peptide, but the suggested tertiary structure of the channel does not strongly suggest the presence of definite binding sites, that is energetic valleys. Such valleys may exist, however, at the ends of the channel, and there is evidence of binding sites in other channels. With these thoughts in the background, we turn back to theory.

It is easy to pick holes (channels?) in the practical approximations derived from the Nernst–Planck equation.

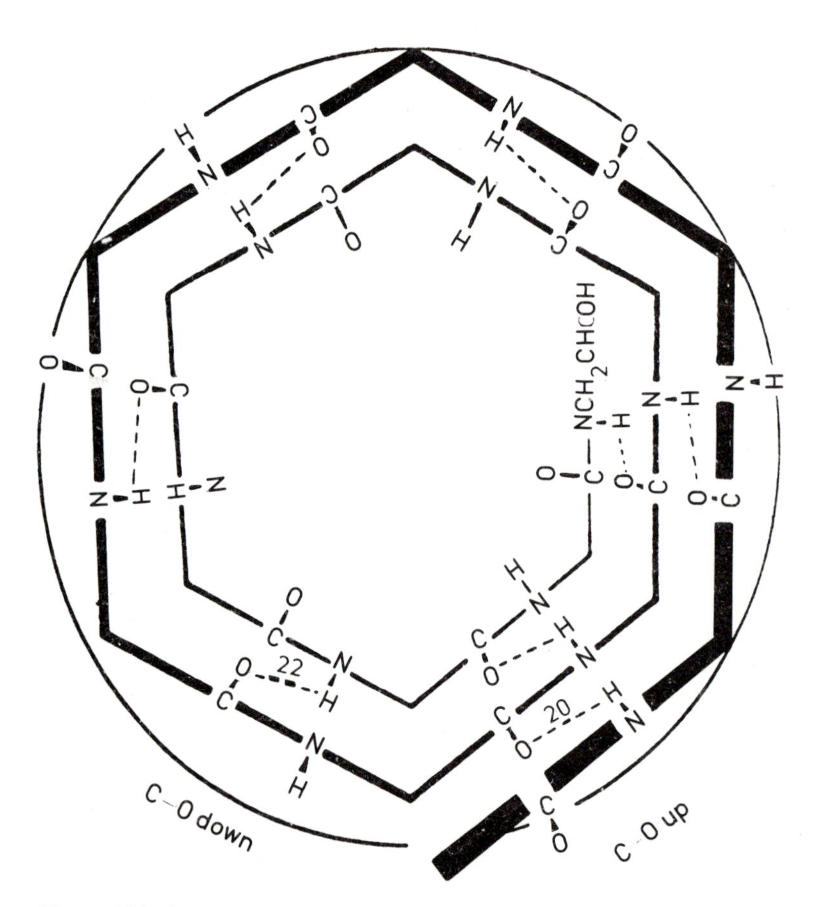

Figure 101. Structure suggested by Urry et al. [*Proc. Nat. Acad. Sci.* 68:672 (1971)] for the π^6 (L,D) helix of gramicidin A. Two of these helices arranged formyl end to formyl end are assumed to form a transmembrane channel. The hydrophobic amino acids on the periphery of the helix are not indicated.

1. The constant field assumption behind the GHK equation is difficult to swallow considering the structural complexity of membranes in general and of known channels. (The linear concentration profile approximation of Henderson is also a compromise with reality.) In fact, it has been shown that the GHK equation can be derived from the Nernst–Planck equation under less restrictive conditions than Goldman's constant field. For example, for monovalent ions, any form of $\phi(x)$ that gives an electric field antisymmetric about the center plane of the bilayer is consistent with the Goldman equation.[215] The relaxation of the original constant-field condition allows a wider range of physical systems to obey, or nearly obey, the GHK equation, but does nothing to help with deviations from the GHK or flux-ratio equations.

2. The GHK equation assumes concentration-independent partition coefficients β. As we saw in Chapter 11, this assumption need not hold for ions. Teorell[216] and Meyer and Sievers[217] constructed a model in which the diffusion potential within the membrane is given by the Nernst–Planck equation or a variation thereof, and the changes in potential associated with the two interfaces are due to Donnan equilibria, so that the total potential change across the membrane due to an ion i is

$$\Delta\phi = -\frac{RT}{z_i F}\ln\frac{c_i(0)}{c_i^{\mathrm{I}}} + \Delta\phi_M + \frac{RT}{z_i F}\ln\frac{c_i(\mathrm{d})}{c_i^{\mathrm{II}}} \tag{13.21}$$

Again this modification leaves much experimental data unexplained.

3. The Nernst–Planck equation is a one-particle equation that says nothing explicit about the possible interactions of diffusing ions. This is probably the major drawback of the theory. Even if we suppose that interionic interactions are contained in the values we take for the mobilities u_i and the form given to $\phi(x)$, this would imply only an *averaging* of interionic interactions, rather than a clear expression of correlation between the movement of different ions. (Theoretical chemists may see an analogy with the absence of explicit reference to electron correlation in Hartree–Fock-type orbitals.) It is significant that at high ionic concentrations, where interionic forces are expected to become noticeable, the GHK equation is often particularly inapplicable.

4. Even at low ionic concentration, the flux-ratio equation, which is consistent with the Nernst–Planck equation, often fails to hold. If transport occurs through narrow channels, a ready qualitative explanation can be hazarded for this failure. The unidirectional flows are measured by effectively holding at zero the concentration of tracer on one side of the membrane. The supposition then is that the tracer crosses the membrane by running down the gradient due to the membrane potential and its own concentration gradient. But if tracer ions are compelled to run through narrow corridors, it is impossible to neglect the flow of

nontracer ions confined to the same path. Thus if there is a large concentration gradient of K^+ ion opposing the flow of $^{42}K^+$, the net flow of nontracer might be expected to slow down the tracer. Such an effect is found in squid giant axons where external K^+ decreases the outward flow of $^{42}K^+$. To complicate matters, an *acceleration* of outward flow was observed in a similar experiment with frog skeletal muscle. We return to these matters later but it is clear that the Nernst–Planck assumption of independent diffusion is not adapted to handle this kind of interference between ions.

We turn to rate theory. In the simple form in which we have examined it there was no provision for molecular interactions; nevertheless, the theory has two apparent advantages. First, from the start it accepts the existence of energy barriers such as almost certainly are seen in nature. Second, the theory can be readily adapted to take ionic interactions into account. A series of energetic valleys, each of which can be occupied by at most one ion, provides a natural model for the correlation of ionic movements, and the mathematical treatment is relatively straightforward compared with the attempt to bring correlation into the Nernst–Planck equation. Thus a model could be envisaged in which there were one or more valleys — binding sites — with concentration-dependent probabilities of being occupied.[218] This simple concept accommodates the existence of multiply occupied channels. If transition probabilities for one site are linked with the occupancy of a second site, we immediately have correlation. Variations are, and have been, easily imagined. Thus "knock-on" models are based on the proposition that an ion leaves a site when another ion enters the site. A twist on this idea is the supposition that a site must be empty before an ion can enter it. Flux equations based on models of this kind have had success in reproducing qualitatively the experimental dependence of flux on ion concentration and in explaining deviations from the flux-ratio equation. Furthermore, the mutual influence of solvent and solute flow is an immediate consequence of a model in which ions or water molecules cannot slip past each other — the "single-file" model[219] — and experimental evidence is becoming available for the interaction of ion and water within channels, a subject into which we delve deeper in Chapter 15. It would be wrong to give the impression that we have even one example of a biological channel for which theory has provided *quantitative* fits to all experiments, but there seems to be no going back from the rate-theory approach.

Can something be salvaged from the Nernst–Planck approach? The answer is reminiscent of the old battle between the valence-bond and molecular orbital theories of bonding. If both are done properly, they must lead to the right answer, but in general molecular orbitals turned out to be easier to handle. Interionic correlation can be fed into the Nernst–Planck equations but the mathematics is not as transparent as the rate-equation formalism. Under conditions in which correlation can perhaps be neglected, say low ionic concentration, the Nernst–Planck equation should hold if we use a realistic potential function for $\phi(x)$, containing, for example, one or more valleys. But if we have valleys, why not use rate theory?

This question is best answered by another: How deep are the valleys? Rate-theory pictures an ion as spending nearly all its life in valleys and very little of its time on the intervening mountains. The theory is best suited to systems in which this physical condition holds. An example of a high-barrier system is the passage of a charged species through the lipid bilayer. This will be our subject in the following chapter. For channels, it is not at all certain that high barriers, or deep valleys, always exist. As mentioned above, gramicidin A does not appear to have any dramatic features associated with the central tunnel. In such cases the ion cannot honestly be supposed to spend most of its time in a small number of locations; its journey could be far closer in nature to the classical continuous diffusion picture, which is best treated by the Nernst–Planck approach. Furthermore, the presence of two ions in a channel *can* be handled by Nernst–Planck theory, if Coulombic repulsion is added.[220]

To summarize the present situation: Both theories are still alive but rate theory is more widely used, especially in dealing with multiple occupancy of channels and high-barrier systems. Simple Nernst–Planck theory fails to explain, even qualitatively, many phenomena that simple rate theory readily accommodates.[221] We end this chapter with an example showing how saturation and other effects are easily explained by one of the many modifications of basic rate theory.

To the double-barrier model analyzed previously we add the condition that the central valley can only be occupied by one ion at a time, that is, we treat the valley as a binding site such as those envisaged in enzyme or surface chemistry.[222] The kinetic scheme for transport is as follows, where X denotes the binding site and i the adsorbed ion:

$$i^{\mathrm{I}} + X \underset{k_1''}{\overset{k_0'}{\rightleftharpoons}} iX \underset{k_2''}{\overset{k_1'}{\rightleftharpoons}} i^{\mathrm{II}} + X \qquad (13.22)$$

The rate constants are those in Figure 100 and they still obey (13.8).

The net flow over barrier 1 is given by

$$J_1 = k_0' c_i^{\mathrm{I}} [X] - k_1'' [iX] \qquad (13.23)$$

and over barrier 2 by

$$J_2 = k_1' [iX] - k_2' c_i^{\mathrm{II}} [X] \qquad (13.24)$$

Under steady-state conditions, $J_1 = J_2 = J_i$, the flow through the system. A conservation equation holds for X:

$$[X]_t = [X] + [iX] \tag{13.25}$$

An enzyme kineticist will solve these equations quickly to obtain

$$J_i = \frac{k_0' k_1' c_i^{\mathrm{I}} - k_1'' k_2'' c_i^{\mathrm{II}}}{k_1'' + k_1' + k_2'' c_i^{\mathrm{II}} + k_0' c_i^{\mathrm{I}}} \cdot [X]_t \tag{13.26}$$

and for the undirectional fluxes:

$$J_i^{\mathrm{I,II}} = \frac{k_0' k_1' c_i^{\mathrm{I}}}{k_1'' + k_1' + k_2'' c_i^{\mathrm{II}} + k_0' c_i^{\mathrm{I}}} [X]_t \tag{13.27}$$

and

$$J_i^{\mathrm{II,I}} = \frac{k_1'' k_2'' c_i^{\mathrm{II}}}{k_1'' + k_1' + k_2'' c_i^{\mathrm{II}} + k_0' c_i^{\mathrm{I}}} \cdot [X]_t \tag{13.28}$$

These expressions are interesting. The unidirectional fluxes here are entirely different in their nature from those coming out of the Nernst–Planck theories:

1 For $c_i^{\mathrm{II}} = 0$, $J_i^{\mathrm{I,II}}$ approaches a *maximum value* given by $k_1' [X]_t$. This is simply Michaelis–Menten behavior; the binding site is *saturated.* Similarly for $c_i^{\mathrm{I}} = 0$, $J_i^{\mathrm{II,I}}$ has a maximum value of $k_1'' [X]_t$.

2. The unidirectional flows depend on the concentrations in both compartments. It is important to realize that in the Nernst–Planck scheme of things the unidirectional flow $J_i^{\mathrm{I,II}}$, say, is *experimentally determined* by putting c_i^{II} at zero, but the flow is supposed to exist, *with the same value*, whatever value c_i^{II} has. Thus at equilibrium, including tracers, we cannot detect any net macroscopic flow, but we can imagine the dynamic microscopic state of the system consisting of two equal and opposite unidirectional flows, of the same value as those measured in standard flux-ratio experiments. For the binding-site model this reasoning is incorrect; not only is the *macroscopic* effect of the flow changed by the concentration in the opposite compartment, but so is its microscopic value. This phenomenon is also typical of certain carrier mechanisms, as we shall see in the following chapter. One consequence of this *transeffect* is the fact that if c_i^{I}, say, is held constant and c_i^{II} is increased, $J_i^{\mathrm{I,II}}$ as measured by tracers *decreases*, a phenomenon that has been observed experimentally in some systems.

3. Returning to (12.53) we see that for $\Delta\phi = 0$ the unidirectional fluxes of the Nernst–Planck model are simply proportional to c_i^{I} or c_i^{II} and are equal if these concentrations are equal, that is, $J_i^{\mathrm{I,II}} = J_i^{\mathrm{II,I}} = p_i c_i$. This symmetry only holds for the binding site model under very restric-

tive conditions. If we put c_i^{I} and c_i^{II} equal to zero in (13.27) and (13.28), respectively, we find for $\Delta\phi = 0$

$$J_i^{\mathrm{I,II}}(c_i^{\mathrm{II}} = 0) = \frac{\bar{k}_0' \bar{k}_1' c_i^{\mathrm{I}}}{\bar{k}_1'' + \bar{k}_1' + \bar{k}_0' c_i^{\mathrm{I}}} \tag{13.29}$$

$$J_i^{\mathrm{II,I}}(c_i^{\mathrm{I}} = 0) = \frac{\bar{k}_1'' \bar{k}_2'' c_i^{\mathrm{II}}}{\bar{k}_1'' + \bar{k}_1' + \bar{k}_2'' c_i^{\mathrm{II}}} \tag{13.30}$$

where we have reverted to the rate constants for zero membrane potential. A high degree of symmetry in the rate constants is required to give these expressions identical functional form, as can be seen by putting $c_i^{\mathrm{I}} = c_i^{\mathrm{II}} = 1$. In general, then, for a fixed value of i the flow I → II when i is only in compartment I differs from the flow II → I when i is only in compartment II. If we were dealing with electric circuits, we would say that the membrane acts as a current rectifier.

4. The flux-ratio equation (12.53) is not obeyed by this system. Because the Nernst–Planck treatment leads to this equation, it can be concluded that a system not obeying the flux-ratio equation is not a simple diffusive system. This conclusion has nothing to do with the form of $\phi(x)$ since the equation is independent of the form of the potential. Deviations from the flux-ratio equation are clearly not to be interpreted as indicating the presence of active transport.

The basic difference between the binding site model and the previous model is that in the former we include a conservation equation for X. This introduces, in the simplest possible way, ionic interaction by mathematically preventing the number of ions occupying the sites from exceeding the number of sites.

Saturation and trans effects are experimentally observable. The fact that channel models can supply these phenomena is no proof of mechanism. In the following chapter we will show that a very different physical model can lead to similar kinetic results. As almost always, kinetics needs and is getting help from other types of observation.

In this chapter we have looked at two simple models based on rate theory. The treatment of channels containing two or more ions of the same or differing species is obviously more mathematically complicated, but much progress has been made in obtaining analytic and numerical solutions to these problems. We do not even hint at the methods or results, but draw attention to two points that are universally relevant in the field of ion transport.

First, we remind the reader of the distinction, made in the previous chapter, between bulk and surface concentrations. Quantitative evidence of the relationship between surface potential and surface ion concentration is contained, for example, in studies on the conductance of artificial membranes containing channels.[223,224] Thus the conductances λ_+^1, λ_+^2 of the cation channel gramicidin at two different surface potentials ϕ_0^1 and ϕ_0^2 obey

$$\frac{\lambda_+^1}{\lambda_+^2} = \exp \frac{-[\phi_0^1 - \phi_0^2]}{RT} \tag{13.31}$$

which follows immediately from the fact that the surface concentration of the ion is proportional to $\exp(-\phi_0/RT)$, provided that saturation effects are absent. The expression (13.31) fits the experimental data well if the surface potentials are calculated by the Gouy–Chapman equation.

A second general point is that the shape of the potential profile depends on the ion being transported. Thus a large ion will generally encounter more frictional opposition than a small ion in negotiating the same channel. The binding energies of different species of ion at a given binding site will usually differ. We have at present no direct way of mapping the potential profile, although attempts are being made.[206,225,226] It is fortunate that the specific form of $\phi(x)$ enters into neither the Nernst–Planck nor the rate-theory expressions for ϕ_M. Looking back at (12.49), we see that the properties of the membrane are all hidden inside the permeabilities P, which, in turn, contain mobilities defined as average qualities of the ion's journey through the membrane. The rate expression (13.14) for P is somewhat more explicit, but in the end again contains quantities (ν, k_0', k_1', k_1'') that are not directly measurable.

Finally, the reader should be aware that the channels that exist in natural membranes are often far more sophisticated in their structure and mode of action than the passive tunnels of our discussion. A common and important example is the control of the opening and closing of certain channels by a *gating* molecule, which operates by changing its conformation as a direct or indirect consequence of the action of a specific stimulus. Thus acetylcholine has been held to bind to specialized receptors, resulting in a change in conformation of the "gate" of the end-plate channel through which cations flow. Since the conformational change also involves a change in dipole moment, the membrane potential can affect the length of time for which the channel is open. None of this is our business here except insofar as it tempts the reader to read elsewhere.

14 Carriers

In the mid-1960s it was discovered that certain antibiotics mediated the transport of alkali metal ions across mitochondrial membranes.[227,228] Subsequently a large number of molecules, in the molecular weight range ~500 to 1000 daltons, have been shown to facilitate ion transport through natural membranes and lipid bilayers. Since these molecules are too small to span a bilayer, they are presumed to act by ferrying ions across the hydrophobic core. Such molecules are known as ion carriers or *ionophores* (which means the same thing). The basic difference between a channel and an ion carrier is that a channel is simultaneously accessible on both sides of the membrane, while a carrier makes contact with its passenger at one surface and chaperones it across to the other where it can be released. This process is what we imply by a "carrier mechanism" in this book. Other operative definitions have been used in the past, mostly based on attempted kinetic distinctions between carriers and channels. Recent work has shown such distinctions to be misleading in some cases, and so we prefer a conceptual rather than an operative definition of carriers.

The primary molecular structure of many carriers is known and the three-dimensional structure of several carriers has been determined by X-ray crystallography.[229,230] Those ionophones that have been characterized have in common a markedly hydrophobic exterior and a cavity or a roughly centrally situated pore the surface of which is hydrophilic, usually due to the presence of carbonyl groups. In many cases the size of the cavity is such as to accommodate exactly a specific unhydrated ion. Thus valinomycin is suited to K^+ but not Na^+ ions and in mitochondria induces the uptake of the former but not the latter. Other carriers are selective for divalent cations.

The chemical nature of carriers is such that their exteriors are designed to be compatible with nonpolar environments while their interiors are designed to accommodate ions. They exhibit these properties not only in membranes, but in solution. Early on it was shown that carriers dissolved in organic solvents were capable of extracting alkali metal ions from aqueous solution.[231] The stability constants for the formation of the valinomycin complex have been measured for all alkali metals from Li to Cs and there is an excellent correlation between the stability constants and the efficiency of the carrier in ion transport as measured by conductivity measurements on bilayers.[232]

Further early evidence for the carrier mechanism came from the fact that carriers such as valinomycin, with a diameter of ~15 Å, increase the conductivity of artificial membranes of far too great a thickness to allow a channel mechanism to be likely. It is also consistent with the carrier hypothesis that the effectiveness of carriers drops dramatically and discontinuously when the membrane is taken to temperatures just below the liquid-crystal to gel-phase transition.[233]

The essential nature of carriers and the fact that they indeed carry now seem beyond doubt. However, their detailed mode of action is still being revealed, mainly by kinetic studies, and the aim of this chapter is an understanding of the broad outlines of carrier-transport kinetics for neutral substrates in terms of formal rate equations and hypothesized energy barriers. We will also be concerned, albeit to a lesser extent, with what are known as *lipophilic ions*. The reason is as follows: Valinomycin is a neutral molecule and so its K^+ complex is positively charged. There are many carriers bearing a carboxyl group that gives a negative charge at pHs over ~4 to 5. The complexes of these charged carriers are therefore either neutral or positively charged. The fact that charged complexes penetrate membranes with comparative ease suggests that large hydrophobic organic ions, *even if charged*, may behave likewise. Such lipophilic ions, both positive and negative, exist (Figure 102) and have been the subject of much experimental and theoretical effort. Being a simpler physical system than a carrier plus its dissociable ion, a lipophilic ion displays somewhat simpler kinetics, and their interpretation has helped in an understanding of carrier mechanisms. For this reason lipophilic ions are often discussed together with carriers, and we will stick to this tradition.

We begin with energetic considerations because general conclusions as to the form of the potential energy profile allow an intelligent choice to be made when we come to decide on a kinetic model. Lipophilic ions are discussed first. The reader is reminded that carriers are presumed to cross the membrane by penetrating the lipid bilayer, not by negotiating polar channels.

The potential profile which the ion is required to negotiate first deviates from constancy well outside the membrane when long-range attractive van der Waals forces come into play. At large distances from the membrane, greater than 200–300 Å, the force acting in the direction of the membrane will be largely determined by the aqueous medium on the far side of the membrane (or the internal contents of a cell). Since the internal and external media differ little in dielectric constant, the force will be small (Chapter 7). At distances less than the membrane thickness, the hydrocarbon core is the determining factor and the force will increase significantly both because of this and the now $1/r^3$ dependence of the force. However, the theory of van der Waals forces for bodies separated by less than atomic dimensions is still being developed and no quantitative estimates have yet been made of the role of van der Waals' forces in attracting lipophilic ions.

Since membranes are almost always charged, there will be an electrostatic force acting on lipophilic ions, a force that becomes significant at two or three Debye lengths. The Gouy–Chapman theory is generally used to estimate surface potential, that is, the difference in electric potential between a point *just outside* the surface of the membrane and a point in the bulk solution. The calculated results usually agree reasonably well with surface potentials estimated from experiment, for example, from zeta potentials.

Many large organic cations and anions are known to be adsorbed to varying extents on membrane surfaces but *not* to pass into the membrane interior. Their presence will locally change the surface potential by changing the net charge at that location. It is not easy to treat this effect rigorously but an approximation that works well is to treat adsorbed charge as if it were uniformly smeared over the surface. The adsorption

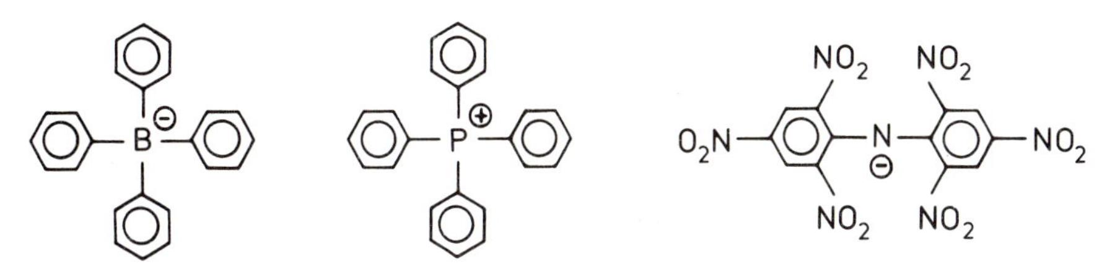

Figure 102. Formula of some lipophilic ions. The borate and phosphonium ions are tetrahedral. [From Langer, P., et al., *Q. Rev. Biophys.* 14:513 (1981).]

process is treated by a Langmuir-type equation:

$$\theta = \frac{1}{K}(\theta_{max} - \theta)[X]_{x=0} \qquad (14.1)$$

where θ is the number of adsorbed molecules per unit area, θ_{max} is the maximum number, K is a dissociation constant, and $[X]_{x=0}$ is the concentration of molecules in the solution adjacent to the surface. $[X]_{x=0}$ can be eliminated by combining this equation with a Boltzmann equation,

$$[X]_{x=0} = [X]\exp[e\phi_0/kT] \qquad (14.2)$$

which relates the bulk and surface concentrations of X to the surface potential ϕ_0. Now ϕ_0 can be eliminated by using the Gouy–Chapman equation and an implicit relationship is obtained for the connection between the number of adsorbed ions, θ, and the bulk concentration $[X]$. There is good experimental reason to suppose that this procedure generally gives reliable results, and it therefore seems that for large ions that are significantly adsorbed on the membrane surface, the concentration profile is effectively determined by electrostatic forces only. This conclusion is of relevance to us since it at least suggests that for lipophilic ions, which *do* penetrate membranes, the concentration profile in solution is likewise dominated by simple electrostatic forces rather than van der Waals or image forces. If this is so, then perhaps we can calculate the concentrations of ions near the membrane surfaces, which are the concentrations that appear in the kinetic rate equations.

Conscience bids us refer to two forces, both repulsive and difficult to evaluate: the image force and hydration force. A charged particle in water should feel a repulsive force near a water/hydrocarbon interface. The difficulties in calculating this force are similar to those encountered in making an estimate of the van der Waals force: The dielectric constant of the head groups is not well defined and the continuum theory of image forces is wobbly at short distances. The hydration force, a structural force arising from the steric repulsion due to surface-adsorbed water molecules, has been seen (Chapter 7) to be the overwhelming short-range force between adjacent bilayers. The response of adsorbed water to the pressure of a compact particle such as a large molecule might be more yielding since there may be little steric hindrance to lateral displacement of water molecules on the surface.

Having reached the head groups, the ion must find a way through and, because of its bulk, it can do this only by causing a major disturbance of a network of hydrogen and electrostatic bonds. Since the ion has a hydrophilic exterior, the process of penetration must surely involve a net rise in the potential energy of the system, which is about all we can say at the moment.

To summarize the ion's odyssey up to now: A cationic ion approaching a negatively charged membrane will slip down a potential slope until it is within a couple of angstroms from the head groups, at which point it has to do some work to penetrate the surface. What forces does it meet inside?

A major contribution to the energy barrier for either a cation or anion is the image force. We have already hinted at the magnitude of this force (Chapter 7), which is, however, not susceptible to accurate estimation. The expression for the image force includes the dielectric constant of the hydrocarbon core, which is usually taken as 2. ESR measurements[234] suggest that ϵ is above 20 as far down as carbon atom 5 in hydrated bilayers, and even at C-12 ϵ is about 5.5. Doubt can be cast on the method, but it is possible that the penetration of water raises ϵ and thereby reduces somewhat the magnitude of the destructive pull on the membrane surface calculated for $\epsilon = 2$ (Chapter 7).

Parsegian[235] has suggested that severe distortion of the membrane could accompany the passage of slowly moving ions. Calculations of image forces thus may be an idle intellectual exercise. Whatever the truth, it seems certain that image forces result in a substantial barrier to transmembrane transport.

The second major contribution to the potential energy profile is a difference in electrostatic potential between the membrane interior and the bulk aqueous solution, which *excludes* Born energies and image forces. Part of this difference is due simply to the double-layer potential; the rest is generally termed a *dipole potential*. Typical respective values for the two contributions are -70 mV and $+300$ mV. The latter value is not as firmly based as the former since potential differences between two phases in principle are not measurable.[236] The value of the potential is *estimated* from electrical measurements on *monolayers* of lipids spread on an air–water interface.[237] The values are always positive; their origin is the subject of hypothesis. The following points help to constrict the theoretician's imagination.

1. That fixed surface charges are not responsible is apparent from the fact that neutral phospholipids such as glycerol monooleate display dipole potentials that tend to be less than but comparable to those of zwitterionic lipids such as DPPE and DPPC and negatively charged lipids such as DPPS.

2. The zeta potential, which gives a rough measure of the surface potential, is effectively zero for neutral and zwitterionic vesicles. Thus the substantial dipole potential of these systems does not show its effect in the neighboring aqueous phase. This implies that the source of the potential is well away from the membrane surface.

3. Replacement of the carbonyl group of phospholipids by an ether

linkage results in a reduction of up to 200 mV in the surface potential of monolayers. This points to the dipole moment of the carbonyl group as a contributing source to the dipole potential. From Figures 6 and 10A it can be seen that the innermost carbonyl group lies about 8 Å below the bilayer surface in the crystal state of DLPE and DMPC. In the fully hydrated liquid crystal state it is possible that the carbonyl group is slightly closer to the surface, perhaps partially hydrogen-bonded to water molecules and oriented such that, on the average, the negative end of the carbonyl dipole, the oxygen atom, is nearer the surface than the positive end. The hydrogen-bonded water molecules likewise might have dipoles so oriented as to contribute to a positive potential within the bilayer. If we replace the parallel arrays of oriented dipoles by four sheets of charge, we see that in crossing the head groups of one surface a positive ion will feel as if it is going through the linear field of a parallel-plate capacitor. In the interior of the membrane the field will be zero, and the potential constant. These approximations give a trapezoidal potential energy barrier for the dipole potential, which should be rounded off a bit at the corners to accommodate reality (Figure 103).

Next in the list of intramembrane forces acting on ions, we need to record the Born energy. As we have hinted, calculations have some degree of associated uncertainty, of which one specific source is the

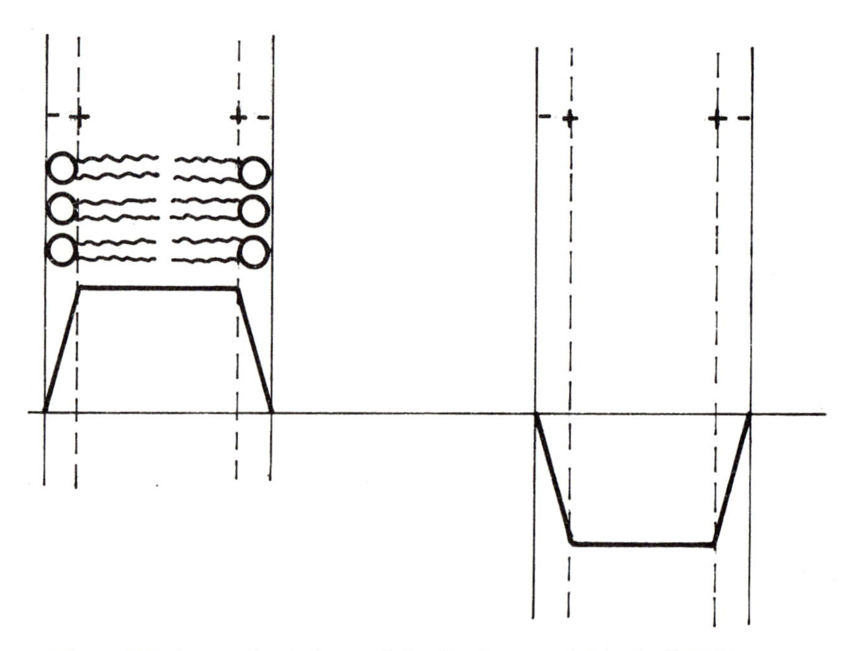

Figure 103. Approximate form of the dipole potential in the lipid bilayer. The energy of an anion in this potential is indicated on the right.

choice of dielectric constant for the hydrocarbon core. Whatever the quantitative unreliability, the qualititative result will always be that a charged ion prefers to be in the aqueous phase (Chapter 7) or in the polar head-group region.

Finally, because we are discussing lipophilic ions, we must take into account the phenomenon that we hesitated to call the "hydrophobic effect" in Chapter 6 (although if "lipophilic" is permissible, why not "hydrophobic"?). Avoiding polemics, we merely state that partition coefficients clearly indicate that in hydrocarbons lipophilic ions have a lower free energy than in water. If they did not, they would be named otherwise.

In Figure 104 there is a qualitative visual summary of the potential energy profiles for lipophilic ions or ion carriers crossing a bilayer. The main difference between cations and anions within the membrane is the effect of the dipole potential; the other forces — Born, image, and "hydrophobic" — will act in the same directions in both cases. The image force is probably the largest of the energetic factors. The size and shape of the barrier are not known even for a one-phospholipid bilayer, but the general form shown in the figure is almost certainly correct in that it shows valleys near the membrane surface and a flattish domelike central area. Note that it is only the image force that, according to our scheme, varies within the bilayer. The reader should appreciate that an externally applied voltage will affect the barrier in the fashion described in Chapter 13, producing an assymmetry.

Before we leave energetic considerations, we pause to note that the preceding analysis is an independent-particle approach; at no point were interactions between ions taken into account. We indicate strategies for attacking this problem. One way is to treat adsorbed ions as a collection of discrete charges and then to attempt either (1) to estimate the effect of the totality of these charges on incoming ions by calculating the change that they produce in the dipole potential, or (2) to take into account the "image" of the charge in the membrane-solution surface. In the second method the charges and their associated images are treated as an array of *dipoles* that influence the total electrostatic barrier surmounted by a diffusing ion. (The distinction between a "diffusing" ion and an "adsorbed" ion is artificial. The effect is due to a mutual interaction between all the ions.) Both models lead to a *nonlinear* dependence of conductance on ion concentration, in contrast to the predictions of independent-particle theories.

A rather different approach from the foregoing is to treat the adsorbed ions as a smeared-out sheet of charge that attracts counter-ions, which themselves, being impermeant, are treated as a smeared-out sheet of charge in the neighboring solution. The two charge distributions and

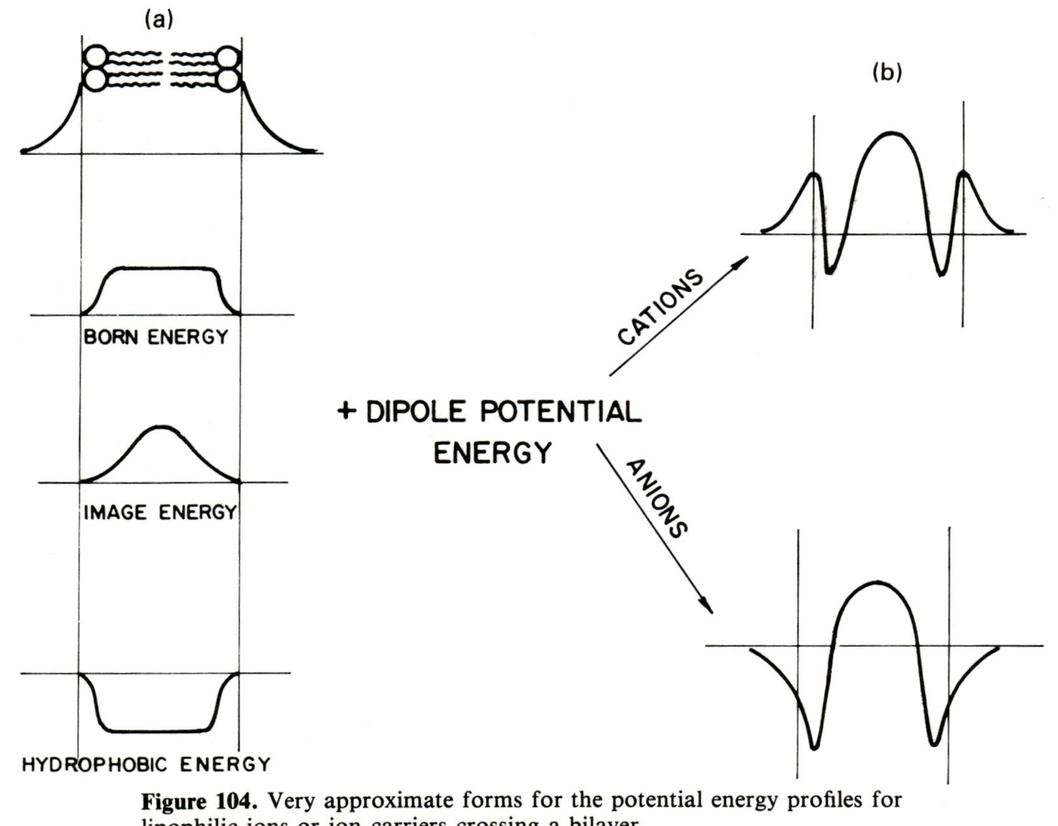

Figure 104. Very approximate forms for the potential energy profiles for lipophilic ions or ion carriers crossing a bilayer.

the intervening head groups are then replaced by a two-plate capacitor and the drop in potential across the surface *due to this effect* calculated from the simple expression $V = Q/C$. Last, the effects of high concentration have been treated by a model in which a finite number of *saturable* adsorption sites is envisaged — a Langmuir-type model. The transition probability for an ion crossing the membrane will be proportional to the fraction of empty "sites" on the other side. The weakness of this approach is that interionic electrostatic forces come into play well below the stage at which site saturation would become significant. In what follows we will avoid high concentrations and associated nonlinear effects.

Switching our attention to kinetics, we must at the very beginning take into account the shape of the potential energy barrier. A constant field approximation would be folly. We have the alternatives of using the Nernst–Planck equation and guessing the form of $\phi(x)$, or of using rate theory. We do what nearly everyone has done and settle for rate theory, partially because a simple version of the theory handles the experimental facts far better than a simple Nernst–Planck approach. Rate theory is applicable because a system with a high barrier between two states spends little time on the top of the mountain and transitions can be modeled as single jumps rather than the discursive meanderings that typify classical diffusion.

We first define our objectives. The many published variations of kinetic scheme and experimental method provide more equations than is seemly in a book of this kind. It is certainly not intended that we batter our way through serried ranks of rate expressions. Rather, we will attempt to understand the main *physical* consequences of some widely used reaction schemes in simple but rigorous terms.

Much of the modern kinetic research is based on electrical measurements. Movements of charged species within a membrane are equivalent to currents. Two types of electrical measurements have been made on bilayers. The older technique is to monitor the steady-state current through a membrane as a function of the applied voltage. The newer methods are based on relaxation kinetics: The equilibrium system of membrane, carrier (or channel), and bathing aqueous solutions is subjected to a sudden change in voltage or temperature and the subsequent development of the current is recorded. An alternative is the charge-pulse method in which a current is passed into the initially equilibrium system over a very short time period and the resulting voltage changes are monitored. Equilibrium and relaxation measurements give different information. For the chemically inclined we give the simple analogy of the equilibrium:

$$\text{keto} \underset{k_{-1}}{\overset{k_1}{\rightleftharpoons}} \text{enol}$$

At equilibrium a measurement of the concentrations of the two forms gives only the *ratio* k_1/k_{-1}. By a temperature jump or some other relaxation experiment, we find the *relaxation time* for the return of the system to equilibrium to be $\tau = 1/(k_1 + k_{-1})$. The two experiments give k_1 and k_{-1} separately. The kinetics of carrier transport or lipophilic ion transport are generally more complicated than this but are similar to our trivial example in the sense that steady-state experiments only give rate-constant ratios.

Again we examine lipophilic ions first. The kinetic scheme, with which it would be hard to quarrel, is shown in Figure 105; solution concentrations are taken as constant. It is not enough to have a large volume of solution to ensure constancy. If the passage of ions into the membrane is faster than the diffusion rate to the surface, the concentration at the surface will be depleted. We discuss this effect later. We assume an independent particle model with no interionic forces. If the system is at

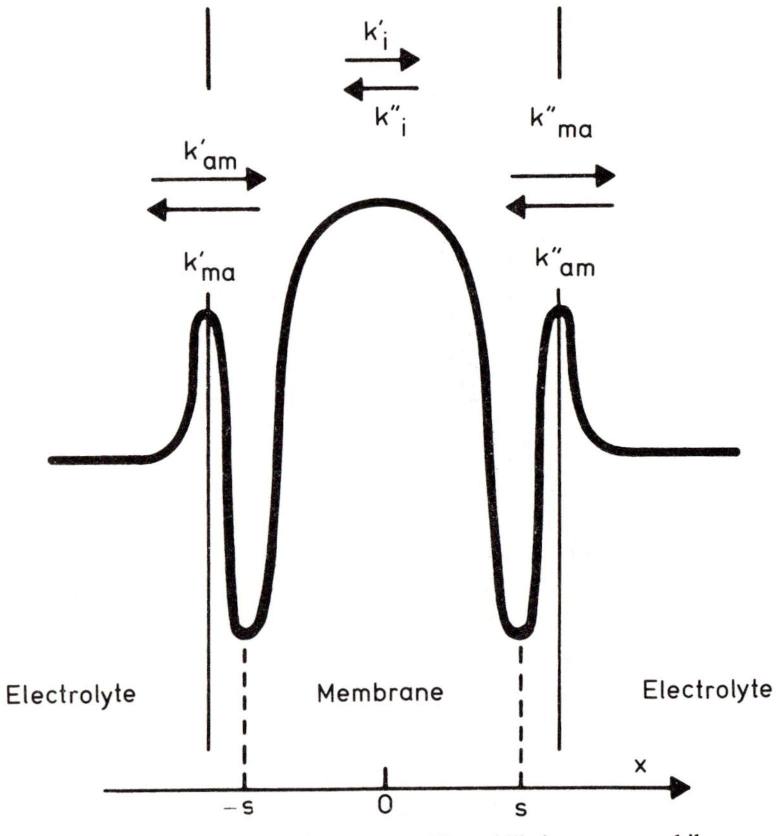

Figure 105. Rate scheme for the passage of lipophilic ions across a bilayer.

equilibrium and a potential gradient is suddenly imposed on the membrane, the distribution of ions will change on a scale far slower than the establishment of the voltage difference, and will reach a new equilibrium distribution determined by the relative energies in the two valleys. This redistribution will cause a current to flow in the membrane and hence through the external measuring circuit. The current is time-dependent since it is a function of the relative populations of the valleys and these change with time until a steady-state population difference is reached in the electric field and a steady current flows. Without going through the derivations, we ask the reader to believe that the general solution for the behavior of the current is given by[238]

$$I(t) = I(\infty)[1 + \Sigma A_n \exp(-t/\tau_n)] \tag{14.3}$$

This sum-of-exponentials expression occurs in many relaxation phenomena. Two examples possibly familiar to chemical readers are nuclear-spin relaxation in a multilevel system and the relaxation of a complex chemical system slightly removed from equilibrium. The mathematics is almost identical. (The voltage-jump case differs from the chemical example in that the latter is usually closed while in the voltage-jump experiments the relaxation is toward a steady *nonequilibrium* state.) It can be shown that the experimental form of $I(t)$ contains the information needed to determine all the rate constants but the practical difficulty of separating exponentials can be a problem. In the single-barrier case illustrated in Figure 104 equation (14.3) reduces to

$$I(t) = I_\infty + (I_0 - I_\infty)\exp(-t/\tau) \tag{14.4}$$

$$I_0 = z^2\alpha^2 e\beta c k_i u \tag{14.5}$$

$$I_\infty = z^2 e\beta c \frac{k_i k_{ma}}{2k_i + k_{ma}} u \tag{14.6}$$

$$\tau = 1/(2k_i + k_{ma}) \tag{14.7}$$

where z is the valency of the ion, α is that fraction of the voltage drop $\Delta\phi_M$ across the membrane that is associated with the intervalley distance, $\beta = k_{am}/k_{ma}$, and $u = \Delta\phi_M e/kT$. Here there is only *one* relaxation time, which recalls the relaxation of a spin 1/2 system that only has two levels so that $N_\alpha + N_\beta$ is constant. For small voltages, $N_1 + N_2$ is also constant.

These equations hold for systems in which the concentration of ions in the solutions adjoining the membrane can be considered constant; c is the bulk concentration of the relevant ion in *both* solutions at equilibrium. We can correct for the effect of surface charge, as we will do in the following example, but if the net adsorption of ions on one side of the barrier or desorption on the other occurs at a rate comparable to solution diffusion, ion concentrations will not be constant. Let us estimate the maximum diffusional rate by supposing that the concentration near the adsorbing surface is zero due to very fast adsorption. The concentration will fall from the bulk value c to zero over a distance δ, the effective width of the depleted layer on the surface. Such a layer is expected even in the presence of stirring and is estimated to be of the order of 0.01 cm thick. Then the ion gradient in this layer is approximately linear and given by c/δ, so that from Fick's first law $J = Dc/\delta$, the flow of course being toward the surface. If this flow is considerably slower than the adsorption of ions at the surface, we cannot assume a constant concentration c in (14.5) and (14.6). From the values of measured or estimated diffusion constants it is certain that diffusion often has to be taken into account, and the necessary rate equations, which are considerably more complicated than those above, have been worked out by Jordan and Stark.[239]

An example of the kind of intimate detail made available by relaxation methods is provided by the voltage-jump results for the tetraphenylborate anion is negatively charged phosphatidylserine membrane.[239] At 25°C, $k_{ma} = 60\ \text{s}^{-1}$, $k_i = 40\ \text{s}^{-1}$, and $\beta = 2 \times 10^{-4}$. This means that the desorption step occurs at a rate comparable to the jump across the barrier. This finding casts doubt on the universality of the common assumption that there is an effective equilibrium at the membrane surface and a rate-determining passage across the membrane. The Theorell–Myers–Sievers theory for membrane potentials and the GHK equation both use equilibrium partition coefficients (Chapter 12). Incidentally the experimentally observed decay of current in this system showed deviations from equation (14.6) but could be analyzed within the framework of the expressions derived for diffusion plus transport.

A complication that we have glossed over is the role of surface charge. The negative charge of phosphatidylserine will reduce the concentration of negatively charged lipophilic ions near the surface. This means that the derived value of the adsorption coefficients k_{am} is too low since the rate of adsorption is given by $k_{am}c_{\text{I}}$ and we have taken c_{I} to be the bulk concentration. Now k_{am} is derived from $k_{ma} = k_{am}/\beta$, giving $1.2 \times 10^{-2}\ \text{cm s}^{-1}$. To correct this apparent value, we have to multiply it by the Boltzmann factor $\exp(-e\phi_0/kT)$, accounting for the repulsion of ions by the surface. A value of $\phi_0 \sim -130$ mV has been estimated by using the Gouy equation (7.47) so that $k_{am}(\text{corr}) \simeq 2.1\ \text{cm s}^{-1}$. Absolute rate theory says that a first-order rate constant can be written in simple form as $k = \nu_0 \exp(-E_a/kT)$ where E_a is the activation energy and ν_0 is an averaged thermal velocity given by a $(kT/m)^{1/2}$. For the tetraphenylborate ion at 300 K, $\nu_0 \simeq 88\ \text{m s}^{-1}$ so that $E_a \simeq 20\ \text{kJ mol}^{-1}$. This is a sensible magnitude for a chemical barrier, and gives us an idea of the hill that the lipophilic ion has to climb to break through the head groups.

It is intuitively obvious that if a strong enough voltage pulse is imposed on a membrane, the difference in energy between the two valleys can be made large enough to ensure that effectively all the ions in the membrane fall into the lower valley. If the pulse length is chosen carefully, the redistribution within the membrane will be over before exchange of ions with the solvent becomes significant. The return to equilibrium gives a current with a total time integral proportional to the number of ions in the membrane. This method of determining the number of adsorbed ions has been used successfully[240,241] with pulses of the order of 300–400 mV. The partioning of lipophilic ions between membranes and solution has also been studied by other spectroscopic methods, including ESR[242] and spectrophotometry.[243] NMR studies indicate that the tetraphenylphosphonium ion and the cetyltriphenylphosphonium ions are adsorbed at the level of the carbonyl groups and C-2 carbons of the lipid chains.[244]

A large number of studies have been made on the effect on rate constants of variations in the head groups, chain length, number of double bonds in the hydrocarbon chain, and so on. Although it is not yet possible to explain all the trends, relaxation methods are providing, and will continue to provide, valuable information at the molecular level on the transport of lipophilic ions.

We will discuss *carriers* from a more conventional standpoint, concentrating on the analysis of a simple *steady-state* system that demonstrates the principal features of carrier-mediated transport. We carry over from our previous discussion the idea of two energy valleys situated just beneath the head groups and containing, at any one time, the vast majority of ions within the membrane. Carriers are assumed to be confined to the membrane. The general kinetic scheme is shown in Figure 106. We neglect completely the role of diffusion to and from the membrane, which often has to be considered to account for experiment, but complicates the mathematics and adds nothing to our understanding of the basic phenomena typical of carrier transport.

The steady-state solution of the kinetic scheme is given by a rather lengthy expression[245] for the flow of "substrate":

$$J = C_t \cdot \frac{k_1 k_3 k_5 k_7 (S_{\mathrm{I}} - S_{\mathrm{II}})}{\left\{ \begin{array}{l} k_1[k_7(k_4+k_5)+k_3(k_5+k_7)]S_{\mathrm{I}} + k_6[k_8(k_2+k_3)+k_4(k_2+k_8)]S_{\mathrm{II}} \\ \qquad + k_1 k_6 (k_3+k_4) S_{\mathrm{I}} S_{\mathrm{II}} + (k_7+k_8)[k_5(k_2+k_3)+k_2 k_4] \end{array} \right\}} \tag{14.8}$$

where $C_t = [C_{\mathrm{I}}] + [C_{\mathrm{II}}] + [CS_{\mathrm{I}}] + [CS_{\mathrm{II}}] \equiv$ the total concentration of carrier. [The enzymatically minded reader may care to derive (14.8) by the King–Altman method.[246]]

The flow vanishes for $S_{\mathrm{I}} = S_{\mathrm{II}}$ as it should for a passive mechanism involving only one kind of molecule. Equation (14.8) hardly shines with immediate physical significance, but a few assumptions will help to bring it closer to earth. Putting

$$k_1,\ k_2,\ k_5,\ k_6 \gg k_3,\ k_4,\ k_7,\ k_8$$

ensures that the equilibrium between carrier and carrier-substrate complex is very fast compared with any jumps across the membrane, and results in

$$J = C_t \cdot \frac{k_1 k_3 k_5 k_7 (S_{\mathrm{I}} - S_{\mathrm{II}})}{k_1 k_5 (k_7 + k_3) S_{\mathrm{I}} + k_2 k_6 (k_8 + k_4) S_{\mathrm{II}} + k_1 k_6 (k_3 + k_4) S_{\mathrm{I}} S_{\mathrm{II}} + k_2 k_5 (k_7 + k_8)} \tag{14.9}$$

It is now convenient to *define*

$$k_2/k_1 = K_{CS_{\mathrm{I}}} \qquad k_5/k_6 = K_{CS_{\mathrm{II}}} \tag{14.10}$$

and to assume that the association constants are the same:

$$K_{CS_{\mathrm{I}}} = K_{CS_{\mathrm{II}}} = K_{CS} \tag{14.11}$$

These definitions lead to

$$J = C_t \cdot \frac{k_3 k_7 (S_{\mathrm{I}} - S_{\mathrm{II}})}{(k_3 + k_7) S_{\mathrm{I}} + (k_4 + k_8) S_{\mathrm{II}} + (k_3 + k_4) S_{\mathrm{I}} S_{\mathrm{II}} / K_{CS} + (k_7 + k_8) K_{CS}} \tag{14.12}$$

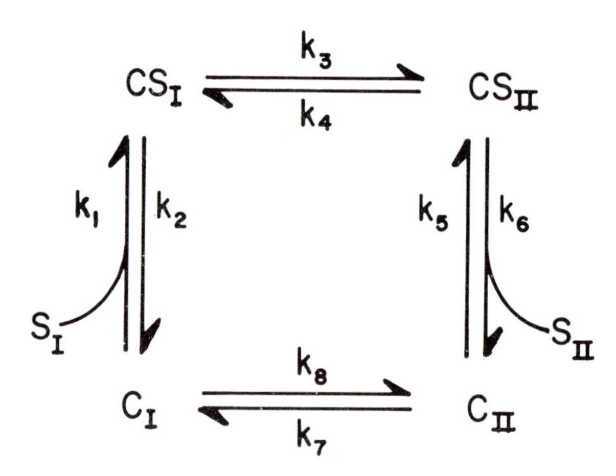

Figure 106. Kinetic scheme for the carrier-mediated transport of ions across a membrane. Subscripts I and II refer to the two sides of the membrane. C = carrier; S = substrate.

Finally, if the membrane is *symmetric*, we can surely write

$$k_3 = k_4 \equiv k_s \quad \text{and} \quad k_7 = k_8 \equiv k_c$$

so obtaining

$$J = C_t \cdot \frac{k_s k_c K_{CS}(S_I - S_{II})}{K_{CS}(k_s + k_c)(S_I + S_{II}) + 2k_s S_I S_{II} + 2k_c K_{CS}^2} \tag{14.13}$$

This is the expression with which we will initially work.

We first consider the flow when $S_{II} = 0$:

$$J = C_t \cdot \frac{k_s k_c K_{CS} S_I}{K_{CS}(k_s + k_c) S_I + 2k_c K_{CS}^2} \tag{14.14}$$

This has the form

$$J = \frac{A S_I}{B S_I + C} \tag{14.15}$$

which is a form familiar in Langmuir-type situations and mirrors the Michaelis–Menten equation for a simple one-substrate enzyme reaction. As S_I is increased, J approaches a maximum value of

$$J_{\max} = A/B = C_t k_s k_c/(k_s + k_c) \qquad \text{for } S_{II} = 0 \tag{14.16}$$

Thus for $S_{II} = 0$, that is, for initial rates of flow in one direction, the flow saturates at high enough values of S_I. This property, which as we will show applies under other conditions, was at one time considered to be the hallmark of carrier transport, distinguishing this mechanism from diffusional mechanisms. However, we have seen that saturation is also a consequence of the simple model of one binding site in a channel (Chapter 13).

Looking at (14.8), we see that for fixed $S_{II}(S_I)$ the flow reaches a maximum independent of $S_I(S_{II})$, so that Michaelis–Menten behavior obtains under quite general conditions. The reason is clear; there is a finite number of carriers, and if they are full of passengers, the total transport rate will not be increased by lengthening the bus queue.

We now look at unidirectional flows. We give, without proof, the equations for the steady state and for a system obeying the conditions used in deriving (14.12):

$$J^{I,II} = C_t \cdot \frac{k_s S_I(k_c K_{CS} + k_s S_{II})}{K_{CS}(k_s + k_c)(S_I + S_{II}) + 2k_s S_I S_{II} + 2k_c K_{CS}^2} \tag{14.17}$$

$$J^{II,I} = C_t \cdot \frac{k_s S_{II}(k_c K_{CS} + k_s S_I)}{K_{CS}(k_s + k_c)(S_I + S_{II}) + 2k_s S_I S_{II} + 2k_c K_{CS}^2} \tag{14.18}$$

The reader should note the following points:

1. $J_{\mathrm{I,II}} - J_{\mathrm{II,I}}$ gives, as it should, the total flow J as expressed by (14.15).

2. In complete contrast to the expressions for the unidirectional flow given by the Nernst–Planck equation, (12.51) and (12.52), the flow in either direction is affected by the substrate concentration on the "receiving" side of the membrane. Thus $J_{\mathrm{I,II}}$ depends on S_{II}. This so-called *trans effect* was once considered to be a prerogative of carrier mechanisms until it was shown that channel models could display the same behavior (Chapter 13).

3. For fixed S_{II}, $J_{\mathrm{I,II}}$ shows a maximum, that is, Michaelis–Menten again appears.

4. Under the special conditions, $k_s = k_c = k$, the flows reduce to

$$J^{\mathrm{I,II}} = C_t \cdot \frac{kS_{\mathrm{I}}}{2(K + S_{\mathrm{I}})} \tag{14.19}$$

$$J^{\mathrm{II,I}} = C_t \cdot \frac{kS_{\mathrm{II}}}{2(K + S_{\mathrm{II}})} \tag{14.20}$$

In this, and only this, case the unidirectional flows depend only on one concentration.

5. When $k_c \gg k_s$,

$$J^{\mathrm{I,II}} \simeq C_t \cdot \frac{k_s S_{\mathrm{I}}}{2K + S_{\mathrm{I}} + S_{\mathrm{II}}} \tag{14.21}$$

In this case, if we keep S_{I} constant, the unidirectional flow from compartment I to compartment II is slowed down if S_{II} increases. This specific transeffect is sometimes called *transinhibition*. Again, simple diffusion mechanisms cannot explain this phenomenon, but the binding-site model of the previous chapter can.

6. A particularly intriguing situation arises when $k_s \gg k_c$. Then

$$J^{\mathrm{I,II}} \simeq \frac{C_t k_s S_{\mathrm{I}}}{S_{\mathrm{I}}[2 + K/S_{\mathrm{II}}] + K} \tag{14.22}$$

Under these conditions, if S_{I} is held constant, the unidirectional flow from compartment I to compartment II is *increased* by increasing S_{II}. This is a case of *transstimulation*. As we mentioned in Chapter 12, this effect has been observed experimentally with frog skeletal muscle but this should not be taken as a proof of mechanism.

7. Finally, consider the possibility that $k_c = 0$, that is, the *uncomplexed* carrier cannot cross the membrane. From (14.13) we see that $J = 0$. The complexed carrier can cross the membrane, but having

released its substrate can only return if it combines with a substrate molecule. This process cannot change the number of molecules in either compartment I or II (apart from microscopic fluctuations). Isotopic exchange *can* be mediated by this mechanism.

It can be seen that the range of effects compatible with the carrier mechanism is large, even for the simple system with which we have dealt. We do not extend our coverage to the multitude of kinetic schemes that have appeared in the research literature. However, one slightly more complex case is worth looking at since it manifests what superficially appears to be an exception to the second law of thermodynamics. The system consists of *two* uncharged substrates that compete for a single carrier. It is not necessary to derive the expressions for the rates to understand the occurrence of two interesting phenomena:

1. Since both substrates, S, S' compete for the same carrier it is not surprising that *competitive inhibition* can be experimentally and theoretically shown to occur; an increase in the concentration of $S_I{}'$ reduces the unidirectional flow $J_S^{I,II}$ of the other substrate. This behavior cannot be emulated by a simple Nernst–Planck system but does follow from the binding-site model presented in Chapter 13. Competitive inhibition is the cause of the phenomenon now to be described.

2. We refer to Figure 107A in which an equilibrium state is depicted. Elderly professors are ferried back and forth across the river Seine, the population of the two banks being equal. The net rate of professorial transport is constant at zero because the flows are equal and opposite. At a certain point in time, a host of go-go girls is deposited on the Left Bank (Figure 107B). They compete for those ferries leaving the Left Bank so that the rate at which professors leave drops below the unchanged rate in the opposite direction. The result is a *net* nonzero flow of professors toward the Left Bank, that on which the go-go girls are based. There are those who will find a psychological reason (but not inhibition) for this flow; the more serious among us will concentrate on the fact that there is a flow of professors in an *absence* of a concentration gradient; $d(\text{prof})/dx = 0$. This appears at first sight to contradict the second law. The system has moved spontaneously away from equilibrium. The solution to the formal paradox is trivial; despite their egocentricity, the professors do not comprise a closed system. The thermodynamic "system" consists of professors *and* go-go girls, and this total system continuously reduces its entropy until there is finally an equal concentration of the two "species" on both banks. The "force" that drives the professors away from equilibrium is the gradient, $d(\text{go-go})/dx$. We have here an example of *coupled flows*, one flow driving another. We can uncouple the flows

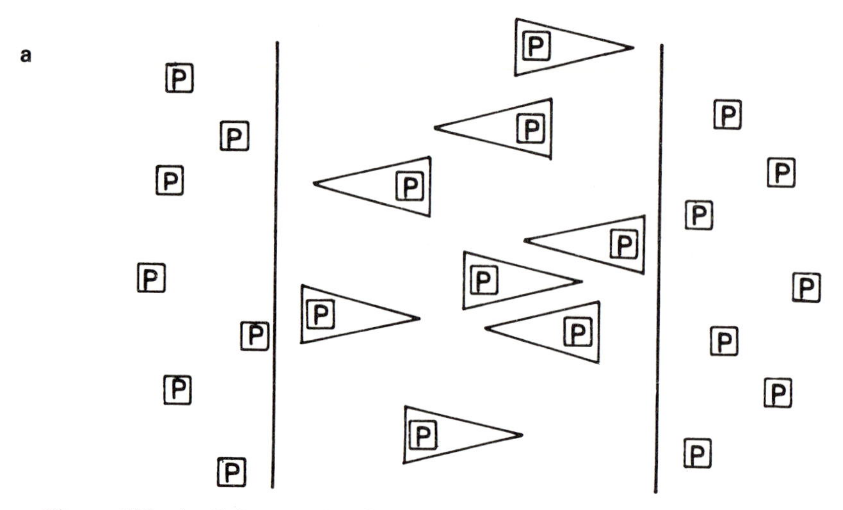

Figure 107. A. Substrate (professors) transported by carriers (ferries) across the membrane (the Seine). The rates from left to right and right to left are equal; the system is at equilibrium.

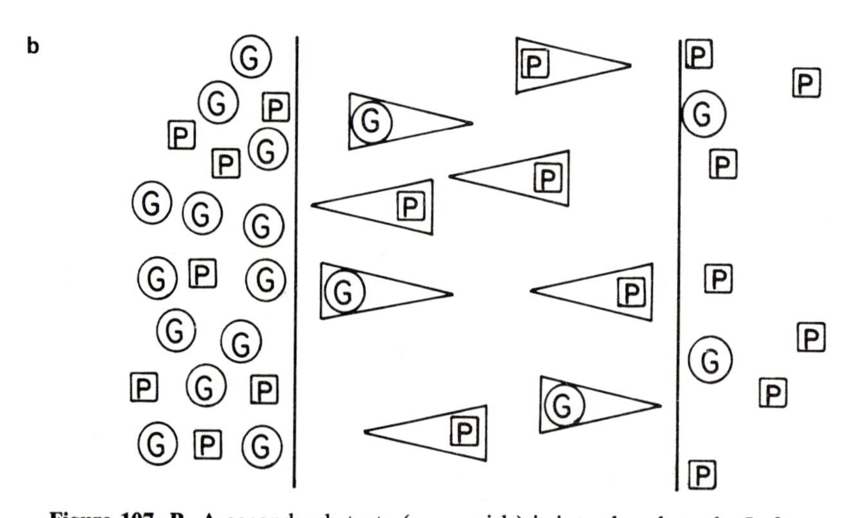

Figure 107. B. A second substrate (go-go girls) is introduced on the Left Bank. The rate of transport of professors from left to right is reduced; that from right to left is initially unaffected. The net flow of professors is now from right to left.

by insisting on there being two types of ferries, one carrying only professors the other only go-go girls. In this case the appearance of the girls leaves the professors' equilibrium undisturbed; the girls equilibrate independently. The formal kinetic analysis shows that substrates can be made to flow "up" a gradient by this mechanism. This phenomenon is reminiscent of that which can occur in a concentrated solution of two kinds of molecule of which one is initially homogeneously distributed while the other displays a concentration gradient. In this system there is often an initial flow of the homogeneously spread molecule induced by the fact that its chemical potential at a given point depends on the concentration of the other species. In both phenomena transport of a substance appears to be influenced by "nonconjugate" forces. We go deeper into this matter in Chapter 15. Here we merely observe that we must be careful not to assume that only metabolic processes can result in flow against a concentration gradient. However, in the absence of metabolic activity the flow of a substance through a membrane against its concentration gradient is a very strong hint indeed (but not unfortunately a proof) that a carrier is involved, which is something to be grateful for in a field in which, as we have seen, different models often lead to identical kinetics.

The carrier-mediated transport of *ions* is associated with the movement of charge and it is natural that electrical measurements, both in the steady state and in relaxation experiments, have provided much of the data on which theory has been built.[247] The subject is too broad and too formally complex to warrant treatment in this book.

We end by going back to the flux-ratio equation (12.53), which, if it holds for a given experimental setup, is a near-guarantee that transport is a simple process involving *only* the gradient in the electrochemical potential of the molecule involved. The reader may care to take the expressions for unidirectional flows derived for the mechanisms listed in this and the previous chapter and be convinced that the flux-ratio equation holds only for the simple two-barrier channel. The deviations from the equation can thus be used to help in identifying the mechanism of a transport process. In fact, the mathematical manipulations we applied to the basic flow expression (14.8) are trivial in the extreme. A more thorough analysis of the equations for flow in carrier-mediated ion transport gives criteria for rejecting certain mechanisms, these criteria being expressed in terms of the relationships between experimental kinetic parameters.[248,249] This degree of sophistication lies beyond our rough-hewn approach, which is designed to show that a reasonable and conceptually economic physical model can "explain" a wide variety of experimentally observable phenomena.

15 Irreversible Thermodynamics

Irreversible thermodynamics is a means of applying the laws of equilibrium thermodynamics, together with a few other "laws", to systems that are not at equilibrium. The quotation marks are there to turn aside the wrath of purists who assure us that while the principles of equilibrium thermodynamics are not to be challenged (in this corner of the universe), the additional principles associated with irreversible thermodynamics all come with a proviso as to their range of applicability. The standard equations of irreversible thermodynamics hold only for systems that are changing slowly, or are at a steady state, or near equilibrium. In this context "slowly" and "near" are usually not easily definable.

Near equilibrium the functions of state familiar from equilibrium thermodynamics are assumed to have a meaning. For example, in a chemically reacting system removed (but not too far!) from equilibrium it is assumed that for each component we can define the enthalpy, entropy, temperature, pressure, and so on, and that *the relationships between these variables are those that hold in equilibrium thermodynamics.* The questions of the range of validity of the theory will hardly concern us. Our aim is to show that if we accept the basic ideas of irreversible thermodynamics, they provide us with a unifying framework capable of providing useful generalizations as to the behavior of systems that are either moving to equilibrium or are in a steady state maintained by forces or flows external to the system. A major area of applicability of the theory in biophysics has been membrane transport. The theory deals with *flows*, those that interest us being flows of matter and charge through membranes. A basic premise that is almost always accepted in biophysical applications is that all flows are linear functions of forces. Nineteenth century science provided the classic example of Ohm's law that states that current (*flow* of charge) is proportional to voltage (electromotive *force*). Fick's first law of diffusion can be similarly interpreted if we are prepared to treat a concentration gradient or, more correctly, a gradient in electrochemical potential as being a force. (Those brought up on the dictum $F = ma$ might feel uncomfortable, but may be reassured by thinking of a force as something that tends to produce a flow. On the other hand, they may see tautology creeping in.) The flow of liquids though a wide pipe is usually taken to be proportional to the difference in pressure across the pipe; the flow of heat energy is often linearly proportional to the temperature gradient. We will not deal with the flow of heat, although this is easily accommodated by the theoretical framework. The flow of matter and charge will be given the units of *quantity* $area^{-1}$ $time^{-1}$. All flows will be with respect to a coordinate frame fixed in the membrane. This may seem too obvious a choice to be worth mentioning, but this is not so. Consider a solution of glucose in which diffusion is taking place. We can define J_s, the flow of solute with respect to the containing vessel, but it is perhaps less apparent that there is an opposing net flow of water molecules. The reader who thinks of a very concentrated "solution" of black spheres in "solvent" white spheres, and follows the process of diffusion into a neighboring region to the right composed entirely of white spheres, will see that there must be flow of white spheres to the left to fill up the holes left by the migrating black spheres. The flow of black *with respect to white*, not to the container, is correctly termed the *diffusional flow* J_D of solute. J_D is clearly not equal to J_s, the flow of solute with respect to the container, but the two flows will converge in value as the solution becomes more dilute. For this reason, since we deal with dilute solutions, we choose the container or membrane as our frame of reference. To make a usable quantitative theory, we must sharpen our definitions.

For linear phenomena of the type mentioned above, we can write $J_i = L_i X_i$, where J_i is the flow of i and X_i the so-called conjugate force acting on i. For Ohm's law, $I = (1/R)V$ so that L_i has the dimensions of reciprocal resistance or conductivity. Other 19th century discoveries complicate but enliven the situation. Thus Seebeck applied a temperature gradient to a bimetallic strip and detected an electric current, showing that I can depend on forces other than a gradient in electric potential. This is one example of a very general phenomenon, the dependence of flow on *nonconjugate* forces. For the flow of charge, Ohm and Seebeck allow us to write an equation in V and ΔT:

$$I = (1/R)V + c\Delta T \tag{15.1}$$

Somewhat more generally we have an equation linear in two forces:

$$J_1 = L_{11}X_1 + L_{12}X_2 \tag{15.2}$$

Completely generally we write, for all the n flows conceivable in a given system,

$$\begin{aligned} J_1 &= L_{11}X_1 + L_{12}X_2 + L_{13}X_3 \cdots L_{1n}X_n \\ J_2 &= L_{21}X_1 + L_{22}X_2 + L_{23}X_3 \cdots L_{2n}X_n \\ &\vdots \\ J_n &= L_{n1}X_1 + L_{n2}X_2 + L_{n3}X_3 \cdots L_{nn}X_n \end{aligned} \tag{15.3}$$

where the L_{ii} are the coefficients of the conjugate forces. The linearity of the above equations in specific applications is an assumption that has to be justified by experiment. There is no doubt that such linear phenomenological relationships apply to a vast number of natural processes. The most familiar flows are those of mass and charge, and in both cases the flow is a vector from which it follows that all the terms $L_{ij}X_j$ in the equation for the flow J_i must also be vectors. Usually this is no problem since the forces X_j themselves have a direction, being expressible as the *spatial* gradient of a potential. Thus the electric force on a charge is given by $e\hat{E}$ where $\hat{E}$ is the electric field, a vector, and is given by $\hat{E} = \text{grad}\phi$, or for a one-dimensional problem $\hat{E} = d\phi(x)/dx$. There is, however, one flow with which we have to deal that is *scalar*, that is, it has the dimensions of a number. This is J_r, the "flow" of a given molecule caused by chemical reaction. Thus if an enzyme produces a certain quantity of ethanol in a unit volume of solution, we can formally express the increase in ethanol concentration per unit volume per unit time as a *flow* of ethanol into the unit volume.

Before continuing with our examination of flows, we mention that we can derive from equations (15.3) expressions for the forces as functions of the flows:

$$\begin{aligned} X_1 &= R_{11}J_1 + R_{12}J_2 + \cdots R_{1n}J_n \\ &\vdots \\ X_n &= R_{n1}J_1 + R_{n2}J_2 + \cdots R_{nn}J_n \end{aligned} \tag{15.4}$$

where the matrix R is the inverse of the matrix L.

Under certain circumstances, a very important relationship holds between the coefficients L_{ij}, namely, $L_{ij} = L_{ji}$. The conditions under which *Onsager's reciprocal relationships*[250] hold will now be discussed.

Our starting point is the second law of thermodynamics because it provides a definition of irreversibility, a criterion for the movement toward equilibrium, which is the driving force behind flows. In any infinitesimal reversible process, the entropy entering the system, dS_{ext}, is given by

$$dS_{ext} = \frac{dQ}{T} \tag{15.5}$$

dQ being the heat entering the system. For an infinitesimal *irreversible* process, the change in entropy of the system is given by

$$dS = dS_{ext} + dS_{int} \tag{15.6}$$

where dS_{int} is the entropy generated *within the system*. We now assume that if we are near equilibrium, dS has a meaning, not only as a difference between initial and final states, but at every stage of a process. Also, even if the system has inhomogeneities and gradients of concentration, temperature, or pressure, within a small volume we can define a local entropy which, when summed over the whole system, gives the total entropy. We assume that all other thermodynamic variables can be similarly treated and that in each small volume the basic equations of equilibrium thermodynamics are valid. Thus we believe that the Gibbs equation holds:

$$TdS = dU + PdV - \sum_i \tilde{\mu}_i dn_i \tag{15.7}$$

where all the variables are *local* variables. Under what circumstances this equation begins to break down is difficult to say, but is is unlikely to be useful during a laboratory explosion. We will now manipulate equation (15.7) to arrive at a useful and suggestive result.

Let us consider a membrane, for convenience of unit area, separating two solutions I and II, of differing concentration, and let us allow dn moles of the solute to pass through the membrane. The total change of entropy is given by

$$dS = dS_I + dS_{II} \tag{15.8}$$

Using (15.7) we find that

$$TdS = dU_I + dU_{II} + P_I dV_I + P_{II} dV_{II} - \tilde{\mu}_I dn_I - \tilde{\mu}_{II} dn_{II} \tag{15.9}$$

From the first law:

$$dQ = dU_I + dU_{II} + P_I dV_I + P_{II} dV_{II} \tag{15.10}$$

Combining (15.9) and (15.10) and realizing that $dn_{II} = -dn_I = dn$, we obtain

$$dS = \frac{dQ}{T} + \frac{(\tilde{\mu}_I - \tilde{\mu}_{II})}{T} dn \tag{15.11}$$

From equations (15.9), (15.10) and (15.11) we find

$$TdS_i = \Delta\tilde{\mu}dn \qquad (15.12)$$

where dS_i is the change in entropy produced in the system by the irreversible process of diffusion. We now boldly divide by dt to reach our objective:

$$\frac{TdS_i}{dt} = \frac{dn}{dt} \cdot \Delta\tilde{\mu} \qquad (15.13)$$

This equation contains the seed of much that we need to apply irreversible thermodynamics. First consider its dimensions: The right-hand side clearly has the dimensions of free energy per unit time and so, therefore has, the left-hand side. Lord Rayleigh named the function $T\,dS_i/(dt)$ the *dissipation function*, Φ, since it measures the rate at which free energy is dissipated by the system. Equation (15.13) was derived for the specific case of mass flow, but if the reader has faith, a generalized expression can be written, taking into account all the flows in the system,

$$\Phi \equiv \text{Dissipation function} = \sum \text{Flow} \times \text{conjugate force} \qquad (15.14)$$

A simple example is the passage of electricity through an ohmic resistance:

$$\Phi = IV$$

We know from our schooldays that $IV = P$ where P is the power dissipated by the circuit, that is, the rate of loss of energy to the surroundings. Onsager found that *if a dissipation function can be written in the form (15.14) and the forces and flows are those used in the linear phenomenological equations (15.3), then* $L_{ij} = L_{ji}$. The theoretical limitations on the validity of Onsager's relationships will not worry us, but they, of course, only apply to *linear* phenomenologic equations. We go straight to an example. Two experimental observations are our points of departure.

1. Referring to Figure 108A, the imposed voltage produces not only a current through the porous partition, but a flow of solution. This is the phenomenon of electroosmosis.
2. Referring to Figure 108B, the applied hydrostatic pressure results in both the expected volume flow and a potential difference across the partition, the so-called streaming potential.

We see that a flow of charge can be induced by a voltage *and* by a pressure difference so that we can write

$$J_{\text{charge}} \equiv I = L_{11}\Delta\psi + L_{12}\Delta P \qquad (15.15)$$

where $\Delta\psi$ and ΔP are the conjugate and nonconjugate forces to the flow

I. Likewise, a flow of volume can be produced by a pressure gradient — the conjugate force — and a nonconjugate force, an electric potential gradient:

$$J_{vol} = L_{21}\Delta\psi + L_{22}\Delta P \qquad (15.16)$$

The indices for L have been chosen to conform with equations (15.3). (Notice that for the flow of current we already have that $\Phi = IV$ for a simple resistance so that I and V are the correct flow and force for Onsager relations to hold. Likewise, J_{vol} and ΔP are a correct choice for flow and force since $J_{vol}\Delta P$ gives the rate at which PV work is done.) Onsager now tells us that $L_{12} = L_{21}$; What is the practical significance of this? It implies that electroosmosis and streaming potential are not

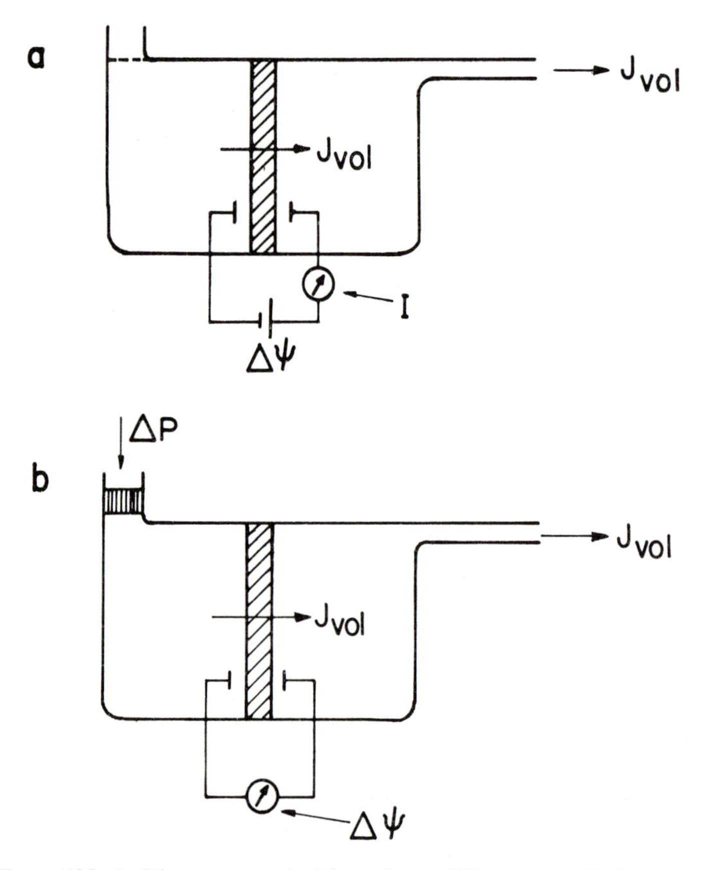

Figure 108. A. Electro-osmosis. The voltage difference results in a current and a volume flow. **B.** Streaming potential. The hydrostatic pressure results in a volume flow and a potential difference across the membrane.

independent phenomena. From (15.15) and (15.16)

$$L_{21} = \left(\frac{J_{\text{vol}}}{I}\right)_{\Delta P=0} \cdot L_{11} \tag{15.17}$$

and from (15.15)

$$L_{12} = -\left(\frac{\Delta\psi}{\Delta P}\right)_{I=0} \cdot L_{11} \tag{15.18}$$

But if $L_{12} = L_{21}$, then

$$\underset{\text{Electroosmosis}}{\left(\frac{J_{\text{vol}}}{I}\right)_{\Delta P=0}} = -\underset{\text{Streaming potential}}{\left(\frac{\Delta\psi}{\Delta P}\right)_{I=0}} \tag{15.19}$$

This equality has been confirmed experimentally for a number of systems. We now manipulate (15.19) to get a convenient expression for membrane transport studies:

$$\Delta\psi = \left(\frac{J_{\text{vol}}}{I}\right)_{\Delta P=0} \cdot \Delta P \tag{15.20}$$

where $\Delta\psi$ is the measured streaming potential and ΔP the pressure difference in the same experiment. Multiplying top and bottom by F, the Faraday, and $\bar{V}_w$, the partial molar volume of water:

$$\Delta\psi = \frac{\bar{V}_w \Delta P}{F}\left(\frac{FJ_{\text{vol}}}{\bar{V}_w I}\right)_{\Delta P\,=\,0} = \frac{\bar{V}_w \Delta P}{F} \cdot N \tag{15.21}$$

The nature of N follows from the fact that $J_{\text{vol}}/\bar{V}_w$ is the number of moles per second of water flowing through the partition. Here we have assumed that $J_{\text{vol}} = J_{H_2O}$, which is a very good approximation for dilute solutions. Also, F/I gives the number of moles of charge flowing per second. Thus N is the number of *molecules* of water crossing the partition per unit of transported charge, in either the electroosmosis or streaming potential experiment. This result, which we now use, is a direct consequence of the Onsager relationship. Rosenberg and Finkelstein[251] incorporated the ion channel gramicidin A into a lipid bilayer separating two solutions, both of which contained 1:1 electrolyte and one of which contained sucrose. Since the bilayer is impermeable to sucrose, an osmotic pressure is created that is the source of ΔP in a streaming potential experiment. For one particular series of experiments using 0.01 and 0.1 M CsCl, KCl, and NaCl, the data were $\Delta P = 24.6$ atmospheres, $\Delta\psi = 3$ mV, $\bar{V}_w = 18 \times 10^{-3}$ liters, giving $N = 6.45$. One simple explanation of this result is based on the picture of a channel wide enough to accommodate a water molecule or an alkali metal cation but allowing only single-file movement. Each cation pushes a row of water molecules in front of it. At higher ionic concentration, 1.0 M, the streaming potential for the same ΔP falls to 2.35 mV, corresponding to $N = 5.1$. The fall in the number of water molecules per transported cation could indicate multiple occupancy of the channel.

The above example illustrates a practical use of irreversible thermodynamics, but the physical system is far simpler than the sophisticated transport systems responsible for active transport in biological membranes. The very complexity of real membranes is an inducement to search for general relationships not dependent on undiscovered details of mechanism. This is the forte of thermodynamics and very general treatments of active transport have been given in terms of irreversible thermodynanics. The results display the drawback of thermodynamics; they tell us nothing direct and little that is indirect about molecular mechanisms. Nevertheless, the analysis of active transport has provided a systematic framework within which to view a large number of experimental data. We will give a simple treatment of a seemingly simple system. We prepare the ground by taking a look at a chemical reaction in solution:

$$A \underset{k_2}{\overset{k_1}{\rightleftharpoons}} B \tag{15.22}$$

where k_1 and k_2 are first-order rate constants. At equilibrium $k_1[A]_{eq} = k_2[B]_{eq}$. Away from equilibrium, the rate of change of the concentration of B is given by

$$\begin{aligned}\frac{d[B]}{dt} &= V = k_1[A] - k_2[B] \\ &= k_1[A]_{eq}\left(\frac{[A]}{[A]_{eq}} - \frac{[B]}{[B]_{eq}}\right)\end{aligned} \tag{15.23}$$

We now assume ideality and write

$$\mu_A = \mu_A^0 + RT\ln[A] \qquad \text{etc.} \tag{15.24}$$

From (15.23) and (15.24):

$$V = k_1[A]_{eq}\left[\exp\left[\frac{\mu_A - \mu_A^0}{RT} - \frac{\mu_{A_{eq}} - \mu_A^0}{RT}\right] - \exp\left[\frac{\mu_B - \mu_B^0}{RT} - \frac{\mu_{B_{eq}} - \mu_B^0}{RT}\right]\right]$$

Therefore:

$$V = k_1[A]_{eq}\left[\exp\left[\frac{\mu_A - \mu_{A_{eq}}}{RT}\right] - \exp\left[\frac{\mu_B - \mu_{B_{eq}}}{RT}\right]\right] \tag{15.25}$$

Near equilibrium $\mu_A - \mu_{A_{eq}}$ and $\mu_B - \mu_{B_{eq}}$ can be much smaller than RT so that the exponentials can be expanded:

$$V = k_1[A]_{\mathrm{eq}}\left[\left(1 + \frac{\mu_A - \mu_{A_{\mathrm{eq}}}}{RT} + \cdots\right) - \left(1 + \frac{\mu_B - \mu_{B_{\mathrm{eq}}}}{RT} + \cdots\right)\right]$$

$$= k_1 \frac{[A]_{\mathrm{eq}}}{RT}(\mu_A - \mu_B) \tag{15.26}$$

since

$$\mu_{A_{\mathrm{eq}}} = \mu_{B_{\mathrm{eq}}}$$

This looks like a flux equation of the form $J = LX$ where the "flow" is the rate of increase of concentration of B, $\mu_A - \mu_B$ is the "force," and $k_1[A]_{\mathrm{eq}}/RT$ is the "straight" coefficient. The requirement that $\mu_B - \mu_{B_{\mathrm{eq}}}$ and $\mu_A - \mu_{A_{\mathrm{eq}}}$ be much smaller than RT implies that $\mu_A - \mu_B \ll RT$. The driving force for the flow is commonly called the affinity A. Since it is usual to define $\Delta\mu = \mu_B - \mu_A$, the affinity is given by $A = -\Delta\mu$. *The condition $A \ll RT$ is the condition for a linear phenomenological equation.* Biochemical processes often involve overall free-energy changes comparable to RT. However, it has been shown[252] that if such a process proceeds by a series of steps each of which obeys the condition $A \ll RT$, the overall reaction can be described by a linear equation.

We have gone through the above example to introduce the idea of affinity and to show a case where the condition for linear phenomenological equations can be clearly stated. For more complex reactions,

$$\nu_A A + \nu_B B + \nu_C C + \cdots \rightarrow \nu_D D + \nu_E E + \nu_F F + \cdots \tag{15.27}$$

the affinity is given by

$$A = -\sum \nu_i \mu_i \tag{15.28}$$

where the ν_i are stochiometric coefficients and are taken as positive for products and negative for reactants.

We now turn to active transport and suppose that the flow of Na^+ ions through a membrane is a consequence of the existence of a gradient in the electrochemical potential and also of "active" transport, which is taken to mean the movement of ions produced by a metabolic process. Formally we have

$$J_{\mathrm{Na}} = L_{\mathrm{Na}}(-\Delta\tilde{\mu}_{\mathrm{Na}}) + L_{\mathrm{Na}r}A \tag{15.29}$$

where A is the affinity of the chemical reaction responsible for active transport. Unfortunately, the nature of this reaction or series of reactions is not known, and at least one risk that we run is that one or more of the reactions does not obey a linear relationship between "flow" and "force." The bilinear form for J_{Na} is complemented by a corresponding form for J_r, the metabolic "flow," or rate:

$$J_r = L_{Nar}(-\Delta\tilde{\mu}_{Na}) + L_r A \qquad (15.30)$$

where we have used Onsager's relation between the coefficients of the nonconjugate forces. Equation (15.30) expresses the fact that the rate of the metabolic process under examination not only is a function of its conjugate "force" A, but is also affected by the transport of Na^+ ions. The off-diagonal coefficient L_{Nar} is one measure of the coupling between transport and metabolism; later we will examine an exact definition of the degree of coupling between flows. If $L_{Nar} = 0$, the flow of Na^+ is purely diffusive and the metabolic process depends only on its affinity.

If we believe in the phenomenological equations (15.29) and (15.30), we can use them to obtain experimental affinities. The transport of Na^+ across frog skin has been studied by following the flow of ions and the rate of oxygen consumption, which serves as a measure of metabolic rate.[253] If the concentration of Na^+ is the same on both sides of the membrane (the frog skin), then the electrochemical potential difference $\Delta\tilde{\mu}_{Na}$ contains only a term in the electric potential drop across the membrane:

$$\Delta\tilde{\mu}_{Na} = F\Delta\psi \qquad (\Delta c_{Na} = 0) \qquad (15.31)$$

For a series of experiments in which $\Delta\psi$ is changed but the chemical composition of the system is kept constant, the affinity remains constant and from (15.30) we find:

$$\left(\frac{dJ_r}{d\Delta\psi}\right) = -FL_{Nar} \qquad \text{(affinity constant)} \qquad (15.32)$$

The left-hand side is the slope of a plot of metabolic rate (say oxygen consumption) against electric potential difference under conditions of fixed affinity. If a series of experiments are carried out with $\Delta\tilde{\mu}_{Na} = 0$, for example if $\Delta\psi = \Delta c_{Na} = 0$, then from (15.29)

$$L_{Nar} = J_{Na}/A \qquad (\Delta\tilde{\mu}_{Na} = 0) \qquad (15.33)$$

From these two equalities it follows that

$$A = -FJ_{Na}/(dJ_r/d\Delta\psi) \qquad \text{(affinity constant)} \qquad (15.34)$$

an equation holding for the restrictive experimental conditions specified. Now FJ_{Na} is equal to the rate of transport of electric charge through the membrane for $\Delta\tilde{\mu}_{Na} = 0$. This flow is called the *short-circuit current*, I_0. Thus

$$A = I_0/(dJ_r/d\Delta\psi) \qquad \text{(affinity constant)} \qquad (15.35)$$

This equation gives an experimental means of measuring the affinity of the reaction driving sodium transport in this biologic system. (Since the affinity is an electrochemical potential difference, its value is a function of the ratio of concentrations of reactant to product in the metabolic reaction supporting transport.) A number of experiments lend weight to this analysis, that is, the results are consistent with the linear phenomenologic equations, together with Onsager's relationships, so that the use of irreversible thermodynamics allows a neat relationship to be written between the experimental variables I_0, J_r, $\Delta\psi$ characterizing the active transport of sodium. We must not ask too much of the theory; it cannot tell us what reaction is responsible for active transport. The use of metabolic inhibitors and the addition of other substances to the system cause changes that can be analyzed by the equations given above to provide hints as to the chemical nature of the sodium pump.

The general bilinear expressions for active transport have been subjected to much mathematical manipulation the main object of which has been to determine the experimental consequences of the relative magnitudes of the straight and cross-coefficients. We present a condensed version of the treatment of Caplan and his collaborators,[254,255] in part to illustrate the meaning of some terms that commonly occur in this field and in part as an introduction to more complex analyses. As a preliminary we derive an important general inequality.

The general equations for a two-flow system are

$$\begin{aligned} J_1 &= L_{11}X_1 + L_{12}X_2 \\ J_2 &= L_{21}X_1 + L_{22}X_2 \end{aligned} \tag{15.36}$$

With a view to the discussion of water transport in the next chapter, we give a specific example of (15.36):

$$\begin{aligned} J_w &= L_{ww}\Delta\mu_w + L_{ws}\Delta\mu_s \\ J_s &= L_{sw}\Delta\mu_w + L_{ss}\Delta\mu_s \end{aligned} \tag{15.37}$$

where the subscripts w and s refer to water and a solute, which, for simplicity, we take to be neutral. The terms $\Delta\mu_w$ and $\Delta\mu_s$ are the differences in chemical potential across the membrane. We take an increase in μ in the x direction to imply a positive $\Delta\mu$ and a flow in the x direction to be positive. In that case, if the equations are to make physical sense, L_{ww} and L_{ss} must be negative or zero. Now the dissipation function (15.14) is always positive since it represents the rate at which free energy is dissipated by the system. Our sign convention then forces us to write

$$\Phi = -J_w\Delta\mu_w - J_s\Delta\mu_s > 0 \tag{15.38}$$

Following Onsager, we write $L_{ws} = L_{sw}$. Substituting from (15.37) into

(15.38), we obtain

$$L_{ss}(\Delta\mu_s)^2 + L_{ww}(\Delta\mu_w)^2 + 2L_{sw}(\Delta\mu_w\Delta\mu_s) < 0 \tag{15.39}$$

Now consider the positive expression:

$$(L_{ss}^{1/2}\Delta\mu_s + L_{ww}^{1/2}\Delta\mu_w)^2 > 0 \tag{15.40}$$

Expanding we find

$$L_{ss}(\Delta\mu_s)^2 + L_{ww}(\Delta\mu_w)^2 + 2L_{ss}^{1/2}L_{ww}^{1/2}\Delta\mu_s\Delta\mu_w > 0 \tag{15.41}$$

Comparison of (15.39) with (15.41) shows that

$$L_{ss}L_{ww} \geqslant L_{sw}^2 \tag{15.42}$$

which is a specific example of a useful general relationship. We now return to active transport and take as our point of departure equations (15.29) and (15.30). We will attempt as far as possible to avoid giving physical interpretations to the terms appearing in the expressions for the flows, but rather will concentrate on the mathematical consequences of the equations. We define the following quantities:

$$Z = (L_r/L_{\mathrm{Na}})^{1/2} \tag{15.43}$$

$$x = \Delta\tilde{\mu}_{\mathrm{Na}}/A \tag{15.44}$$

$$j = J_{\mathrm{Na}}/J_r \tag{15.45}$$

$$q = L_{\mathrm{Na}r}/(L_{\mathrm{Na}}L_r)^{1/2} \tag{15.46}$$

where q is termed the *degree of coupling*. From (15.42) and (15.46), it follows that $-1 \leqslant q \leqslant 1$. Direct substitution verifies the useful relationship:

$$j/Z = [L_{\mathrm{Na}r}/(L_{\mathrm{Na}}L_r)^{1/2} + Zx]/[1 + L_{\mathrm{Na}r}Zx/(L_{\mathrm{Na}r}L_r)^{1/2}]$$

or

$$J_{\mathrm{Na}}/J_rZ = [q + Z\tilde{\mu}_{\mathrm{Na}}/A]/[1 + qZ\tilde{\mu}_{\mathrm{Na}}/A] \tag{15.47}$$

which is of the form $Y = (q + X)/(1 + qX)$ where $X = Z\tilde{\mu}_{\mathrm{Na}}/A$ and $Y = J_{\mathrm{Na}}/J_rZ$. We consider the nature of the system for various values of q.

A major distinction can be made between systems having negative and positive values of q. If q is positive, then $L_{\mathrm{Na}r}$ is positive and the flow of Na^+ ions is aided by the coupled chemical reaction. Another possible example of this effect is the apparent entrainment of a solute molecule in the osmotic flow of water through a membrane. On the other hand, as we noted in Chapter 12, the diffusion of given molecules in one direction may result, by an exchange-type process, in their places being taken by a second species. In this case the flow of one component is opposed to that of the other and the cross-coefficient L_{12} is negative, as is q. The most uninteresting situation is when $q = 0$, as then the cross-coefficient vanishes and the flows are completely uncoupled and independent of each other, presumably because the two flows occur in physically or chemically distinct spaces. The uncoupling of linked processes within the cell often can be accomplished by the addition of specific inhibitors. The two special cases, $q = \pm 1$, are important. From (15.47) we find

$$J_{\mathrm{Na}}/J_r = \pm Z \qquad (q = \pm 1) \tag{15.48}$$

Thus in this case the flows bear a constant ratio to each other independant of the forces. The flows are completely coupled and Z gives the stoichiometry of the system. In the example of active transport, this means that if we increase the concentration gradient of Na ions across a membrane, we may affect the flow of ions through the membrane but the rate of the reaction responsible for the active pumping of sodium will change in such a way as to leave J_{Na}/J_r unchanged. In such systems we see from (15.14) and (15.48) that the dissipation function can be written as

$$\Phi = J_2(X_2 \pm ZX_1) \tag{15.49}$$

which in the case of our simple example of active transport becomes

$$\Phi = J_{\mathrm{Na}}(-\Delta\tilde{\mu}_{\mathrm{Na}} \pm ZA) \tag{15.50}$$

It is clear that by adjusting the forces X_1 and X_2 we can make Φ vanish. Physically this means that no free energy is dissipated by the working of the system, a condition that applies only for a reversible reaction proceeding infinitely slowly. This limiting case can occur only for a completely coupled system in which the possibility of complete conversion of one form of energy into another can be envisaged at least. For general values of q it is instructive to plot Y $(= J_{\mathrm{Na}}/J_r Z)$ against X $(= Z\Delta\tilde{\mu}_{\mathrm{Na}}/A)$ and curves for some values of q are shown in Figure 109. The intercepts of the curves with the X-axis represent systems in which $J_{\mathrm{Na}} = 0$. Now if $q \neq 0$, then at the intercept $\Delta\tilde{\mu}_{\mathrm{Na}} \neq 0$ and we have a system in which the conjugate force to J_{Na} produces no flow. A mechanical example of such a system could be a pump maintaining a constant head of water. The initial action of the pump will result in a rise in the level of the pumped water, but eventually the weight of the water will produce a pressure, that exactly balances the driving force of the pump. A very important biological example of a *static head* is the maintenance of a time-independent gradient of ions across a membrane by the operation of an ion pump. In this case the force opposing the pump could be the electrochemical gradient, which results in back-diffusion. This "leakage" of ions may be via a passive pathway such as a lipid bilayer area of a membrane, or *via the pump.* In the static head situation no work is done on the external world, as can be seen from the water-pump example. The energy put into the system by the pump is spent in keeping the water level at a constant height.

Returning to Figure 109, we consider the intercepts of the Y versus X plots on the Y-axis. These points represent systems for which $J_{Na} \neq 0$ but $\Delta\tilde{\mu}_{Na} = 0$. Since there is a flow of ions in the absence of an electrochemical gradient, the driving force must be the nonconjugate force, namely, the chemical pump. Coupled systems in which the flow of a component occurs in the absence of a conjugate potential gradient are said to exhibit *level flow*. A mechanical example is the pumping of water along an infinite horizontal canal. An important biologic example is the

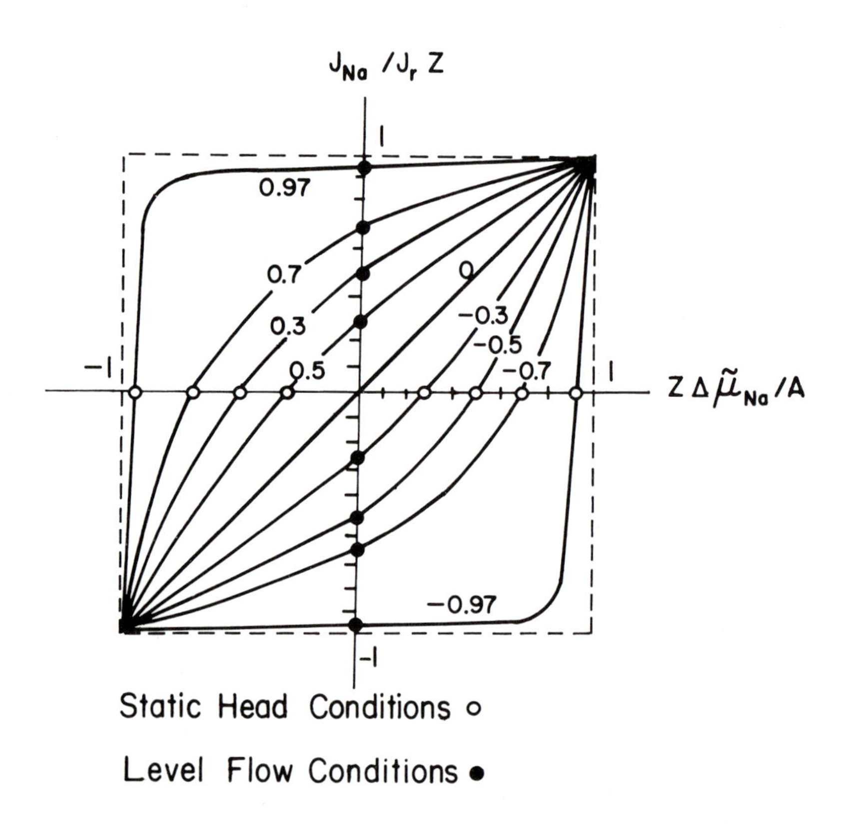

Figure 109. Plots of J_{Na}/J_rZ against $Z\Delta\tilde{\mu}_{Na}/A$ for a two-flow/two-force active transport system. See text for explanation.

active transport of Na^+ in renal proximal tubules for which the ionic concentrations are very similar on both sides of the membrane. The use of level flow conditions in the determination of active transport was exploited by Ussing,[256] who compensated for the membrane potential by using an external current source. When the solutions on both sides of the membrane are identical, the current that must be passed across the membrane in order to result in $\Delta\phi = 0$ is a direct measure of the rate of active transport of ions across the membrane, as we shall see in Chapter 18.

As in the case of static head conditions, systems showing level flow do no work. In one case no net transport occurs and in the other the flow is against a zero potential energy gradient. In general, for a two-flow system there will be a flow of either mass or charge, or both, and it usually will be a flow that is opposed by a force resulting in the performance of work. The efficiency of such a system can be expressed as

$$\text{Efficiency} = -\frac{\text{Power output}}{\text{Power input}} \tag{15.51}$$

which, for a two-flow system, is given by $-J_1X_1/J_2X_2$. Here we are assuming that the flow J_2 takes place in the direction that nature decides to be "spontaneous." For example, process 2 could be a chemical reaction driving ions across a membrane. If the spontaneous reaction is causing work to be done, it must be causing a flow against an opposing force. This means that J_1 has an opposite sign from the conjugate force X_1, and for the efficiency to be positive, we need the negative sign in (15.51). This is also consistent with the usual thermodynamic convention in which work done on the environment is negative. From the generalized form of (15.47) it follows that

$$\text{Efficiency} = \frac{-J_1X_1}{J_2X_2} = -[q + Z(X_1/X_2)]/[q + 1/Z(X_1/X_2)] \tag{15.52}$$

The maximum efficiency can readily be shown[254] to be given by

$$\text{Maximum efficiency} = q^2/[1 + (1 - q^2)^{1/2}]^2 \tag{15.53}$$

The value of the above expressions is that they provide a framework for the analysis of experimental results. Since one of the principal objectives of studies on membrane transport is to understand the nature of the coupling between the metabolic activity of the membrane and the associated transport of ions and neutral molecules, theoretical expressions for the degree of coupling and the efficiency are clearly desirable.

The theory and illustrations given in this chapter were chosen to demonstrate the way in which irreversible thermodynamics has found a use in biophysics. A further important example appears in the following chapter. Authoritative summaries will be found in the references.[257-260]

A closing note: The theory deals with systems removed from equilibrium but it is unusual to apply it to systems that are not in a steady state. Thus in our examples all the flows are assumed to be constant. For a system in which the incoming and outgoing fluxes of mass or charge are not equal the dissipation function (15.14) will clearly have to include more terms.

16 Water Transport

The passage of water through artificial or biological membranes is, considering its apparent simplicity, a process that has received and is receiving a great deal of theoretical attention and yet remains poorly understood. A basic example that is of great historic importance is that of a membrane permeable to water but impermeable to a solute. The facts are familiar: Water passes spontaneously across the membrane from the less to the more concentrated solution. This flow was, in the 19th century, attributed to a so-called *osmotic pressure* that was taken to be equal to the magnitude of the external pressure, which, when applied to the more concentrated solution, just prevents the flow of water. A quantitative theoretical relationship between the *equilibrium* osmotic pressure and the composition of the solutions on either side of the membrane is easily derived. The chemical potential of water is written:

$$\mu_w = \mu_w^0 + \bar{V}_w P + RT \ln x_w \tag{16.1}$$

where the symbols have the meanings they had in (11.5) and we have assumed ideality. At equilibrium the chemical potential is the same on both sides of the membrane:

$$\mu_w^0 + \bar{V}_w P(1) + RT \ln x_w(1) = \mu_w^0 + \bar{V}_w P(2) + RT \ln x_w(2) \tag{16.2}$$

where 1 and 2 label the two solutions. Rearranging we find:

$$P(2) - P(1) = \Delta\pi = \frac{RT}{\bar{V}_w} \ln(x_w(1)/x_w(2)) \tag{16.3}$$

In dilute solution well-worn approximations, accessible in most undergraduate textbooks of physical chemistry, yield the equation originally due to van't Hoff:

$$\Delta\pi = RT(c_s(2) - c_s(1)) = RT\Delta c_s \tag{16.4}$$

where c_s is the concentration of solute.

The derivation given above leads to a description of an equilibrium state, but, of course, says absolutely nothing about the rate at which that state is approached from a nonequilibrium situation. This problem, which is the one that will concern us, has been tackled theoretically in two ways:

1. By the method of irreversible thermodynamics, which, by its nature, is limited to predicting the *relationships* between the linear coefficients in the phenomenologic equations. As we shall see, the theory has often been sullied by the attempted grafting of mechanical or molecular concepts onto the thermodynamic equations — the mighty may not have fallen in these endeavors, but they have certainly staggered.
2. By kinetic or hydrodynamic theories often involving the concept of friction and occasionally based on geometric models of the pore.

The point of departure for the thermodynamic treatment of water transport is the dissipation function:

$$\phi = -J_w\Delta\mu_w - J_s\Delta\mu_s \tag{15.38}$$

The inclusion of a term in J_s is a recognition of the fact that many membranes are permeable to a greater or lesser extent to a variety of molecules. The chemical potential of the solute is given by

$$\mu_s = \mu_s^0 + RT \ln c_s + P\bar{V}_s \tag{16.5}$$

Now μ_s and μ_w are not independant but are related by the Gibbs–Duhem equation for a two-component system:

$$d\mu_s = -(c_w/c_s)d\mu_w \tag{16.6}$$

which, together with (16.5), leads to

$$\mu_w = \mu_w^0 + P\bar{V}_w - RTc_s/c_w \tag{16.7}$$

We have used the ideal solution forms for the chemical potentials, which can be justified if we are dealing with dilute solutions. If the solute concentrations are small, we can also take c_w to be the same on both sides of the membrane and then (16.4) and (16.7) give

$$\Delta\mu_w = \bar{V}_w\Delta P - RT\Delta c_s/c_w = \bar{V}_w\Delta P - \Delta\pi/c_w \tag{16.8}$$

To forestall inconvenience as a result of the logarithmic term in (16.5), we use the quantity $\bar{c}_s$ defined by Kedem and Katchalsky[261] as

$$\bar{c}_s = c_s/\ln[c_s(2)/c_s(1)] \tag{16.9}$$

which, for smallish Δc_s, allows us to write

$$\Delta\mu_s = \bar{V}_s\Delta P + RT\Delta c_s/\bar{c}_s = \bar{V}_s\Delta P + \Delta\pi/\bar{c}_s \tag{16.10}$$

Substituting from (16.8) and (16.10) into (15.38):

$$\begin{aligned}\Phi &= -(J_w\bar{V}_w + J_s\bar{V}_s)\Delta P - (J_s/\bar{c}_s - J_w/c_w)\Delta\pi \\ &= -J_v\Delta P - J_D\Delta\pi\end{aligned} \tag{16.11}$$

This form for the dissipation function defines new forces (ΔP and $\Delta\pi$) and new flows. J_v is usually termed the *volume flow* but can be seen to be the sum of the *velocities* of solvent and water just as J_D, usually termed the *exchange flow*, is in fact the difference in the average *velocities* of water and solute. We can now set up the phenomenological equations:

$$J_v = L_p\Delta P + L_{pD}\Delta\pi \tag{16.12a}$$

$$J_D = L_{Dp}\Delta P + L_D\Delta\pi \tag{16.12b}$$

The approximations involved in the assumptions of ideality and linearity are acceptable in dilute solution and within these limits there seems no reason to doubt the validity of the phenomenological equations. In contrast to the form (15.37), ΔP and $\Delta\pi$ can be varied independently of each other while $\Delta\mu_w$ and $\Delta\mu_s$ cannot, a fact that allows us to attach meanings to the linear coefficients. Thus L_{pD} gives the volume flow for a unit osmotic gradient when the hydrostatic pressure gradient is zero. L_{Dp} gives a measure of the difference in velocity of solute and water when flow is produced solely by hydrostatic pressure. If Onsager's relationships are applied, then we must put $L_{Dp} = L_{pD}$. The reader is warned that this equality is not accepted universally. We will soon look at this controversial question, which is discussed at length in a review by Hill[262] that is recommended to all who wish to understand more about osmosis than we can enter into here. The mathematical manipulations that the phenomenological equations have had to suffer are long drawn out and ghastly in their complexity, and many of the results are of very doubtful value. We will give some indication of the directions that have been taken and also comment on some of the inconsistencies in the literature.

With respect to a solution of a given solute, membranes can be classified according to their degree of permeability to the solute. Thus a certain membrane may be permeable only to water, in which case we can label it a *semipermeable membrane.* At the other extreme, a piece of wire gauze would certainly not hinder the flow of solute with respect to that of water and a membrane displaying similar properties would be termed *nonselective.* Such a membrane would not be expected to display the phenomenon of *ultrafiltration*, which means that the application of pressure to a solution would result in its passing unchanged in composition through the membrane. It has been stated[263] — and the statement has been perpetuated — that for a nonselective membrane neither osmotic flow nor ultrafiltration takes place, and thus in this case $L_{pD} = 0$. This is untrue. If a solution is placed on one side of a wire gauze and pure water on the other, diffusion of both solute and solvent will occur across the gauze but the rates of diffusion will differ and the partial molar volumes of the two components usually will also differ. It is intuitively clear that there will be a volume flow proportional to $(D_s\bar{V}_s - D_w\bar{V}_w)$ even in the absence of an external pressure gradient, that is, $J_v \neq 0$. But if it is true that in this case $L_{pD} = 0$, then, because ΔP has been taken as zero, it follows from (16.12a) that $J_v = 0$. The inconsistency can be formally resolved by accepting that $L_{pD} \neq 0$ for nonselective membranes. But now another problem arises. For a gauze separating two identical solutions, $\Delta\pi$ is zero and the application of a mechanical pressure gradient ΔP will force unchanged solution through the gauze. No ultrafiltration occurs, that is $J_D = 0$. But if J_D and $\Delta\pi$ are both zero, then for $\Delta P \neq 0$ (16.12b) can only be satisfied for $L_{Dp} = 0$. Thus we are forced to the conclusion that for this system $L_{Dp} \neq L_{pD}$: Onsager reciprocity does not apply. It could be argued that we have compared differing systems and that the linear coefficients are concentration dependant. This could save the theory but it is extremely discouraging for those hoping to determine the coefficients experimentally.

Many membranes are neither semipermeable nor nonselective, thus permitting the passage of both solvent and solute and exhibiting ultrafiltration. In the absence of osmotic pressure, the effect of mechanical pressure will be to cause both volume and exchange flow and the ratio of L_{Dp} to L_p is equal to the ratio of the flows:

$$\frac{J_D}{J_v} = \frac{L_{Dp}}{L_p} = \frac{J_s/\bar{c}_s - J_w/c_w}{J_v} = \frac{J_s(1 + \bar{c}_s\bar{V}_s)}{J_v\bar{c}_s} - 1 \qquad (16.13)$$

For dilute solutions, $c_s\bar{V}_s \ll 1$ so that

$$-\frac{L_{Dp}}{L_p} = 1 - J_s/(J_v\bar{c}_s) = \sigma_f \qquad (16.14)$$

This can be rearranged to give

$$J_s = (1 - \sigma_f)\bar{c}_s J_v \qquad (16.15)$$

which can be regarded as an expression for the flow of solute produced by pressure-induced solvent flow. Thus, without committing ourselves to specifying a mechanism for this phenomenon, we can term σ_f a *drag reflection coefficient*. Another way of looking at σ_f emerges from a consideration of its value for semipermeable and nonselective membranes. For a semipermeable membrane $J_s = 0$, from which it follows that $\sigma_f = 1$. For a nonselective membrane (when $\Delta\pi = 0$) $L_{Dp} = 0$ and thus $\sigma_f = 0$. This leads to the dubbing of σ_f as a reflection coefficient since the solute is "reflected" completely when the membrane is semipermeable, corresponding to σ_f being unity, and is not "reflected" at all when the membrane is nonselective, corresponding to the vanishing of σ_f. In fact, back in 1951 Staverman[264] defined a reflection coefficient that was intended to measure the ability of the membrane to discriminate between solute and water:

$$\sigma_s = 1 - (c_s^{\text{filtrate}}/c_s^{\text{filtrand}}) \tag{16.16}$$

It can be seen that σ_s has the same values for semipermeable and nonselective membranes as has σ_f. Staverman showed that for a membrane that was permeable to solute the measured osmotic pressure is given by

$$\Delta P = \sigma_s \Delta\pi = \sigma_s RT \Delta c_s \tag{16.17}$$

On the other hand, turning to our theoretical expressions, we find that the external pressure needed to balance the osmotic pressure in this case is, from (16.12a) putting $J_v = 0$,

$$\Delta P = -\frac{L_{pD}}{L_p}\Delta\pi \tag{16.18}$$

Comparison of (16.17) with (16.18) shows that

$$\sigma_s = -\frac{L_{pD}}{L_p} \tag{16.19}$$

Staverman assumed that $L_{pD} = L_{Dp}$, in which case his reflection coefficient given by (16.19) is identical with the *drag reflection* coefficient given by (16.14). We prefer to leave our options open and reserve the term *reflection* coefficient for L_{pD}/L_p, which, as we have seen, is basically a ratio of conductivities. It has yet to be shown experimentally that L_{pD} is equal to or different from L_{Dp} in a specific system.

An experimental factor relevant to our theoretical juggling is the fact, pointed out by Kedem and Katchalsky,[265] that in most experiments it is easier to measure the flow of solute and volume, J_s and J_v, rather than J_D and J_w. They derived an equation for J_s, which can be obtained most simply by taking the flow of solute to be compounded of the solvent drag term (16.15), and a diffusion term, which can be written $\omega_s \Delta\pi$. This form implies that *solute* diffusion is a linear function of osmotic pressure. The resulting expression and (16.12a) are sometimes referred to as equations for practical calculation:

$$J_v = L_p \Delta P + \sigma_s L_p \Delta\pi \tag{16.20a}$$

$$J_s = (1 - \sigma_f)\bar{c}_s J_v = \omega_s \Delta\pi \tag{16.20b}$$

where (16.19) has been used to substitute for L_{pD}. In the original derivation it was assumed that $L_{pD} = L_{Dp}$ and thus that $\sigma_s = \sigma_f$, but we again leave room for doubt.

The object of theory is to rationalize experiment and so we pause to consider the experimental relevance of the flow equations. First it must be emphasized that the linear coefficients cannot be taken to be constants, even for a given membrane. In Chapter 13 we saw that for diffusion through channels there are a variety of situations in which we can expect a nonlinear dependance of rates on the concentrations of the transported species. It is not difficult to believe, for example, that the rate at which water is driven through a channel by mechanical pressure can depend on the presence and concentration of solute molecules in the same channel. Whatever the molecular details, we cannot a priori take the linear coefficients to be independent of the composition of the system, and so the value determined for L_{pD}, say, only applies for a particular membrane under particular conditions. If, in addition, as there is good reason to suppose, we cannot put L_{Dp} equal to L_{pD} but have to determine its value independently, we might begin to question the utility of irreversible thermodynamics to water transport. After all, with a little physical insight the phenomenological equations could have been written down without the benefit of the dissipation function, and if Onsager reciprocity does not hold, then thermodynamics seems almost a side issue. Moreover, even accepting the validity of the linear equations, thermodynamics can never provide direct evidence of molecular mechanisms and the realization of this has prompted attempts to force mechanistic significance onto the parameters appearing in the linear equations. Others abandoned thermodynamics completely and attempted to explain water transport in terms of purely mechanical and hydrodynamic concepts. We now have a look at some extra-thermodynamic approaches, concentrating particularly on the concept of friction. Our base for the expedition will be equations (16.20), which we need not regard as taking their validity from irreversible thermodynamics. Equation (16.20a) merely states that volume flow is dependent on both mechanical and osmotic pressure while (16.20b) states that solute flow depends partly on volume (bulk) flow and partly on osmotic pressure. One might have guessed as much. The assumption of linearity is common. As for the coefficients, they certainly cannot be determined by thermodynamics. The derivation of σ_s and σ_f suggests that they are equal if Onsager reciprocity holds between L_{pD} and L_{Dp}, and this is unlikely. If theory is to advance, we need to understand and find ways of estimating the coefficients in (16.20).

In steady-state mechanical motion, the driving force is exactly balanced by the opposing frictional force, which is usually assumed to be proportional to the velocity. Equating these forces for the case of water and solute transport along a given axis, we find

$$-\frac{d\mu_w}{dx}=f_{wm}v_w+f_{ws}(v_w-v_s) \tag{16.21a}$$

$$-\frac{d\mu_s}{dx}=f_{sm}v_s+f_{sw}(v_s-v_w) \tag{16.21b}$$

where the f's are partial molar frictional coefficients and the v's are velocities. The terms have the following origin.

$f_{wm}v_w$ is the force due to the friction between a mole of water and the membrane. The corresponding term for the solute is $f_{sm}v_s$.

$f_{ws}(v_w-v_s)$ gives the force acting on a mole of water due to its motion relative to the solute. The corresponding term for the solute is $f_{sw}(v_s-v_w)$.

The number of coefficients to be determined can be reduced by using the relationship

$$f_{sw}/c_w=f_{ws}/c_s \tag{16.22}$$

which was first derived by Spiegler.[266] We now show that the simple equations (16.21), which have been used as a basis for the treatment of water transport,[267] are too simple. Consider a system consisting of a semipermeable membrane and pure water. A mechanical pressure gradient will cause a volume flow $J_v=J_w\bar{V}_w$. In the absence of solute, (16.21a) gives

$$-\frac{d\mu_w}{dx}=f_{wm}v_w=f_{wm}J_w\bar{V}_w/c_w\bar{V}_w=f_{wm}J_v/c_w\bar{V}_w \tag{16.23}$$

From the expression $\mu_w=\mu_w^0+RT\ln c_w+P\bar{V}_w$ we find, for c_w constant,

$$d\mu_w=\bar{V}_w dP \tag{16.24}$$

From (16.23) and (16.24):

$$\bar{V}_w dP=(f_{wm}J_v/c_w\bar{V}_w)dx \tag{16.25}$$

Integrating across the membrane and rearranging the result gives

$$J_v = \Delta P c_w V_w^2 / f_{wm} \Delta x \tag{16.26}$$

Comparison with (16.12a) shows that

$$L_p = c_w V_w^2 / f_{wm} \Delta x \tag{16.27}$$

We now change the system to a semipermeable membrane separating two solutions of differing composition but identical mechanical pressure. For dilute solutions, in which $c_w \bar{V}_w \simeq 1$, we find that $d\mu_w = RT\bar{V}_w dc_w$, which, together with (16.23), yields

$$J_w = RTc_w \bar{V}_w \Delta c_w / f_{wm} \Delta x \tag{16.28}$$

This has the form $J_w = P_w \Delta c_w$ used to define membrane permeability and gives the value

$$P_w = RTc_w \bar{V}_w / f_{wm} \Delta x \tag{16.29}$$

for the permeability constant. From (16.27) and (16.29), it follows that

$$L_p / (P_w \bar{V}_w / RT) \equiv \beta = 1 \tag{16.30}$$

where the ratio β is usually termed the bulk-to-diffusive flow ratio. Theory thus predicts that, within the experimental range in which linearity and dilute solution approximations hold, the value of β is always unity. Theory disappoints: Bulk flow caused by a given pressure gradient ΔP is often far larger than the flow caused by the "equivalent" osmotic pressure. We use quotation marks because mechanical pressure and so-called osmotic pressure are two very different beasts. Certainly the destruction of concentration gradients is usually to be attributed predominantly to entropy rather than to conventional force. This is the root of the theoretical failure, a failure that we assured by using the same value for f_{wm} for both bulk and diffusive flow. This is clearly nonsensical for a wide channel through which bulk flow can be far greater than the diffusive flow of water. For narrow pores, on the other hand, bulk flow is effectively impossible, and even flow induced by pressure proceeds by a diffusive mechanism so that β approaches unity. We return to this point in the following chapter. In intermediate cases the value of the frictional "constant" f_{wm} will depend on the geometry of the channel. We have to accept that we need a minimum of two constants to describe the frictional force between water and the membrane. The relative weights of these two constants in the observed constant f_{wm} will depend on the dimensions, and possibly the chemical nature, of the pore. The present state of theory is not such that we can either estimate or experimentally determine these constants. Upon reflecting that for any but semipermeable membranes we also need to know f_{sm} and f_{sw}, one is hardly filled with joy. With faith we can approximate f_{sw} from the observed diffusion constant in dilute solution by using the relationship $f_{sw} = RT/D_{sw}$. However, f_{sm}, the interaction between the solute molecules and the membrane, is not accessible.

Frictional concepts provide but an illusory explanation of water transport. We are left with equations that we cannot use fruitfully because they contain parameters that elude evaluation. In the light of this impasse, there seems little point in elaborating here the many mathematical manipulations of the basic equations, for example, the treatment of transport through heterogeneous membranes.[268] It is a pity that textbooks have tended to present equations (16.12) and (16.20) as if they were end points rather than tentative, and faulty, formulations of an as yet unsolved problem.

Hydrodynamics may in the end prove to be the most effective avenue of approach to the understanding of water transport. The parameters entering mechanically flavored theories usually include the ratios of pore diameter to solute and water molecule size. These ratios are presumed to determine the reflection coefficient. It would not be appropriate for us to enter into a detailed comparison of the theories; none has yet proved satisfactory and we again recommend Hill's review[262] as a starting point for those wishing to tackle the original literature. One remark is perhaps necessary, and that is that some of the theories have ignored diffusional flow altogether, surely a fatal omission.

Despite the poor impression given by irreversible thermodynamics in the present chapter, the edifice is not in danger of tumbling. Nevertheless, the reader's faith may well be shaken by the fact that there exists even one example of a probable violation of the Onsager relationship. The practical moral is that it is advisable to demonstrate reciprocity experimentally, especially when dealing with two processes as disparate as diffusion and bulk flow. Another lesson is to be learned from the attempts to attach physical meaning to the terms in the phenomenologic equations. Thus it is not difficult to show[269] that for a model in which solute and water cross the membrane via *different* channels, the solute flow for $\Delta c_s = 0$ is proportional to ΔP. This would normally be taken to prove that solute is being dragged by the solvent flow, in contradiction to the premises of the model. The example merely reaffirms the old truth that different models can often lead to the same equations. A similar but subtler problem arises in connection with the equations (15.29) and (15.30) for active transport, from which we find that

$$(J_{\mathrm{Na}})_{\Delta\mu=0} = L_{\mathrm{Na}r}A \qquad (J_{\mathrm{Na}})_{A=0} = L_{\mathrm{Na}}(-\Delta\tilde{\mu}_{\mathrm{Na}}) \qquad (16.31)$$

These equalities have direct, unambiguous physical interpretations. Thus the first equation means that in the absence of an electrochemical gradient ion flow is to be entirely attributed to the action of an "ion pump." It is then tempting to suppose that in the presence of an electrochemical gradient the term $L_{Nar}A$ gives the contribution to the net ion flow due to a unidirectional pump dependent only on the affinity A and operating independently of $\Delta\tilde{\mu}$. This conclusion is incorrect. As Essig has stated:[270] "An active transport mechanism of finite free energy permits the movement of Na^+ in either direction depending only on the orientation of the *total* force acting on the particular test species." (Emphasis added.) A "pump," being basically a coupled or sequential series of chemical and physical processes, can permit flow in either direction and the measured rate of pumping in a two-flow system *in the active pathway* is a net rate dependent on both the affinity and $\Delta\tilde{\mu}$. Again the subject is too specialized for this book but we send the concerned reader to a book[271] and an article[272] by Terrell Hill.

The initial enthusiasm that greeted the introduction of irreversible thermodynamics into biophysics has cooled somewhat. The basic theory itself is still in a state of flux (see, for example, Jayne[273] on the subject of minimum entropy production) and the applicability and meaning of the phenomenological equations have, as we have seen, sometimes been questioned. We make no claim to having presented a deep analysis here — the specialists in this field differ. Our purpose is merely to erect a warning notice for those approaching the mine field.

17 Fluctuations, Channels, and Lipids

Theory unifies. Four superficially differing physical phenomena mentioned in previous chapters are NMR relaxation times, van der Waals forces, diffusion, and number density fluctuations. Underlying these four subjects is a common denominator — the microscopic time-dependent fluctuations of molecular properties. Spin relaxation is normally caused by randomly fluctuating magnetic fields sensed by nuclei or electrons. Van der Waals forces are a consequence of fluctuating dipoles. A diffusing molecule is subject to random forces due to collisions with its neighbors. Number density fluctuations are due to random molecular translations.

Random, or *stochastic*, processes are all around us, in the air we breathe, and the analysis of such processes attempts to relate microscopic fluctuations to macroscopic observations so that laboratory observations of samples containing huge numbers of molecules can be used to reveal details of the chaotic microscopic world. The measurement and analysis of fluctuations of physical properties are a day-to-day occupation in many laboratories. In this chapter we will present a simple example of a fluctuating system and carry through an elementary analysis of its behavior. We focus on a problem that is of major importance in understanding the working of membranes, namely, the transport properties of channels, a subject examined from a rather different point of view in Chapter 12. Our final objective is to show how experimental observations on multichannel membranes can lead to information on the behavior of single channels and how the more difficult to obtain observations on single channels are statistically analyzed. Fluctuation analysis, known as noise analysis in certain fields, is basically a branch of probability theory. Its magnificent generality is a weakness reminiscent of other wide-ranging theories, such as thermodynamics, which give no direct mechanistic information, but its power lies in the constraints it imposes on our model making and in its ability literally to turn chaos into order, allowing the characterization of a random process by a handful of parameters. We will progress from simple to more complex ideas.

Figure 110 shows the time dependence of two fluctuating properties (1) the electric current I passing through a single channel in a membrane across which a constant voltage has been applied and (2) the angle θ between the director and a C—H bond in the hydrocarbon chain of a phospholipid in the liquid crystal state. The current jumps discontinuously between *discrete* conductance states, and the angle varies over a *continuum* of values (we cannot at present directly observe the fluctuations of bonds). Two important parameters used to characterise such

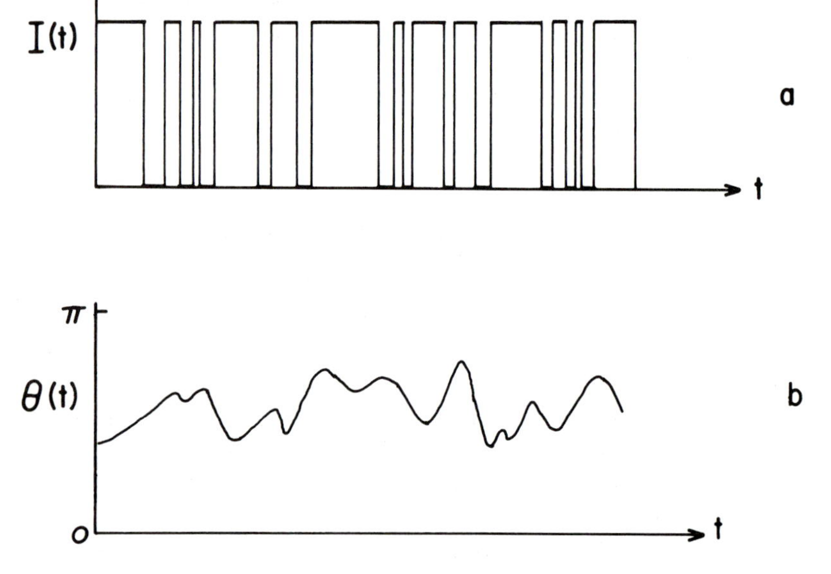

Figure 110. A. Stochastic process involving transitions between two discrete states. **B.** Stochastic process involving a continuum of states.

fluctuations are their amplitude and their frequency. We concentrate first on defining "amplitude." We will need the language of statistics. For the processes in Figure 110 a *mean* value can be defined by the following operation. The time axis is divided into intervals Δt and a value A_i of the variable $A(t)$ is assigned to each interval i, as shown. The mean of the stochastic variable A is then given by

$$\langle A \rangle \equiv \bar{A} = \lim_{n \to \infty} \left[\sum_{i=1}^{n} A_i/n \right] \tag{17.1}$$

As $\Delta t \to 0$, the mean calculated by (17.1) approaches the mean defined by integration, but the summation procedure is often preferable since most data-processing devices are programmed to collect observations at discrete time intervals; in any case, for the processes concerning us, we do not have analytical functions to integrate. Of course, averages other than the mean can be calculated from the basic data. In the case of the angle θ, the time average value of $\langle 3 \cos^2 \theta - 1 \rangle/2$ gives the order parameter S_{33} (Chapter 5). An important average is the *mean-square deviation* or variance, σ^2, obtained by taking each deviation from the mean, squaring it, and averaging the results:

$$\sigma^2 = \lim_{n \to \infty} \frac{1}{n} \sum_{i=1}^{n} (A_i - \bar{A})^2 = \langle A_i^2 \rangle - \bar{A}^2 \tag{17.2}$$

The standard deviation, together with the mean, is often enough to characterize a random variable. A common case is when the deviations obey a Gaussian (normal) distribution, such as is obtained by solving the problem of the one-dimensional random walk for a large number of steps. If the probability for a step to the right is p_R and to the left is p_L, then after n steps the variance is given by

$$\sigma^2 = np_L p_R = np_L(1 - p_L) = np_R(1 - p_R) \tag{17.3}$$

and the distribution curve for the probability that in n steps there have been n_1 positive steps is given by

$$P(n_1;\, n) = (\pi\sigma^2)^{-1/2}\, e^{(n_1 - \bar{n}_1)/2\sigma^2} \tag{17.4}$$

where $\bar{n}_1$ is the average number of positive steps. Thus the distribution is completely determined by $\bar{n}_1$ (which fixes the mean) and σ^2. The mean is given by the mean number of steps to the right minus the mean number of steps to the left $= n(p_R - p_L)$. The expression for the mean and the variance hold for a binomial distribution and therefore for the one-dimensional random walk. The Gaussian curve is the large number limit of the binomial. In what follows the expressions for the mean and variance will appear, but some readers may prefer to obtain the results given below from first principles.

We can already apply the scant definitions made above to a real problem. Consider a channel that can exist in one of two states, which we call "closed" and "open." We label these states 1 and 2 for later convenience. In the closed state the channel is supposed to be nonconducting; in the open state the current is given by $I = \gamma V$ where V is an externally applied voltage and γ a conductance. If the probability of the channel being closed is P_1 and of being open is P_2, then for this one channel the mean conductance is given by

$$\bar{\gamma} = P_1 \cdot 0 + P_2 \cdot \gamma = P_2\gamma \tag{17.5}$$

The mean here is taken over a time that is very large compared with the average lifetimes in each state. The reader should realize that the definition of $\bar{\gamma}$ given above is a disguised version of (17.1). For two channels the possible states are (1, 1), (1, 2), (2, 1), and (2, 2), which have the conductance 0, γ, γ, and 2γ respectively. Note that the distinct *conductance states* are given by the binomial distribution for $n = 2$ (i.e., 1 : 2 : 1) and the probabilities of these states are familiar binomial terms, P_1^2, P_1P_2, P_2^2. Using $P_1 + P_2 = 1$, the mean is found to be $2P_2\gamma$. Either by working through a few more cases or by going to the formal treatment of a binomial distribution, the reader will come to the conclusion that the mean for n channels is

$$\bar{\gamma} = nP_2\gamma \tag{17.6}$$

The variance of $\bar{\gamma}$ is given by (17.3), which for the one-channel case gives $\sigma_\gamma^2 = \gamma_\gamma^2 P_2(1 - P_2)$ and for the n-channel case

$$\sigma_\gamma^2 = n\gamma^2 P_2(1 - P_2) \tag{17.7}$$

Experimentally both $\bar{\gamma}$ and σ_n^2 can be determined from traces of the kind shown in Figure 110A. The ratio of the two quantities is given by

$$\sigma_\gamma^2/\bar{\gamma} = \gamma(1 - P_2) \tag{17.8}$$

There are cases where the probability of a channel being open can be controlled. For example, one model for the channel operation in the so-called end-plate membrane is the two-state model presented here. It is known that for low concentrations of acetylcholine the conductance is very small, that is, the probability P_2 of the channel being open is much smaller than unity. In this case we find from (17.8)

$$\frac{\sigma_\gamma^2}{\bar{\gamma}} = \gamma \tag{17.9}$$

This result means that we can determine the conductance of a *single* channel from the mean and the variance of the fluctuating conductance of a membrane containing many channels, the number of which we do not have to know. The uses of statistics begin to become apparent.

The preceding analysis tells us nothing about the time scale on which channels open and close. At no point did time enter into the data or the model. The probabilities P_1 and P_2 are static probabilities, not transition probabilities.

For time-dependent stochastic variables, the probability distribution of the deviations from the mean is clearly not enough to characterize the fluctuations. The plots in Figures 111A and 111B have the same probability distribution for deviations from the average but the fluctuations have different temporal distributions. The data in Figure 110B allow a calculation of S_{33} but the order parameter does not give us the *rate* of reorientation and the meaning of the term "rate" is not obvious for a random variable. To bring time into the analysis of fluctuations, we use for illustrative purposes a specific but very widely occurring class of stochastic processes, namely, *Markovian processes.* The basic distinguishing mark of a Markovian process is exemplified by the fact that the result of tossing an unbiased coin is completely independent of its past history. For Markovian processes we cannot predict the result of

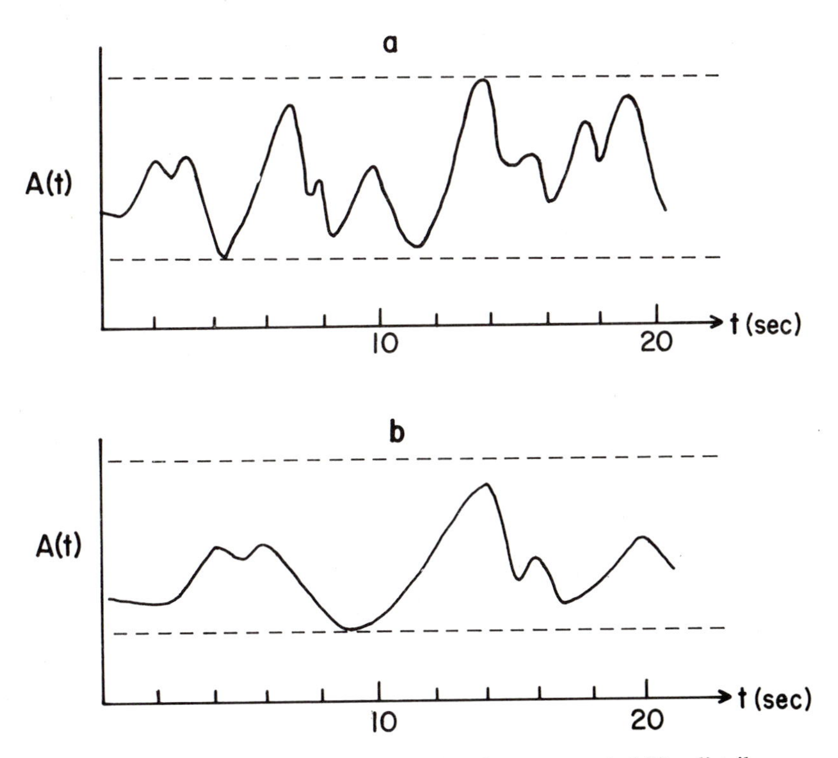

Figure 111. Two random processes having the same probability distribution for deviations from the mean but different time scales.

a "trial" by a knowledge of the history of the system. Only *now* matters, not *then*. For the skeptical: If a coin has been tossed 19 times and has given heads ten times, the next toss of the coin will give either 10 : 10 or 11 : 9 as the overall balance of heads to tails. Thus the 20th toss of a coin gives an overall result that falls within narrow limits set by the entire previous history of the experiment. However, this does *not* mean that coin tossing is a non-Markovian process, it is the result of an *individual* event, in this case the 20th throw, that is required to be independent of the past. A large number of naturally occurring stochastic processes are almost certainly Markovian, and many are presumed to be. In our two examples, the change in the angle θ at a given time is presumed to be independent of orientation at that time. The change of state of a membrane channel from "closed" to "open" was similarly supposed to be completely divorced from its past history. (Note that we sneaked in a model here. There is no reason to suppose that all channels have only two states, or even if they have, that one of the states is nonconducting.) We now follow up the consequences of defining a process as Markovian.

To determine completely the temporal behavior of a Markovian system, we need to specify the measurable properties of every state and the time-dependent transition probabilities, $\pi_{i,j}(t)$, giving the probability that if the system is in state i at a certain time, it will be in the state j after a time t. In our example of a two-state channel, i and j can be 1 or 2. [$\pi_{i,j}(t)$ is a *conditional probability* in the language of probability theory.] Notice that $\pi_{1,1}(t)$ gives the probability that if the channel is in state 1 at a certain time, it will be in the *same* state at a time t later. For our system there are four transition probabilities, and theory proclaims them to obey the following equations:

$$\frac{d\pi_{i,j}(t)}{dt} = \sum_{k=1}^{2} \pi_{i,k}(t) a_{ik} \tag{17.10}$$

where the a_{ik} are given in our case by

$$a_{12} = -a_{11} = \alpha \quad \text{and} \quad a_{21} = -a_{22} = \beta \tag{17.11}$$

α and β are *rate constants* in that the transition probability for the system jumping from say, 1 to 2, in an infinitesimal time Δt is given by $\alpha \Delta t$. [Equation (17.10), the *Kolmogorov equation* for the transition probabilities, is well known in the theory of stochastic processes. An application perhaps familiar to readers concerned with magnetic resonance theory is Anderson's treatment of spin relaxation.[274]] Obviously the larger the α and β, the faster the system fluctuates. To solve (17.10) we need the initial conditions

$$\pi_{i,j}(0) = \delta_{i,j} \tag{17.12}$$

where $\delta_{i,j}$ is the Kronecker delta:

$$\delta_{i,i} = 1; \; \delta_{i,j} = 0 \qquad \text{if } i \neq j \tag{17.13}$$

Equation (17.12) makes the common-sense statement that if a system is in a given state, it will certainly be in the same state no time later. We write (17.10) in full for the two-state case:

$$\begin{aligned}
\frac{d\pi_{11}}{dt} &= \pi_{11}a_{11} + \pi_{12}a_{21} = -\alpha\pi_{11} + \beta\pi_{12} \\
\frac{d\pi_{12}}{dt} &= \pi_{11}a_{12} + \pi_{12}a_{22} = \alpha\pi_{11} - \beta\pi_{12} \\
\frac{d\pi_{21}}{dt} &= \pi_{21}a_{11} + \pi_{22}a_{21} = -\alpha\pi_{21} + \beta\pi_{22} \\
\frac{d\pi_{22}}{dt} &= \pi_{21}a_{12} + \pi_{22}a_{22} = \alpha\pi_{21} - \beta\pi_{22}
\end{aligned} \tag{17.14}$$

The solution, taking into account the initial conditions, is

$$\pi_{i,j}(t) = P_j + (\delta_{i,j} - P_j)e^{-t/\tau_c} \tag{17.15}$$

where

$$P_1 = \frac{\beta}{\alpha + \beta}, \qquad P_2 = \frac{\alpha}{\alpha + \beta}, \qquad \tau_c = \frac{1}{\alpha + \beta}$$

τ_c will be called the *correlation time*, or decay constant of the fluctuations. P_1, P_2 are just the *stationary equilibrium* probabilities of finding the channels in state 1 or 2. The chemically educated reader may care to think of another two-state system, the enol–keto equilibrium.

If we write

$$\text{keto} \underset{\beta}{\overset{\alpha}{\rightleftharpoons}} \text{enol}$$

then $d[\text{keto}]/(dt) = -\alpha[\text{keto}] + \beta[\text{enol}]$ and at equilibrium $\alpha/\beta = [\text{enol}]/[\text{keto}]$ so that the probability of finding a molecule in the enol form is $[\text{enol}]/([\text{enol}] + [\text{keto}]) = \alpha/(\alpha + \beta)$.

We are now ready to introduce the *time autocorrelation function*, a pivotal concept in much of modern spectroscopy and statistical mechanics. The autocorrelation function $\Phi_I(\tau)$ for the current is defined by:

$$\Phi_I(\tau) = \langle I(t + \tau)I(t) \rangle - \bar{I}^2 \tag{17.16}$$

The bracketed term requires us to measure the current at a time t and multiply the result $I(t)$ by the value of the current at a later time $t + \tau$.

We repeat this for a large number of different initial times, always maintaining the same time interval τ, and take the average of the products. For stochastic processes this average always converges as we increase the number of products. The limiting value is what is meant by $\langle I(t+\tau)I(t)\rangle$. Perhaps a study of Figure 112 will clarify the procedure. The value of $\Phi_i(\tau)$ versus τ for the two processes in Figure 111 are shown in Figure 113.

We will now derive an explicit form for the time correlation of the current through a single channel, which will form a link between theory and experiment. We again work with a simple "open-or-closed" channel.

Since $I = \gamma\nu$ or 0, we can write

$$\Phi_i(\tau) = (\gamma\nu)^2[\langle Y(\tau)Y(0)\rangle - \langle Y\rangle^2] \tag{17.17}$$

where Y is a stochastic variable equal to 0 in state 1 and unity in state 2:

$$y_1 = 0, \qquad y_2 = 1 \tag{17.18}$$

To calculate the average value of $Y(\tau)Y(0)$, we only have to take all possible products of the form $y_i(\tau)y_j(0)$ multiplied by the probability of the product. [This procedure is just a variation of (17.1) for a finite number of observations.] The probability of a term $y_i(\tau)y_j(0)$ is the probability of the channel being in the state j at time zero multiplied by the probability that if the channel is in this state, it will be found in the state i at time τ. In symbols we then have

$$\begin{aligned}\langle Y(\tau)Y(0)\rangle = {} & y_1(\tau)y_1(0)\pi_{11}(\tau)P_1(0) + y_2(\tau)y_1(0)\pi_{12}(\tau)P_1(0) \\ & + y_1(\tau)y_2(0)\pi_{21}(\tau)P_2(0) + y_2(\tau)y_2(0)\pi_{22}(\tau)P_2(0)\end{aligned} \tag{17.19}$$

Since the probabilities $P_1(0)$, $P_2(0)$ are time-independent, we revert to the nomenclature P_1, P_2. Substituting (17.15) and making use of the relationship $P_1 + P_2 = 1$, it takes three or four lines of trivial algebra to obtain the needed result:

$$\Phi_I(\tau) = \langle(\gamma VY)^2\rangle \varrho e^{-\tau/\tau_c} - \bar{I}^2. \tag{17.20}$$

where $\langle(\gamma VY)^2\rangle = \langle I^2\rangle$ and $\varrho = \bar{Y}(1-\bar{Y})/\bar{Y}^2$.

For the particularly simple case of a channel that has an equal probability of being open or shut, $\alpha = \beta$, which gives $\bar{Y} = P_2 = 1/2$, and therefore $\varrho = 1$ and $\Phi_I(\tau) = \langle I^2\rangle e^{-\tau/\tau_c} - \bar{I}^2$. The exponential decay of the correlation function is typical of a Markovian process involving two states, and also of some other problems involving a continuum of states, for example, isotropic rotational diffusion. What the correlation function measures is the time dependence of the correlation between the value of a variable at a given time and its value at a time τ later. For slow fluctuations the value of a variable changes slowly, and if, for example, a

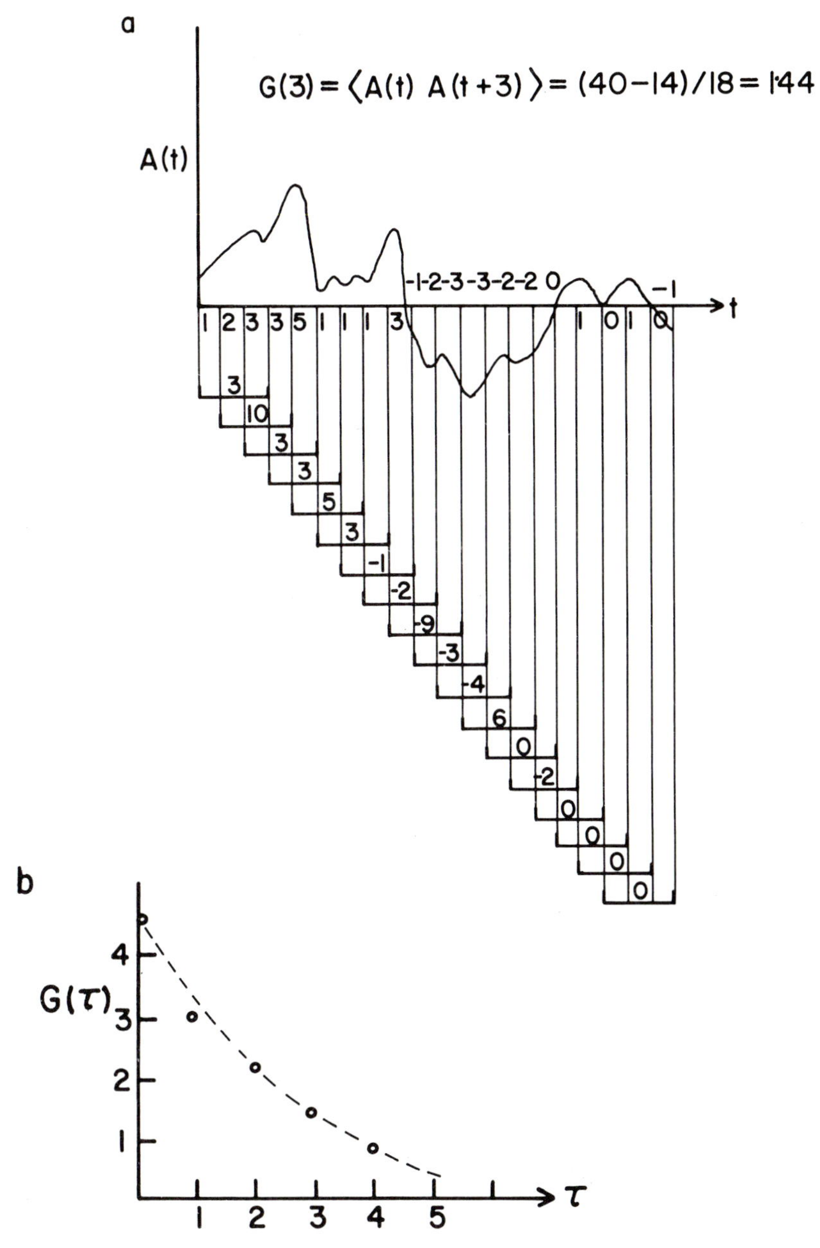

Figure 112. A. Construction of the correlation function $G(\tau)$ for a random process. In the figure, $G(3)$, the correlation between times differing by three units, is derived by taking the average of $A(t)$ times $A(t+3)$. The values of $A(t)$ are indicated on the t-axis. **B.** Values of $G(\tau)$ plotted against τ give the correlation function.

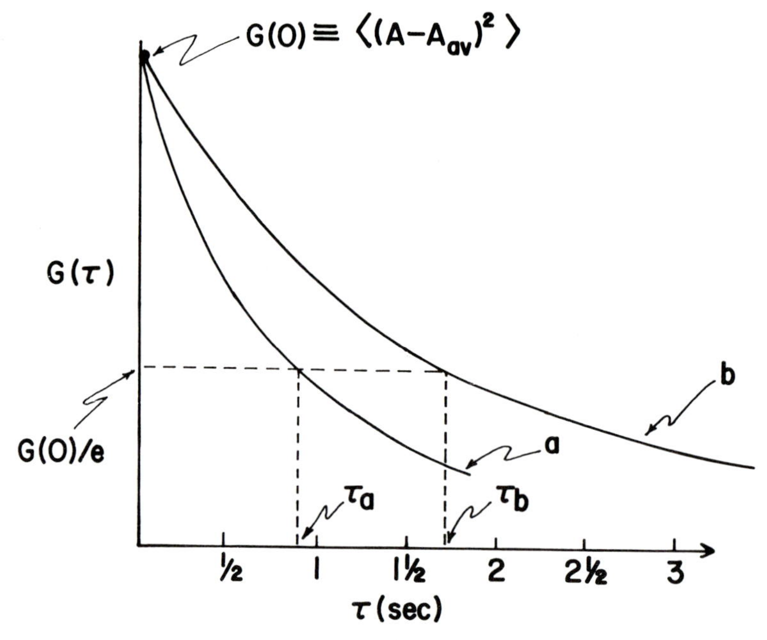

Figure 113. Correlation functions for the two processes shown in Figure 111. The correlation function for the faster process, 111a, decays more rapidly and has a shorter correlation time.

high positive deviation is observed at a given time, there is a good chance that a short time later the observed value will still be large and positive. The decay of correlation is measured by τ_c. The forms of $\Phi(\tau)$ for many stochastic processes have certain common features shared by $\Phi_I(\tau)$ in our example. First:

$$\Phi_I(0) = \langle I(0)I(0) \rangle - \bar{I}^2 = \langle I^2 \rangle - \bar{I}^2 \tag{17.21}$$

which is the variance or mean square deviation. (There is an implicit assumption here — that the process is a *stationary* random process, but Markovian processes are.)

Second, the limiting value of $\Phi_I(\tau)$ as τ goes to infinity is

$$\Phi(\infty) = \lim_{\tau \to \infty} \langle I(\tau)I(0) \rangle - \bar{I}^2 \tag{17.22}$$

Now for $\tau \to \infty$, $I(\tau)$ and $I(0)$ become *independent* variables. This statement can be made in other ways. It can be said that we cannot predict anything about $I(\tau)$ from $I(0)$ or that $I(\tau)$ and $I(0)$ are not *correlated*. There is no correlation between the angle of a C—H bond in a lipid at noon and the angle at midnight. The average $\langle AB \rangle$ over the product of uncorrelated or independent variables A and B is given by $\langle AB \rangle = \langle A \rangle \langle B \rangle$ so that $\Phi(\tau)$ as $\tau \to \infty$ is given by

$$\Phi(\tau)_{(\tau \to \infty)} = \langle I(\tau) \rangle \langle I(0) \rangle - \bar{I}^2 = \bar{I}^2 - \bar{I}^2 = 0 \tag{17.23}$$

The correlation function dies after enough time. For an exponential correlation function, $\Phi(\tau)$ goes to zero for times longer than a few multiples of the correlation time τ_c. Thus τ_c is a statistical measure of how long a system remembers its past. Correlations last longer in slowly moving systems; τ_c is longer. This point indicates the diagnostic nature of the correlation function — it is an indicator of the rapidity of fluctuations.

The concept of the correlation function and the above deviation of its form perhaps seem more entertaining than useful. After all, if we have the record of the behavior of a *single* channel, such as shown in Figure 110A, we have all the statistical information we need. It is a trivial task to estimate from the experimental data the average lifetime of the channel in the "open" or "shut" state. The distribution of lifetimes and other statistical information can be extracted directly from the trace of $I(t)$. However, the observation of single-channel conductances is not easy, and at present most experiments on biological membranes are limited to the observation of a large number of channels. The current fluctuations for such a system are not susceptible to easy visual analysis, but require the essentially statistical treatment embodied in the determination of a correlation function. The observed properties of a collection of randomly fluctuating independent units are the sums of the individual properties of those units. The fluctuating conductance of a membrane can be treated by the same methods that we used for a single channel. Assuming n identical channels to be independent of one another, it is intuitively clear that the average current will be n times the average current for one channel:

$$I_n = n\bar{I} \tag{17.24}$$

Probability theory shows that the autocorrelation function of the current is n times that of a single channel:

$$\Phi_{I_n}(\tau) = n\Phi_I(\tau) = n\langle(\gamma V\bar{Y})^2\rangle \varrho e^{-\tau/\tau_c} - \bar{I}^2 \tag{17.25}$$

where $\bar{I}$ now refers to the total current. This expression supplies an answer to the question: "What use is the autocorrelation function?" First, let it be clear that experimentally the function can be obtained automatically from the channel noise by simple data processing. The form of $\Phi_I(\tau)$ or $\Phi_{I_n}(\tau)$ for *the model assumed* should be exponential, both functions having the *same* decay constant $1/\tau_c$. In Figure 114 are shown the experimental autocorrelation functions of a single-channel

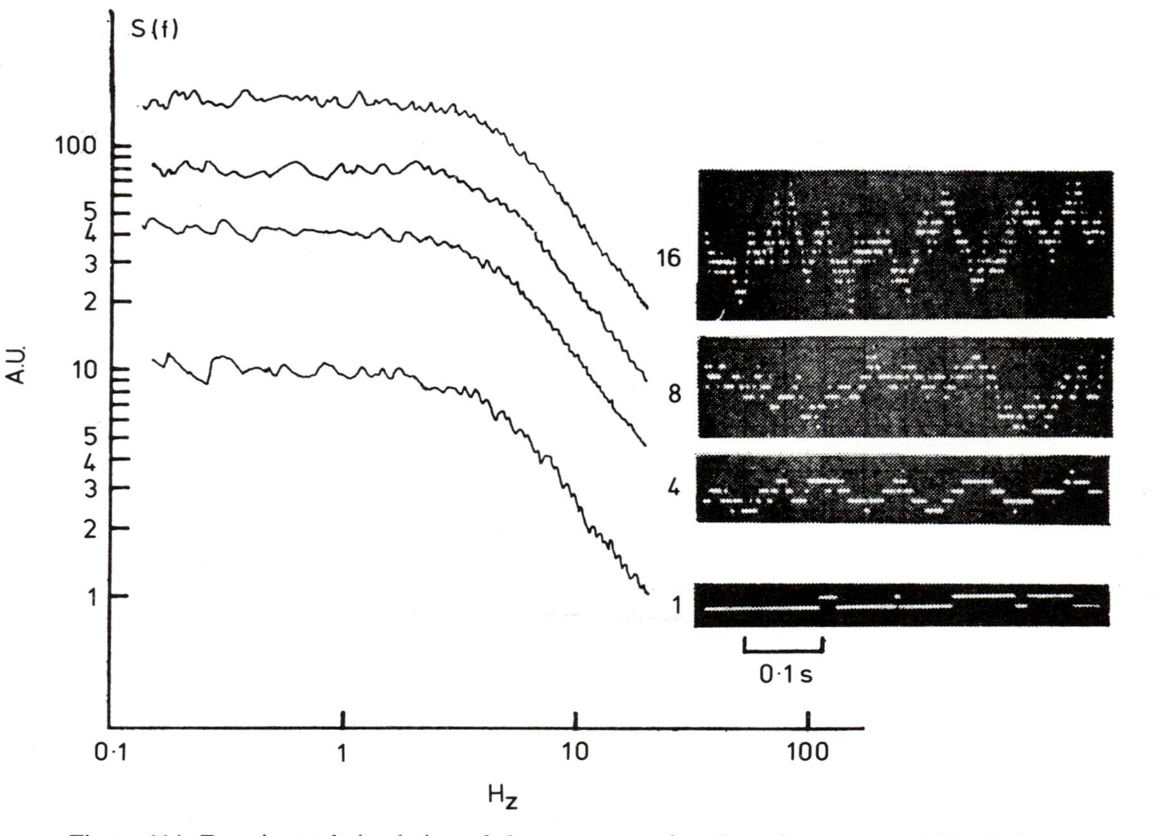

Figure 114. Experimental simulation of the current passing through a system of identical channels having only open and closed states with single-channel rate constants of ~20 sec^{-1}. To the right are shown the random currents obtained from one, four, eight, and 16 channels. On the left are the respective power spectra that can be seen to differ in magnitude but to have the same frequency distribution. [From Conti, F., and Wanke, E., *Q. Rev. Biophys.* 8:451 (1975).]

model system and a many-channel system containing the same channel. The values of τ_c are identical within experimental error which encourages the belief that we can extract meaningful results from the fluctuation analysis of multichannel preparations. Of the microscopic parameters P_2, γ, n, and τ_c that characterize the system, we have obtained τ_c from the correlation function and γ from our previous analysis, which we presume to hold here. Since $\bar{I} = \gamma V \bar{Y}$, we can obtain $\bar{Y}$, and therefore P_2 (because $\bar{Y} = P_2$) and n.

It is true that we have worked within the limits of a particularly simple model but that model probably applies to some channels. The reader is certainly familiar with the problem of results that are consistent with more than one model (compare Nernst–Planck and rate theories for transport), and this is true in fluctuation analysis, as in most fields. Among the documented traps awaiting the experimentalist are multiple conductance states and voltage-dependent conductances. However, it is not our object to enter into the complications apparently inherent in any natural system, but rather to illustrate the use of a certain technique, hint at its possibilities, and possibly whet a few appetites.

The interpretation of ^{13}C NMR relaxation times and van der Waals forces leads naturally to the idea of the *power spectrum* of a random process.

A tuning fork giving low C on the piano vibrates effectively only when subjected to low C sound waves and the amplitude of its vibrations can be taken as a measure of the intensity of the monochromatic sound. A record of Stravinsky's *Rite of Spring* (or of the Who) is hardly monochromatic and a seismographic plot of the sound would certainly not be a sine wave. Nevertheless, the tuning fork, if placed in front of a speaker, from time to time vibrates in response to the sound. The fractional contribution of low C to the total sound would be reflected by the behavior of the tuning fork. The ^{13}C nucleus has a spin of 1/2 and in an external magnetic field has two energy levels. Radiation of only one frequency, obeying the condition $h\nu = \Delta E$, will induce transitions between the two energy levels of the nucleus. The more intense the appropriate radiation, the more effective it is in inducing transitions, that is, in shortening the lifetime of the nucleus in its different energy levels. This lifetime can be measured by standard NMR techniques and the results are listed as "spin-lattice relaxation times," T_1. The term *lattice* carries the implication that the radiation producing relaxation comes from the molecular environment, not from an external source of radiation.

In the case of hydrocarbon chains in phospholipids, "the lattice" is shorthand for the protons bonded to the carbon atom. The proton has a magnetic moment that gives a magnetic field at the ^{13}C nucleus, a field dependent on the relative orientation of the two nuclei. Since molecular motion, assumedly stochastic, results in a time-dependent orientation, the field felt by ^{13}C is also time-dependent, and we are involved with a stochastic process with a variable having a continuum of allowed values. The ^{13}C nucleus is exposed to this randomly varying field and, like the tuning fork, it will respond to only that component of the field at the right frequency, $\nu = \Delta E/h$. The relaxation time T_1 of the ^{13}C nucleus is thus controlled by the weight of molecular motion of frequency ω_0 contained in the random molecular motion, where ω_0 is the frequency in cycles per second (or hertz), the usual units used in this field. This weight is called the *spectral density* at ω_0, $J(\omega_0)$. It is an experimental fact that if we increase the rate of molecular motion, say, by heating the lipid, T_1 *increases*. This implies that the spectral density $J(\omega_0)$ at the resonance frequency has *decreased*. [A rough analogy is contained in the fact that if the Stravinsky record is played at 78 rpm instead of 33 rpm, the tone of all the instruments will be raised and the weight of the high frequencies will increase while the contribution of lower frequencies will be reduced. The weight of low C will be cut down (check the score) and the tuning fork will be disturbed less.] In vesicles[275,276] T_1 for ^{13}C nuclei in lipid chains *increases* toward the end of the chain. This is consistent with a decrease in $J(\omega_0)$, the weight of molecular motion at the ^{13}C resonance frequency, and is interpreted as meaning that the methyl groups, which easily have the longest T_1, are moving very fast compared with, say, segment C_3 in the chain. This conclusion is supported by the fact that all the T_1's decrease with increasing temperature, that is, with faster motion. It can be said fairly safely that the correlation time for the orientation of a C—H bond toward the end of the chain is much shorter than that at the beginning. The general relation between T_1 and motion is shown in Figure 115.

We have oversimplified in the previous paragraphs. Motion in liquid crystals is complex, certainly not isotropic, and it may be better to use two correlation times $\tau_\perp$ and $\tau_\parallel$ associated with different modes of motion perpendicular and parallel to the director. Nevertheless, the general conclusions are unaltered by these refinements. (For the afficionados, the reported NMR results are all for $\omega_0^2\tau_c^2 \gg 1$.)

Since theory connects T_1, via $J(\omega)$, to the detailed molecular motion of a molecule or segment thereof, the reader should not be surprised to be told that the random motion of molecules, or the random variation of any stochastic variable, can be analyzed in terms of spectral densities, just as light was analyzed by Newton's prism. What we need is the Weiner–Khinchin theorem, which states that the spectral density at frequency ω is the time Fourier transform at frequency ω of the correlation function for the random process. We go immediately to an example. For the two-state Markovian channel, we found the form of the correlation function $\Phi_1(t)$; the Weiner–Khinchin theorem says that the *power spectrum* is given by

$$J(\omega) = 4\int_0^\infty \Phi_I(t)\cos 2\pi\omega t dt \tag{17.26}$$

On substituting from (17.25) for $\Phi_I(t)$, we find

$$J(\omega) = 4\bar{I}^2\varrho\,\frac{\tau_c}{1+(2\pi\omega\tau_c)^2} \tag{17.27}$$

A plot of this function is shown in Figure 116. A spectrum of this form is termed a Lorentzian spectrum. For any value of ω, $J(\omega)$ is proportional to the weight of the frequency ω in the mixture of frequencies contributing to fluctuations. Power spectra of the Lorenztian and other forms have been obtained from biological membranes and curve fitting used to obtain values for ϱ and τ_c.

The Weiner–Khinchin theorem is the bridge between model and experiment in many branches of spectroscopy and chemical physics, but often the power spectrum is not extracted from the correlation function. Both contain the same information, as they are Fourier transforms of each other.

As a final example of a correlation function, we consider the use of fluorescence correlation spectroscopy in determining the diffusion constants of molecules in bilayers. The method is based on the fact that if a certain number of molecules in a system are labeled in some way, then if we concentrate on one part of the system, the number of labeled

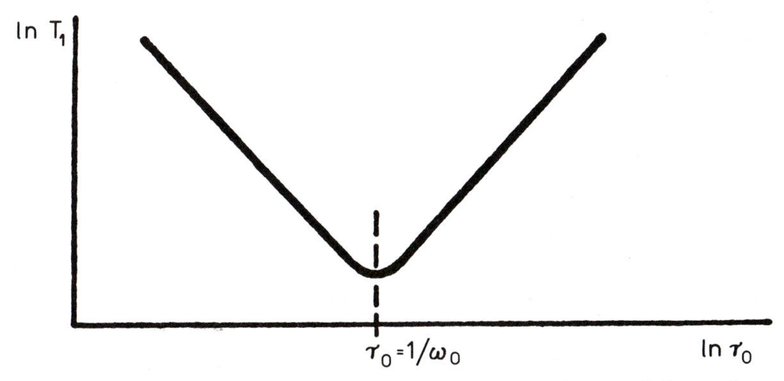

Figure 115. The dependence of T_1 on the correlation time τ_0 of the random motions producing relaxation. T_1 is shortest—relaxation most effective—for $\tau_0 = 1/\omega_0$ where ω_0 is the correct frequency to induce nuclear transitions. For faster or slower motions, T_1 increases—transitions are less frequent. (From Slichter, C. P., *Principles of Magnetic Resonance*, Harper & Row, New York, 1963.)

molecules can be expected to fluctuate as a result of random translational motion, provided, of course, that diffusion can occur. The smaller the volume under observation, the more pronounced we expect the fractional fluctuation to be. The fractional fluctuation in the number of oil sheiks standing on a given carpet in Mecca is more dramatic than that in the number of sheiks in Saudi Arabia. For this reason, fluctuation measurements are usually performed on samples containing small numbers of probes. In practice, fluorescent molecules are incorporated into a membrane and the preparation is irradiated by a laser beam of constant intensity. The intensity of the emitted light varies due to fluctuations in the number of fluorescent probes in the illuminated area. If the sole reason for the fluctuations is diffusion, then we can describe the rate of change of the probe concentration, or better the time variation of the deviation in concentration from the average concentration $\bar{C}$ by

$$\frac{\partial \Delta C(r, t)}{\partial t} = D\nabla^2 \Delta C(r, t) \tag{17.28}$$

Here $\Delta C(r, t) = C(r, t) - \bar{C}$ where $C(r, t)$ is the concentration of the probe at a given point and a given time. Compare (12.9). Here D is, for

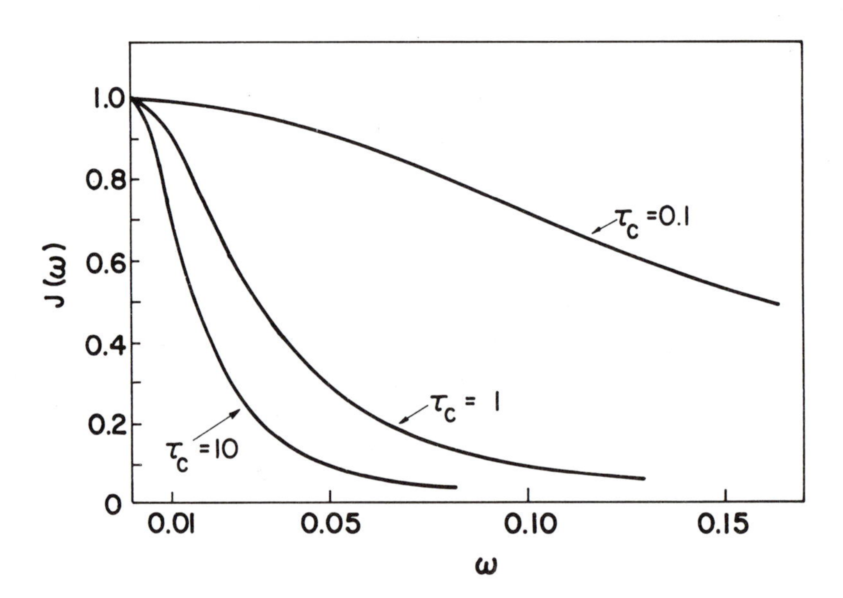

Figure 116. Plots of $J(\omega)$ as defined in equation (17.27). For slower motions, that is, for longer τ_c, the contribution of high-frequency motions, large ω's, is smaller.

the most general case, a three-dimensional diffusion constant. In examining a single bilayer, however, we can assume that there is no diffusion in the direction of the bilayer normal and D is then a two-dimensional constant. For a system displaying lateral anisotropy, two diffusion constants are necessary. The correlation function at a given point in the sample is given by

$$\begin{aligned} G_C(r, \tau) &= \langle \Delta C(r, t)\Delta C(r, t+\tau)\rangle \\ &= \langle \Delta C(r, 0)\Delta C(r, \tau)\rangle \end{aligned} \tag{17.29}$$

where the second equality follows because the system is assumed to be stationary. The probe concentration can be related to the experimental observable, which is usually a photomultiplier current:

$$i(t) = g\epsilon Q\int I(r)C(r, t)d^3r \tag{17.30}$$

where $i(t)$ is the current; g includes the electrical, optical, and geometrical characteristics of the apparatus, and ϵ and Q are the extinction coefficient and fluorescence quantum yield of the probe at the experimental excitation and emission frequencies. $I(r)$ is the intensity of the illuminating source, usually a laser, at the point r. The correlation function of the photomultiplier current is then

$$G_i(\tau) = g^2\epsilon^2 Q\int I^2(r)G_C(r, \tau)d^3r \tag{17.31}$$

The integration is over the whole illuminated area of the sample. All now depends on obtaining an expression for $G_C(r, \tau)$ from the diffusion equation (17.28). This is a mathematical rather than a physical problem, and the final form of G_C, and therefore of G_i, will obviously include the diffusion coefficient, which thus is experimentally accessible from (17.31). As we saw earlier, (17.21), $G(0)$ gives the variance of a random variable, that is, the mean-square deviation σ^2. Our Arabian analogy suggests that for the fluorescence correlation experiment, σ^2 will bear some relationship to the average number of probes. In fact, for a Poisson[277] process, which is what we are dealing with, the normalized correlation function for $\tau = 0$ gives the average number of probes[278]:

$$\beta = \langle i(t)^2\rangle/\langle i(t)\rangle^2 = 1/\langle N\rangle \tag{17.32}$$

where $\langle N\rangle$ is the average number of probes in the supposedly uniformly illuminated area. In practice the laser beam usually has a Gaussian profile. If the $1/e^2$ radius is R, then $\langle N \rangle$ refers to the probes within this radius. Since $\langle N \rangle$ can be estimated directly from the composition of the system, the value of β serves as a means of assuring that the probes in the system retain their chemical and physical integrity.

The power of fluctuation analysis is evident. In the following chapter we comment on some experimental achievements and limitations.

18 Physical Chemistry and the Cell Membrane

The simplicity of lipid bilayers as compared with natural membranes has determined the emphasis of this book. The physical chemist traditionally works, both experimentally and theoretically, on systems of defined composition, and there is not a single natural membrane for which the structure is known to the (albeit still limited) precision attainable for single lipid bilayers. Nevertheless, membranes have been examined by physical techniques and the results are heartening in that they demonstrate unequivocally that the structure, dynamic behavior, and general properties of the lipid bilayer are carried over in modified form to the living membrane. The purpose of this chapter is both to justify this last statement and to place before the reader some examples of the relevance of physical chemistry to the study of the cell membrane. Of course, the selected research represents a minute fraction of the literature and is not intended to be in the nature of a review.

We start with structure, and in particular we return to our basic model, the lipid bilayer, and ask what evidence there is to support the existence of a bilayer in biological membranes. X-rays again are the method of choice. The irregular lateral distribution of proteins and the variety of proteins associated with one membrane are not conducive to well-resolved diffraction patterns. Another relevant factor in weighing the utility of X rays is the relative electron densities of protein and lipid. Most proteins have a density of 0.42–0.44 electron/Å^3 while lipid head groups have minimum values of around 0.47 electrons/Å^3 and maximum values of well over 0.6 electrons/Å^3. The center of the lipid core in the liquid crystal state has a density that may fall below that of liquid water, 0.334 electrons/Å^3. We see that an intrinsic protein should raise considerably the electron density at the bilayer core. In Figure 117 we see that electron-density profile of the purple membrane of *Halobacterium halobium,* a membrane containing 75 wt% of protein. The two profiles reflect the difficulties of choosing phase angles in the analysis of the diffraction patterns. Nevertheless, both profiles are very reminiscent of those obtained for lipid bilayers (compare Figure 22) so that our first tentative conclusion is that a bilayer structure exists in this cell membrane. In addition, both profiles are asymmetric and show a higher density at the core than does the profile obtained from the extracted lipids alone. Furthermore, fragments of purple membrane in dehydrated stacks are, on X-ray evidence, very closely packed. The conclusion is that the protein, bacteriorhodopsin, is an intrinsic protein situated within the purple membrane and projecting very little, if at all, above the membrane surface. Numerous cases of bilayer-type diffraction patterns have been obtained from aqueous dispersions of biologic membranes including those of *Escherichia coli, Acholeplasma laidlawii,*[279,280] and red blood cell ghosts.[281] An interesting profile is that shown in Figure 118 which was obtained from a virus and shows a deep bilayer-suggestive minimum located somewhat below the surface of the virus. Outside this minimum the profile indicates a layer of high electron density, which may be due to a layer of protein around the virus. The presence of protein on the outer surface of a cell membrane is also indicated by the profile of the complete *H. halobium* envelope shown in Figure 119. The bilayer signature is again clear. In addition, there are two high-density areas, both of which are assigned to protein. The 'periplasmic space' may be filled with an aqueous medium, the spacing conceivably being determined by a balance between attractive and repulsive forces rather than by specific chemical structures.

Accepting the existence of bilayer structures in natural membranes, the next question is whether the properties of these bilayers are similar to those in artificial systems. We consider first the gel-to-liquid-crystal transition, which, if it occurs in natural membranes, we expect to be broadened by the presence of proteins, cholesterol, and other membrane

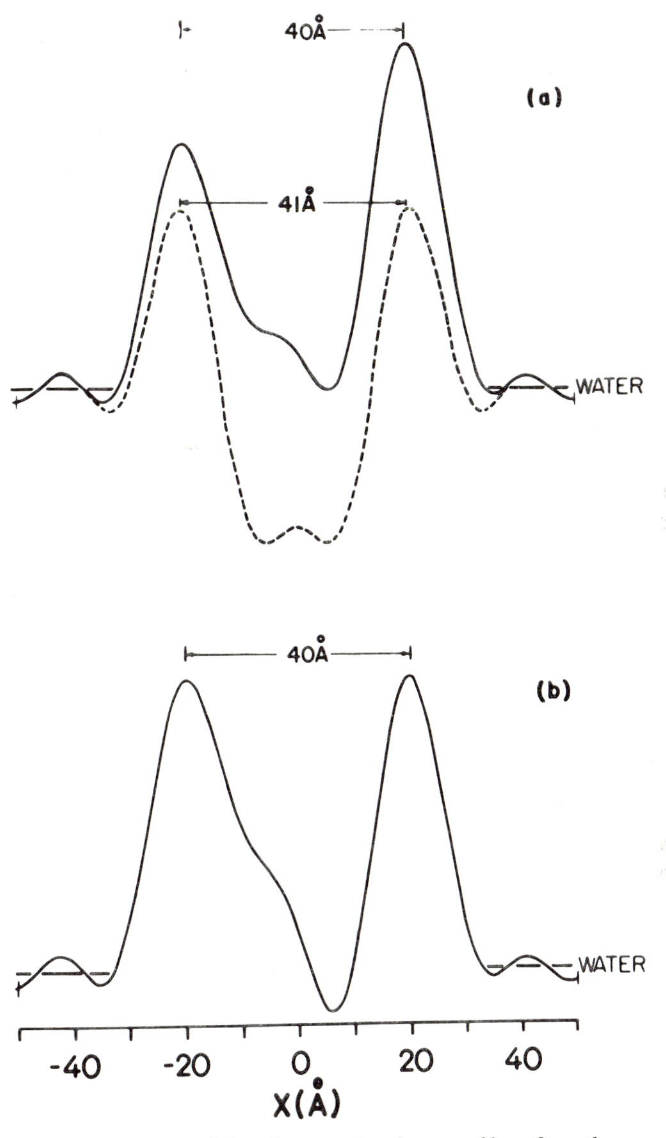

Figure 117. Two possible electron-density profiles for the purple membrane of *H. halobium*. (A) and (B) are the results of differing assumptions about phase angles in analyzing the diffraction patterns. The asymmetry of both the (solid-line) profiles and the relatively high central valley indicates the presence of protein in the lipid bilayer. The dotted lines indicate the profile of the *bilayer*, which is not plotted on the same vertical scale as the solid line. [From Blaurock, A. E., and King, G. I., *Science* 196:1101 (1977).]

components (Chapter 9). Many membrane fragments display thermotropic transitions as initially revealed by the DSC technique. Some curves are shown in Figure 120. The nature of the transition bears much in common with that of simple bilayers. Thus X-ray studies on fragments of *A. laidlawii* membrane showed that cooling produces changes in the diffraction patterns of exactly the kind associated with lipid chain crystallization in bilayers. The patterns indicate a bilayer structure both above and below the transition.[280] For *A. laidlawii,* chain melting is accompanied by an increase in membrane volume of ~0.7%.[282] Since the membrane is about two thirds protein, the increase in the lipid bilayer volume is about 2%, which can be compared with the figure of ~4% for DPPC bilayer. On melting *A. laidlawii* membranes, a 17% decrease in membrane thickness is observed[280] (see Figure 27).

The broad DSC peaks characteristic of membranes are explicable partly in terms of the protein and partly as a consequence of the heterogeneous lipid composition of the bilayer. If *A. laidlawii* is grown in lipid-containing media, the membrane can be enriched in specific lipids—a process that changes the temperature and shape of the DSC

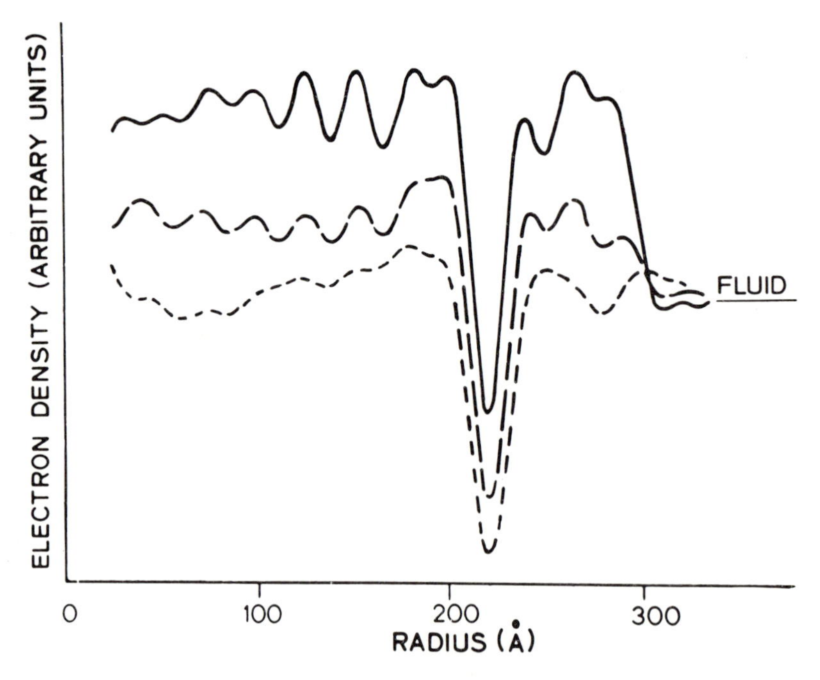

Figure 118. Electron-density profile of a PM2 virus particle showing a deep minimum below the surface of the virus. (The different profiles are for media of differing density.). [From Blaurock, A. E., *Biochim. Biophys. Acta* 650:167 (1982).]

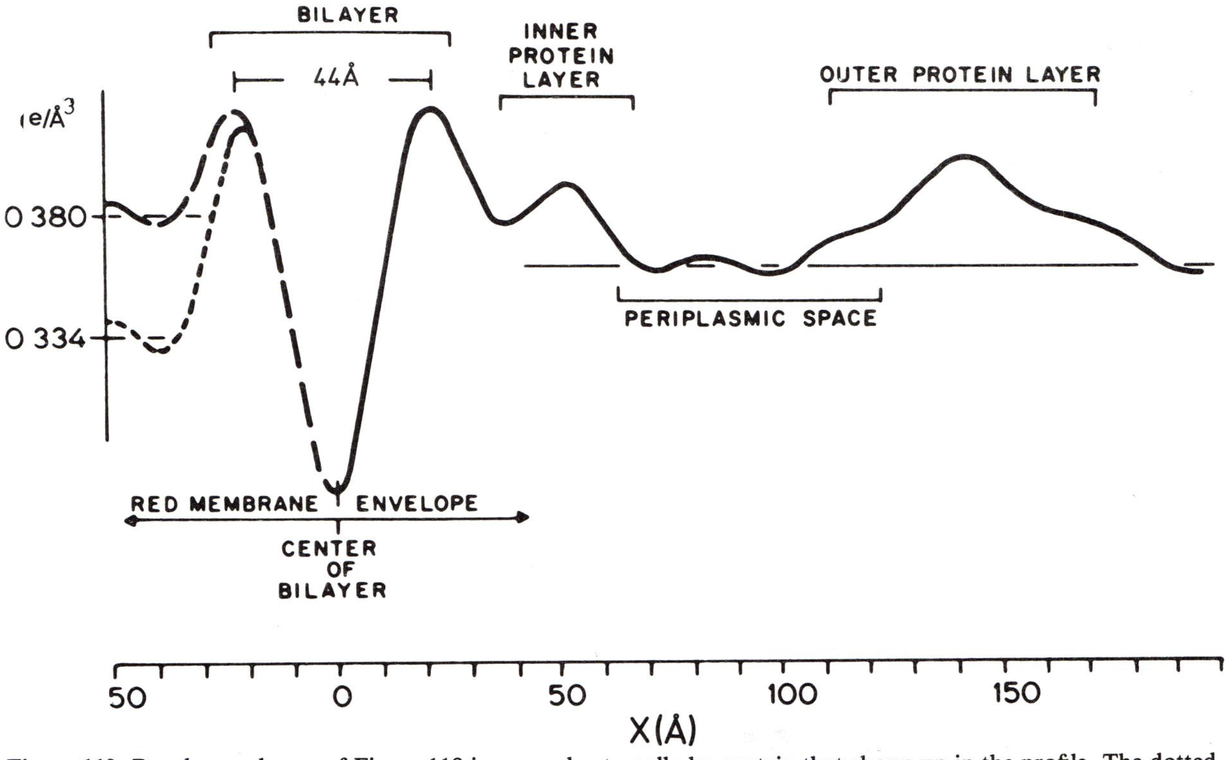

Figure 119. Purple membrane of Figure 118 is covered externally by protein that shows up in the profile. The dotted lines to the left of the bilayer center are the expected lipid bilayer electron-density profiles constructed with respect to a basal salt medium (— —) and a water medium (- - - -). [From Blaurock, A. E., et al., *J. Cell Biol.* 71:1 (1976).]

peaks. An example is shown in Figure 120 in which there appears the trace for a membrane enriched to ~80% in oleate chains. The peak is considerably sharper and falls at ~70°C lower than that for the unmodified membrane. The melting transition has been shown to be reversible in *live A. laidlawii* cells,[283] confirming the relevance to biology of studies on model bilayers.

The order-parameter profiles described for bilayers (Chapter 5) are echoed in the results of deuterium NMR studies on membranes of *E. coli* and *A. laidlawii*. Perdeuterated lipids have been incorporated into *E. coli* membranes and the ^{2}H NMR spectra recorded[284] as a function of temperature (Figure 121). Recalling that the splitting $\Delta\nu_Q$ is a measure of the order parameter, we can interpret the steep falloff of the spectrum

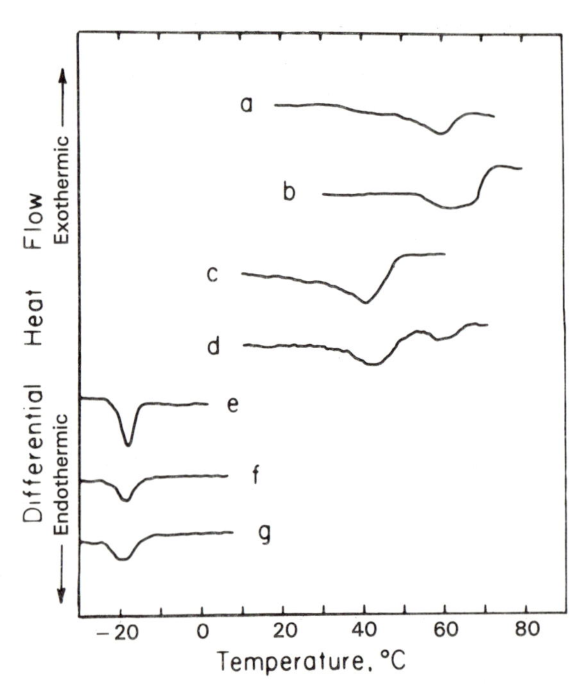

Figure 120. Differential scanning calorimetry curves for *A. laidlawii. a* and *b* are obtained, respectively, from extracted lipids and membranes of stearate-enriched membranes. Curves *c* and *d* are for extracted lipids and membranes of cells grown in normal growth medium. Curves *e*, *f*, and *g* are from extracted lipids, membranes, and *whole cells* of oleate-enriched membranes. The higher temperature peak in curve *d* is due to protein denaturation. The curves show that the transition for the membrane occurs at lower temperatures as the lipids are enriched in fatty acids of decreasing melting point.
[From Steim, J. M., et al., *Proc. Nat. Acad. Sci.* 63:104 (1969).]

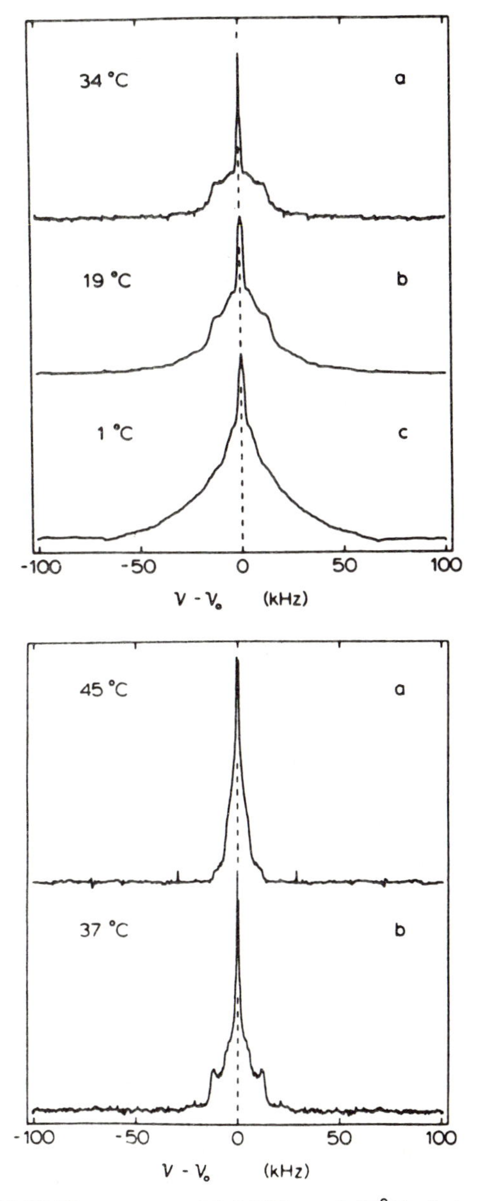

Figure 121. ^{2}H NMR spectra at 34.4 MHz and 37°C of the cytoplasmic membrane and lipid extract of *E. coli* cells grown on a medium containing perdeuterated palmitic acid. The upper three spectra are of the membrane at different temperatures and the lower two of lipids extracted from the membrane. The central peak is from HDO. The broad line in the low-temperature spectrum originates in the gel phase of the lipid bilayer.
[From Davis, J. H., et al., *Biochem.* 18:2103 (1979).]

at ~34°C as indicating a similar value of the order parameter for most of the lipid chain, reminiscent of the order-parameter profile for pure lipids. As the temperature is lowered, the broad spectrum of the gel state is seen to grow but to coexist with the liquid crystal spectrum over a wide range of temperature, consistent with the broad thermal transition observed calorimetrically. The changes in the second moment of the NMR spectrum closely parallel those for pure DPPC bilayer. Lipids with a palmitic chain deuterated at the terminal methyl group have been incorporated into *E. coli* membranes.[285] Again, gel-like and liquid-crystal-like spectra coexist over a wide range of temperature. As we noted in Chapter 9, proteins reduce the average value of $\Delta\nu_Q$ in lipid bilayers, that is, they appear to disorder the chains. The same phenomenon is found when the spectrum of the *E. coli* membranes is compared with that of the extracted lipids, the values of $\Delta\nu_Q$ being ~1 kHz and ~3.4 kHz respectively. On the other hand, *A. laidlawii* membranes containing chains labeled in the 4 or 8 position show very similar membrane and extracted lipid spectra, indicating that the disordering effect of the protein at these positions is much smaller than at the methyl group.[286]

On the basis of the above and many other studies, it can be confidantly concluded that the lipid portions of biological membranes are organized and behave in much the same way as in pure lipid bilayers and protein–lipid preparations. This provides a framework in which to analyze the results of spectroscopic and thermal studies on biological membranes. It probably would be a mistake, however, to dismiss non-bilayer structures from consideration.[287] The fact that the evidence for such structures in both artificial and natural membranes is far from convincing does not exclude their occurrence as transient species, difficult to detect spectroscopically but playing an essential role in processes such as membrane fusion. The experimental observations in this field are fragmentary and their interpretation open to discussion, so it would be premature to attempt a summary at present.

Since the measurement of chain order parameters is experimentally demanding, it has become fashionable to compromise and to use spin-label or fluorescent probes to measure "fluidity". This concept and its "measurement" have documented drawbacks (see Chapters 5, 9 and 10). Nevertheless, the experiments are comparatively simple and qualitative conclusions so easy to arrive at that a bulky literature has accumulated that is devoted to attempts to measure fluidity differences caused by chemical or physical factors and to correlate these changes with biochemical or physiological processes. Many investigators have been content to chronicle the steady-state fluorescence anisotropy as a function of the experimental parameters without attempting to interpret their results in terms of microviscosity. This is a wise course since the evaluation of viscosity from steady-state measurements is unreliable.[288] Time-resolved fluorescence anisotropy decay measurements provide far more information but are more time-consuming. Typical examples of the use of fluorescent probes are given in the following.

A sizable proportion of studies has been aimed at detecting gross differences in "fluidity" between membranes subjected to different chemical or physical environments. Typical is the finding that rats fed on alcohol show more rigid whole-brain membranes than do normal rats, while hibernation appears to soften the membranes.[289] In vitro studies of the interaction between ions or molecules and membranes are more common. The binding of fibrinogen to human blood platelets is accompanied by a decrease in the membrane lipid fluidity but augmented mobility of the membrane proteins.[290] The increased rigidity of the lipids results in a greater exposure of proteins above the membrane surface, which can be crudely attributed to a squeezing-out effect as the lipids move toward the gel state. Molecular linkage between platelets could be facilitated by the increased protein exposure. This study underlines the danger of attempting to correlate microviscosity with protein diffusion.

The germination of spores can be triggered by a variety of molecules. Using diphenylhexatriene as a fluorescent probe, it was found that compounds triggering spore germination in *Bacillus megaterium* in vivo caused an increase in the fluorescence anisotropy of the isolated membranes in vitro. After germination, the same compounds failed to affect the anisotropy. Furthermore, chemical or genetic blocking of the in vivo triggering by L-proline also blocked the in vitro change in anisotropy caused by L-proline. The authors[291] wisely decline to translate the anisotropy changes into specific changes in molecular organization of the membrane, but the link between germination and membrane modification has been established.

Although fluorescence studies are the most common in this field, spin labeling runs a close second (and NMR a poor third). Using a series of spin labels situated at different depths in the membrane bilayer, an order-parameter profile similar to that observed in pure bilayers was observed in virus particles.[292, 293] The similarity is, incidentally, strong circumstantial evidence for the existence of a bilayer in the virus. Additional support comes from the measurement of ^{13}C relaxation times for different carbon atoms in the lipid chains in the virus membrane.[294] Again, the results parallel those for pure lipid bilayers.

Observations on simple bilayers have helped to explain apparently complex biological effects. For example, it is known that Ca^{++} decreases the fluidity of lipid bilayers (Chapter 4). Noradrenaline increases the fluidity of rat liver plasma membranes when it acts in the presence of Ca^{++}–mobilizing hormones, that is, substances that increase the Ca^{++}

concentration in the cytosol of the cell, in this case by releasing Ca^{++} from the membrane.[295] The desorption of Ca^{++} from the lipid head groups is the cause of the increase in fluidity. Care must be taken, nevertheless, in drawing parallels between bilayers and natural membranes. Thus by using spinlabels it has been found that Ca^{++} decreases fluidity in the afore-mentioned liver membrane, and also produces changes in the activity and Arrhenius plots of the membrane-bound enzyme adenylate cyclase.[296] It is tempting to ascribe the modification of enzymatic behavior to the alteration in lipid organization. In fact, the activity change is almost certainly due to binding of Ca^{++} to the enzyme. Another salutory tale is that of *Paramecium tetraurelia.*[297] This organism, when grown at 25°C, moves away from an environment held at 42°C. Paramecia lose this thermal avoidance capability if they are incubated for four hours or longer at 35°C. Furthermore, these incubated paramecia show an increase in membrane fluidity when they are returned rapidly to 25°C. We noted an example of bacterial response to heat in Chapter 9, the observed changes in fluidity being associated with marked alterations in the ratio of saturated to unsaturated lipid chains. In contrast, the paramecium varies the fluidity of its membrane not by changing the fraction of unsaturated lipid, but by changing the overall composition of the membrane without altering the double-bond population. We are still a long way from understanding the connection between fluidity and membrane lipid composition.

The electrostatic, electrodynamic (dispersion), and structural forces that control the interactions of simple bilayers must be involved in the behavior of natural membranes but their estimation is not always easy.

Charges in vivo originate primarily on simple ions or on ionized groups of proteins and lipids. The surface and bulk charges and potentials are related—in a simple treatment—by the Poisson–Boltzmann and Gouy–Chapman equations. To measure the range of validity and the applicability of these equations in a given system, it is necessary to have experimental means of measuring surface charge σ and potential ϕ_0. In Chapter 7 we mentioned electrophoretic, spin-label, and fluorescence methods of determining surface potential. The electrophoretic approach yields the ξ potential, which is lower than the surface potential and generally leads to smaller values of σ than other methods. Dyes and spinlabels cannot be guaranteed to bind uniformly on real membranes, and therefore may give local information rather than the averaged picture provided by electrophoresis. A less used method, which can certainly furnish more or less local information, relies on the influence of electric potential on the distribution of charged species near the membrane.[298] (This effect was invoked in discussing the transport of lipophilic ions in Chapter 14.) The rate of a reaction at a membrane site will depend, if one of the mobile reactants is charged, on the concentration of that reactant at the surface, which in turn is a function of the electric potential. The theory is trivial: if the reaction depends linearly on the charged species A in the solution, then formally the rate can be written:

$$\text{Rate} = k[A][\ ][\ ]\cdots \tag{18.1}$$

where k is the apparent rate constant and $[A]$ is bulk concentration that is related to the concentration at the surface, $[A]_0$, by

$$[A]_0 = [A]\ \exp(-zF\phi_0/RT) \tag{18.2}$$

where we have accepted the usual approximations of the conventional Poisson–Boltzmann method. The rate can also be written:

$$\text{Rate} = k_0[A]_0[\ ][\ ]\cdots \tag{18.3}$$

where k_0 is the true rate constant. Equating (18.1) with (18.3) and using (18.2):

$$k = k_0\ \exp(-zF\phi_0/RT)$$

or

$$lnk = lnk_0 - zF\phi_0/RT \tag{18.4}$$

Now the Gouy–Chapman equation relates the surface potential to the surface charge, and for smallish potentials, say 50 mV or less, we can use the linear form of the equation to obtain

$$lnk = lnk_0 - C\sigma[A]^{-1/2} \tag{18.5}$$

where $C = (2\pi/RT\epsilon)^{1/2}$ and activity corrections are ignored. The slope of the plot of lnk versus $[A]^{-1/2}$ gives σ, which we take to be a good approximation to the charge density at the reaction site.

A natural membrane that has received a good deal of attention is the thylakoid membrane of the chloroplast, which contains photosystems I and II (PSI and PSII). It is in this membrane that the initial photosynthetic electron-transfer reactions occur under the stimulus of light. At pH 7.0 the charge density determined by electrophoresis[299] for the outer surface of the membrane is about 10^{-2} coulomb of negative charge per square meter or one negative charge on 1600 Å^2. Spectroscopic probe methods give two to three times this value[300, 301]. An important point to note is that the charge probably originates largely on ionized groups on proteins since the membrane contains predominantly neutral lipids such as monogalactosyl and digalactosylglycerides.[302] By employing charged redox reactants, the reaction rate method has been used to estimate the charge density near the PSI and PSII proteins. At pH 8.0 the charge density on the outer part of PSII was found[303] to be −2.1 to -2.2×10^{-2} C m^{-2}, which can be compared with -1.2×10^{-2} found by electrophoresis for the outside of the total membrane[304]. The values are consistent with the supposed low charge density on the lipid bilayer. A knowledge of charge densities on thylakoid membranes and of the ionic composition of the ambient aqueous medium allows an estimation of the electrostatic forces between stacked membranes.

It is found, both theoretically and experimentally, that divalent cations play a dominant role in determining the electrostatic force between membranes. This is not surprising considering the presence of Ca^{++} and Mg^{++} in all cells and the preferential concentration of divalent ions near negatively charged surfaces (see Chapter 7). If stacked thylakoid membranes are isolated under conditions in which they are not in contact with media having a high salt concentration, they remain stacked when exposed to cation-free solution. Mg^{++} appears to be the ion mainly responsible for reducing the repulsion between the surfaces.[305] If divalent ions are replaced by prewashing the membranes in a solution of monovalent cations, the stacks disintegrate when they subsequently are suspended in salt-free media.[306] Presumably in this case osmotic pressure is sufficient to force water between the surfaces. The stability of stacked layers in vivo suggests that the equilibrium distance of ~40 Å is the result of a balance between electrostatic and dispersion forces such as described in Chapter 7.

Dispersion forces between natural membranes are not directly measurable but reliable order-of-magnitude calculations can be made. We need to determine or estimate polarizabilities. For lipid bilayers the dielectric properties can be well approximated by those of a slab of hydrocarbon. For proteins the polarizabilities are less certain but reasonable guesses can be made[307] and the range of these guesses cannot be much in error. Calculations have been carried out by Sculley et al.[307] in which membrane proteins are treated as cylinders. Using the fact that about 50% of the thylakoid membrane consists of protein, the dispersion force between two membranes was estimated as a function of distance, assuming that the intervening medium is water. The calculated dispersion force is easily outweighed by the electrostatic repulsion for distances up to ~80 Å, and probably for far greater distances. This discomforting result, which suggests that stacked thylakoid membranes cannot be stable, illustrates the role of theory in stimulating hypotheses. The search for an explanation has resulted in interesting and provocative experiments. Since the dispersion forces *for randomly arranged protein* are almost certainly fairly close to the theoretical estimate, we must look to the electrostatic forces for help. (An unlikely suggestion is that there is a polypeptide, of molecular weight ~2000, which forms a structural bridge between membranes.[308]) Two types of explanation have been brought forward. It has been supposed[307] that cations are bound to the membranes, thus drastically reducing the surface charge and, of course, the electrostatic repulsion. A more interesting hypothesis is that the proteins are not randomly laterally distributed in the membrane. This suggestion has some experimental backing[309]. It is proposed that apressed areas of the membranes are depleted in relatively highly charged proteins such as photosystem I with respect to nonapressed areas (Figure 122). Taking into account the low charge density associated with thylakoid lipids, the removal of protein charge from apressed areas will result in a strong reduction in the electrostatic repulsion and could account for stacking. If stacks were unstacked, it might be expected that diffusion could act to bring about a more homogeneous spread of protein in the plane of the membrane, especially as the fluidity of the membrane is known to be high. The interested reader will find evidence for such a rearrangement in a paper by Barber[310].

The quantitative estimation of forces between natural membranes is becoming possible and will find a place in the description of biological systems. The same theory allows a discussion of forces within membranes, such as dispersion forces between proteins[311]. On the other hand, a useful thermodynamic treatment of membrane organization is a long way off, and perhaps may rarely be profitable in view of the structural complexity of most natural membranes. It is true that classical thermodynamics has occasionally been quantitatively applied to real membranes, as in the demonstration that the *A. laidlawii* membrane obeys the Clausius–Clapeyron equation with respect to the variation of T_t with pressure.[312] Nevertheless, attempts to apply thermodynamics to phase separation and equilibria in much simpler systems, for example, mixed lipid bilayers, are still meeting with difficulty. Real membranes contain many types of lipid, proteins with unknown structures, and molecules that, perhaps by using covalent linkages, control the lateral arrangement of other membrane components. If we now add the presence of a complex environment on both sides of the membrane, we are forced to admit that equilibrium thermodynamics is hardly likely to be a useful means of explaining membrane structure.

The confident application of theory to natural membranes is not yet possible in the field of lateral and rotational diffusion. There are, as we saw in Chapter 10, shortcomings in the theory of lateral diffusion even when applied to simple systems. Of course, the physical techniques by which diffusion is measured have been used on cells and much valuable data have accumulated on the mobility of membrane components. Some experiments have been made on lipids,[313] it generally being found that the diffusion constants in the cell membrane and in the bilayer made

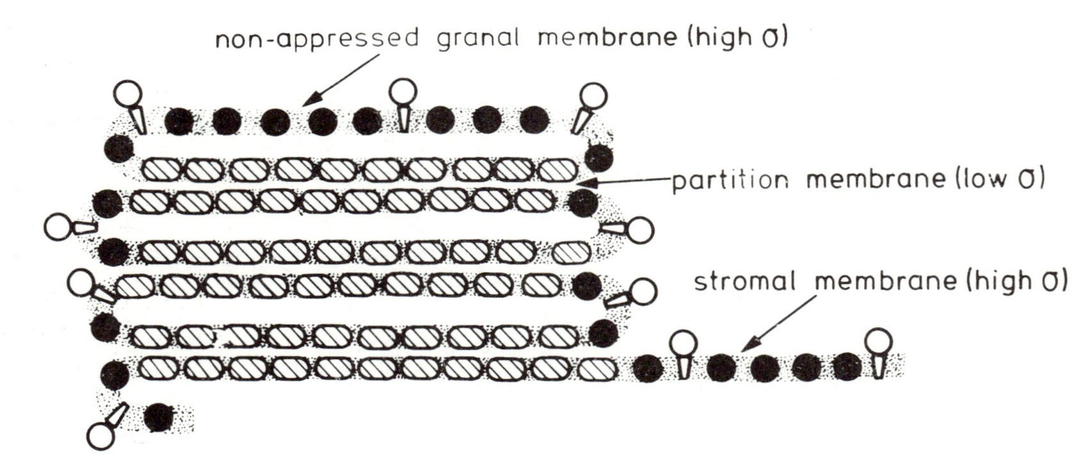

Figure 122. Model for the distribution of proteins in stacked thylakoid membranes. [From Barber, J., *Ann. Rev. Plant Physiol.* 33:261 (1982).]

from the total extracted lipids are very similar in value. Of much greater biological interest is the lateral diffusion of proteins, which has been demonstrated by a range of methods that include electron microscopy of freeze-fractured or ferritin-particle-labeled membranes, fluorescent probe techniques, and, indirectly, immunological techniques. A large proportion of these experiments has shown that movement of proteins occurs without attempting to measure diffusional rates. Those theoretical studies that have been concerned with the estimation of rates must be regarded as interesting rather than convincing. The specter of "fluidity" rears its head in these regions, but seldom contributes more than self-evident generalizations.

In contrast to lateral diffusion of membrane components, the transport of ions and molecules through the membrane is a field in which theory has had very considerable success; the reason for this is that theory has concentrated on accounting for rate laws rather than rate constants. A survey of experimental studies of transport across natural membranes would require a volume in itself. We limit ourselves to showing that the theory given in earlier chapters predicts phenomena and rate expressions that are exemplified in the behavior of cell membranes.

We consider first simple diffusion of an uncharged molecule, through the important example of water. Following our discussion in Chapter 16, we focus on the two permeabilities P_w, the diffusive conductivity (16.29), and RTL_p/V_w, which we write as P_f and which is a measure of the hydraulic conductivity. P_f is chosen in this way since simple theory gives $P_f/P_w = \beta = 1$, as we see from (16.30). The ratio β has been determined for a variety of natural membranes under conditions for which the transport rate is not limited by bulk diffusion, that is, by unstirred layers of solution near the membrane. The vital importance of thorough stirring was demonstrated by Cass and Finkelstein,[314] who found that for water transport through artificial bilayers the experimental value of P_w, the diffusive constant, rose and β approached unity as stirring of the bulk solution increased in vigor. The value of β indicates that water passes through lipid bilayers by an essentially diffusive process even when the driving force is mechanical pressure. The effect of increasing pressure in this case is is not to increase real bulk flow, but to raise the solubility of water in the bilayer. Membranes taken from animal cells show values of β well above unity. For the membrane of the toad urinary bladder $\beta = 4.8$ at 25°C,[315] a value similar to that found for frog skin[316] and for the cortical collecting tubule[317].

We found in Chapter 16 that values of β greater than unity are explicable in terms of pores large enough to permit bulk flow, but another possible explanation is that water is crossing the membrane in single file through a narrow channel. It has been suggested[251] on theoretical grounds that for a single-file mechanism $P_f/P_w = n_w$, the number of water molecules in the channel. Water transport through artificial bilayers is effected by diffusion through closely packed lipid chains in which there are no specialized structures acting as channels and so single-file diffusion does not occur. For gramicidin channels in bilayers, n_w was found to be 6.5 by streaming potential measurements (Chapter 15). The value obtained by direct determination of P_f/P_w was 5.3. Taken together with the internal diameter of the gramicidin channel, ~ 4 Å, the two values of n_w form a consistent set of data and the similarity of the β value for this system and for natural membranes strongly suggests the involvement of single channels in water transport in some cell membranes. We see that theory is helping in the understanding of water transport even though—or perhaps specifically because—no attempt has been made to calculate rates. An interesting review is presented by Hebert and Andreoli.[318]

Moving a step upward in mechanistic sophistication, we consider the transport of a neutral molecule by a carrier and show that phenomena such as countertransport, predicted by theory in Chapter 14, can occur in biological membranes.

Corticosterone acts on its target cells in human placenta by a process involving preliminary binding to a cytoplasmic protein receptor and the subsequent binding of the resultant complex onto the nucleus of the cell. The entry into the cell appears to be mediated by a carrier,[319] with the main evidence being kinetic:

1. The uptake of corticosterone by placental membrane vesicles is a saturable process. Although this is characteristic of carrier mechanisms (Chapter 14), it is also a feature of transport through a channel (Chapter 13).

2. The efflux from membrane vesicles containing tritium-labeled corticosterone was enhanced by the addition of steroids to the extra-vesicular solution. This countertransport is a typical trans effect associated with a carrier mechanism. In this particular system, other nonkinetic evidence supports the carrier hypothesis.

The classic, and most quoted, case of countertransport concerns the transport of glucose through the membrane of the red blood cell[320]. Using isotopically labeled glucose, it was shown that the efflux of glucose from the cell was stimulated by the addition of unlabeled glucose to the external solution. Mannose has a similar effect on glucose transport, implying that the two sugars are conveyed by the same carrier. (The "uphill" transport of a molecule against its concentration gradient had previously been observed for phosphate and arsenate ions crossing a bacterial cell wall.[321]) By turning back to Chapter 14 and glancing at our fable concerning the countertransport of professors, the reader will see that the addition of go-go girls to the Left Bank stimulates an *apparent* efflux of professors from the other Bank by reducing the flow of their colleagues from left to right. The *unidirectional* flow from right to left was in fact not initially affected. If this tale is a correct analogy for glucose transport, the unidirectional rate of efflux of labeled glucose from a cell, as distinct from the net efflux, should be independent of changes in the glucose concentration of the extracellular fluid. Experiment shows otherwise.[322] An increase in extracellular glucose concentration results in an increase in the unidirectional efflux, which approaches a maximum at high (> 15 mM) glucose concentration. These results are not consistent with equations (14.20) and (14.21), which hold for $k_s = k_c$—that is when the rate at which the carrier crosses the membrane is equal to the rate at which the carrier–substrate complex migrates. However, for $k_s \neq k_c$ the unidirectional flow $J^{I,II}$ does depend on S_{II}, as equation (14.23) for $k_s \gg k_c$ shows. We have an example of transstimulation. The carrier–substrate complex shuttles inwards faster than the empty carrier and thus, when the glucose concentration is raised, the rate of arrival of carriers at the inner surface of the red blood cell membrane is also raised.

The transport of sugars, and of other molecules such as amino acids, is not always effected by simple carrier diffusion. The biochemists have amply demonstrated the operation of active transport in the conveyance of many molecules and ions through biological membranes. We will later look at some theoretical aspects of active transport, but first we focus on the simple diffusion of ions across membranes.

The passage of ions across membranes is inextricably linked with the membrane potential, and in Chapter 12 we derived expressions connecting the diffusion potential and the permeabilities of ions. The most famous of these relationships, the GHK equation, is limited in its range of applicability, but at rather low ion concentrations proves satisfactory for many real membranes. The original papers of Hodgkin, Katz, and Keynes are still one of the best introductions to the subject of diffusional ion transport. They worked on the giant squid axon, a long tubular structure wide enough to allow the insertion of a microelectrode, and thus the measurement of the potential difference $\Delta\phi_D$, the diffusion potential. Employing nine different solutions of Na^+, K^+, and Cl^- ions, $\Delta\phi_D$ was recorded using microelectrodes filled with seawater.[323] An experimental difficulty is the junction potential between the microelectrode solution and the axoplasm. Hodgkin and Katz proceeded by assuming the validity of the GHK equation and then finding best-fit values for the junction potential and the permeability ratios P_{Na}/P_K, and so on. The other parameters appearing on the right-hand side of the GHK equation are the ionic concentrations, which can be determined experimentally. The value found for the best-fit junction potential, 11 mV, is reasonable on other evidence. An excellent fit to all the experimental data could be obtained with one set of permeability ratios. This can be regarded as an experimental "proof" of the validity of the equation, at least for the system examined. (In fact, the Henderson and Planck equations also fit the data just as well[324] but all three equations are broadly based on the same physical principles. The GHK equation is simplest to apply.) The breakdown of the GHK equation, discussed in Chapters 12–14, has been attributed, among other things, to the presence of channels in membranes, and much work has been done on channels such as gramicidin incorporated into lipid bilayers.

What evidence is there for channels in natural membranes? Work over the last decade has firmly established the existence of channels in several types of membrane.[325] It is not within the terms of reference of this book to enter into the biochemical and chemical characterization of channels. Perhaps the strongest evidence comes from the isolation of membrane proteins that, when incorporated into lipid bilayers or vesicles, enormously increase the permeability of the bilayer to one ion or a small number of ions, and which exhibit kinetic behavior characteristic of that predicted for transport via channels.

Separate channels for Na^+ and K^+ appear to coexist in nerve membranes,[326] although it is claimed that the acetylcholine-activated channel in frog neuromuscular junction is a common path for both cations.[327] Furthermore, the experimental observation of the inconstancy of the permeability ratio P_{Na}/P_K, not consistent with the simple Nernst–Planck formalism, receives at least a qualitative explanation in channel models. Thus in the case of squid axon, P_{Na}/P_K has been altered by a factor of over three by changing the internal K^+ concentration[328] and similar observations have been made on the P_K/P_{Tl} ratio for echinoderm eggs bathed in K^+/Tl^+ solutions.[329] A two-barrier model with barriers of differing height on either side of a binding site can reproduce much of the experimental data if a knock-on mechanism is assumed; that is, one ion can knock another ion off the binding site.[330] The mathematical treatment is a simple extension of that used in Chapter 13. Of course, the fact that a model can be fitted to experiment is not a proof of the model's correctness, but rate theory is certainly more readily fitted to the facts than classical diffusion theory.

Our knowledge of the physical and chemical nature of channels has been broadened by studies of the permeability and electrical properties of natural membranes and of channels incorporated into lipid bilayers. We make no attempt to review this field, but some of the difficulties facing the theoretician may be appreciated from the permeability data for the K^+ and Na^+ channels in frog-nerve axon as obtained by

Hille.[326,331] For the *sodium* channel, the ratio P_M^+/P_{Na}^+ for the alkali metal ions falls from 1.11 for Li^+ to 0.086 for K^+ and 0.010–0.015 for Rb^+, Cs^+, and NH_4^+. On their own, these figures suggest a geometrical interpretation in terms of the size of the hydrated or partially hydrated ions. Consider, however, the corresponding data for P_M^+/P_K^+ in the *potassium* channel: Li^+, 0.02; Na^+, 0.01; Rb^+, 0.91; Cs^+, 0.08; NH_4^+, 0.13. We see that the Na^+ ion has a very low relative permeability in the K^+ channel while the K^+ ion does not traverse the Na^+ channel with ease. These facts are not immediately explicable in terms of simple size considerations. By studying permeability as a function of pH[332] it has been deduced that there is a single ionizable group within the Na^+ channel with $pK_a = 5.4$. This might be a carboxyl group. As mentioned earlier some channels are guarded by so-called *gates*. The most famous case is that of the Na^+ and K^+ channels in nerve axons for which the opening and closing of the gates is dependent on the potential across the membrane.[333] Biochemical evidence indicates that gates are protein-like molecular moieties. The fact that gates exist leads to the fluctuations in membrane conductivity discussed in Chapter 17. Such fluctuations have been observed for channels incorporated into bilayers and for real membranes (Figures 123 and 124). In certain cases it is possible to observe single-channel conductances, which simplifies the extraction of rate constants. Some gates have been shown to have charges localized on them, and since the movement of such a gate when it opens or closes results in charge movement, an associated *gating current* is observed[334]. The fact that a gate carries charge, either as net charge or as a dipole, explains why gating is influenced by the charge state of the membrane as a whole.

Experiments on conducting channels incorporated into lipid bilayers have contributed, and continue to contribute, greatly to our understanding of the behavior of channels in natural membranes. Particularly helpful have been studies on EIM, excitability-inducing material. This protein, extracted from Aerobacter bacteria, endow lipid bilayers with a voltage-dependent conductance of the type found in nerve cells.[335] A change in applied voltage across this channel results in a noninstantaneous change of conductivity characterized by a relaxation time that is itself voltage-dependent. This behavior is qualitatively similar to that observed by Hodgkin and Huxley[333] in their studies on axons. Again we see the value of the lipid bilayer as an experimental model. Interestingly, the open-state, but not the closed-state, conductance is independent of the lipid used in constructing the bilayer.[335]

The transport of ions across membranes is often "coupled" with chemical processes and/or the passage of other molecules. The quotation marks are there to remind us that two processes may be interdependent but not share a common chemical intermediate (see the following).

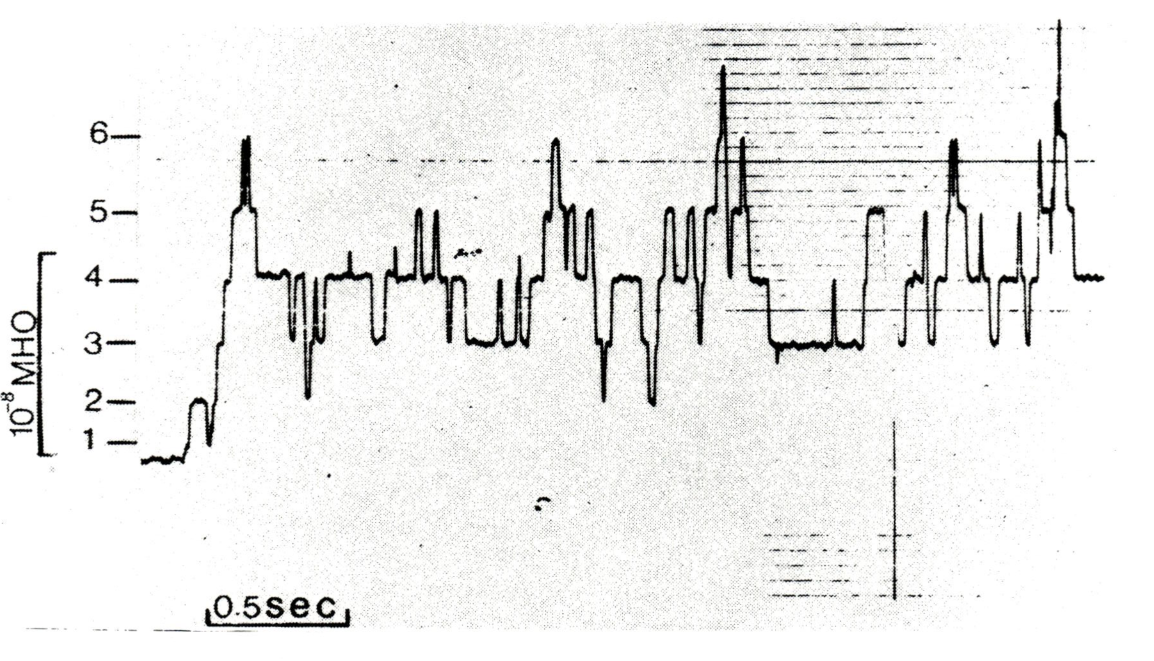

Figure 123. Single-channel conductance fluctuations from a lecithin bilayer in the presence of alamethicin and KCl. A potential of 200 mV was applied to a 10-μm-diameter membrane patch. [From Mueller, P., *Ann. N.Y. Acad. Sci.* 264:247 (1975).]

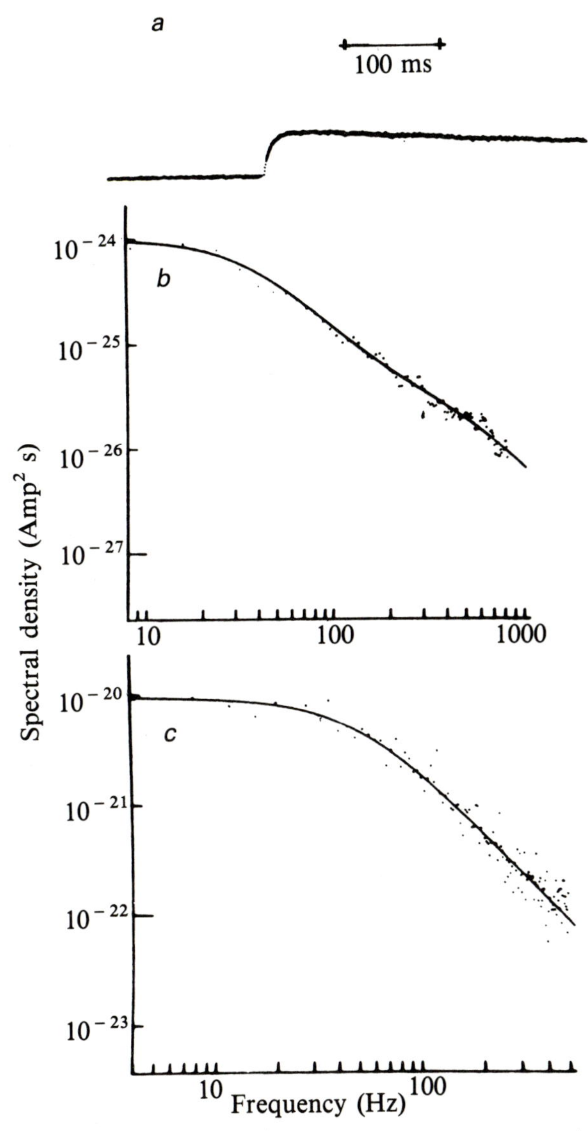

Figure 124. A. Current fluctuations—to the right of the straight segment of baseline—from a voltage-clamped frog node of Ranvier. The lower trace is the mean current showing the rise after a voltage step. **B**. A frequency spectrum corresponding to fluctuations of the kind shown in **A**. **C**. A frequency spectrum for acetylcholine-activated channels at the frog nerve–muscle junction. [From Stevens, C. F., *Nature* 270:391 (1977).]

Examples include the pumping of Na^+ and K^+ ions at the expense of ATP breakdown and the coupling of sugar and amino-acid transport to the flow of Na^+ ions. The details of these processes are still under biochemical investigation, and since we are ignorant of the full molecular mechanisms, there is a temptation to turn to thermodynamics for help in defining the expected general behavior of, and the limitations on, these coupled systems. We have twice looked at a two-flow system through the lens of irreversible thermodynamics. Water transport did not profit spectacularly from this scrutiny but active transport fares better. For an ion pump, assuming the affinity to be a constant, equation (15.35) implies that we will obtain a linear plot for J_r, the rate of the metabolic process, as a function of $\Delta\phi$, the potential difference across the membrane. Vieira et al.[336] tested this prediction for the sodium pump of frog skin by using oxygen consumption as a measure of the metabolic rate. An excellent linear plot was obtained, strengthening one's belief in the correctness of the irreversible thermodynamics approach to the coupled system. While this is encouraging, it should be appreciated that the application of the theory to more complex cases can result in lengthy expressions overloaded with parameters that are, at present, experimentally inaccessible.[337] Furthermore, respectable authors who shall be nameless have misunderstood the nature of the terms appearing in the phenomenological equations.

The standard treatment of metabolism and photosynthesis was dominated for many years by the concept of *chemically* coupled reactions. In particular, oxidative and photosynthetic production of ATP was supposed to involve the participation of intermediate compounds containing so-called high-energy phosphate bonds. Putting to one side the justified criticism of this nomenclature[338] and its implications, we highlight the basic tenet that there is a chain of reactions, successive pairs of reactions being linked by a common chemical intermediate, for examples,

$$X + P \rightleftharpoons XP$$

$$XP + YZ \rightleftharpoons YP + XZ$$

This general idea is relevant to us since many of the enzymes participating in the synthesis of ATP have been shown to be membrane-bound and many of the steps certainly involve the transport of molecules and inorganic ions across membranes. An alternative to the chemical coupling hypothesis was put forward by Peter Mitchell in the late 1950s and early 1960s.[339] In the *chemiosmotic hypothesis* the center of the stage is occupied by an intermediate state rather than by intermediate compounds. In briefest precis: It is proposed that in photosynthesis say, the absorption of radiation by the membranal photosystem results in the production of a "protonmotive" potential across the (thylakoid) membrane:

$$\Delta p = \Delta\phi - Z\Delta pH \tag{18.6}$$

The term $\Delta\phi$ is the electrostatic potential difference across the membrane and $-Z\Delta pH$ is just the disguised Nernst potential caused by a proton gradient across the membrane. Thus radiant energy is supposed to be the immediate driving force for proton translocation through the membrane and Δp, a combination of electrostatic and entropic forces, is presumed to be the energy source for the subsequent chain of chemical reactions that ends in the formation of ATP. The complete process is symbolized in Figure 125. At site I a proton-translocating pump is driven by an oxidation-reduction chain, resulting in the transfer of electrons to O_2. The operation of this pump creates or increases a proton concentration gradient across the membrane. The back flow of protons induced by this gradient and $\Delta\phi$ is the source of free energy for the synthesis of ATP at side II, which is the location of Mitchell's "proton-translocating ATPase system."

The chemical coupling and the chemiosmotic hypotheses both have their proponents but the weight of evidence appears to favor Mitchell at present. Among the early experiments that gave a strong impetus to the theory was that a Jagendorf and Uribe, who equilibrated chloroplasts in aqueous medium at pH 4.0 and then quickly raised the external pH to 8.0. As Mitchell predicted, the resulting efflux of protons resulted in the onset of phosphorylation.[340] In this case most of the "protonmotive force" comes from the pH gradient, but there may well be small diffusion

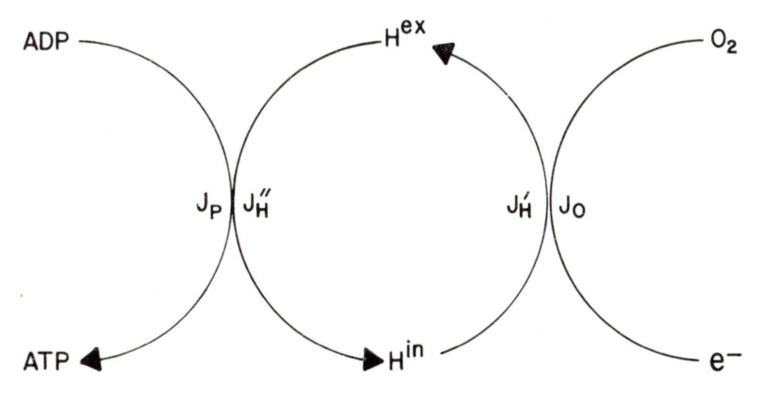

Figure 125. Basic scheme for Mitchell's chemiosmotic hypothesis.

potential contributing to $\Delta\phi$ and arising from cation diffusion associated with proton translocation. (The cation concerned, at least in thylakoids in vivo, appears to be Mg^{++}.[341,342]) The qualitative aspects of the chemiosmotic theory in the main are satisfactory in that the size of the "force" Δp needed to drive ATP synthesis falls within the range of values of Δp found experimentally. It might seem that there is some ambiguity in defining Δp since, if the membrane carries a surface charge, we know that the bulk pH and electric potential differ from the corresponding values near the surface. Fortunately, thermodynamics removes the difficulty. It is clear from its definition, (18.6), that Δp is a disguised electrochemical potential gradient. If we now assume that despite ion flow the bulk solutions are effectively at equilibrium, then it must be true that the electrochemical potential of all species in the bulk must be independent of their location. The relative weights of $\Delta\phi$ and $Z\Delta pH$ will vary with the distance from the membrane but their sum must remain constant. This means that ΔpH and $\Delta\phi$ can be taken as differences between bulk properties.

References

1. Overton, E., *Visch. Naturforsch. Ges. Zurich* 41:383 (1896).
2. Guidotti, G., *Ann. Rev. Biochem.* 41:731 (1972).
3. Gorter, E., and Grendel, F., *J. Exp. Med.* 41:439 (1925).
4. Danielli, J. F., and Davson, H., *J. Cell. Comp. Physiol.* 5:459 (1935).
5. Hitchcock, P. B., Mason, R., Thomas, K. M., and Shipley, G. G., *Proc. Nat. Acad. Sci., U.S.A.* 71:3036 (1974); Elder, M., Hitchcock, P. B., Mason, R., and Shipley, G. G., *Proc. Roy. Soc. London.* A354:157 (1977).
6. Abrahamsson, S., Dahlen, B., Lofgren, H., and Pascher, I., *Prog. Chem. Fats Lipids* 16:125 (1978).
7. Dorset, D. L., *Biochem. Biophys. Acta* 424:396 (1976).
8. Pearson, R. H., and Pascher, I., *Nature* 281:499 (1979).
9. Mushayakarara, E., Albon, N., and Levin, I. W., *Biochem. Biophys. Acta* 686:153 (1982).
10. Hauser, H., Pascher, I., and Sundell, S., *J. Mol. Biol.* 137:249 (1980).
11. Pascher, I., and Sundell, S., *Chem. Phys. Lipids* 20:175 (1977).
12. "Lyotropic Liquid Crystals and the Structure of Biomembranes," *Advances in Chemistry Series*, vol. 152, S. Friberg, Ed., American Chemical Society, Washington, D.C., 1976.
13. Chapman, D., and Salsbury, N. J., *Trans. Far. Soc.* 62:2607 (1966); Chapman, D., Byrne, P., and Shipley, G. G., *Proc. Roy. Soc.* A290:115 (1966).
14. Finean, J. B., *Q. Rev. Biophys.* 2:1 (1969); Charvolin, J., and Tardieu, A., *Solid State Physics*, Suppl. 14, J. Liebert, Ed., Academic Press, New York, 1978.
15. Janiak, M. J., Small, D. M., and Shipley, G. G., *J. Biol. Chem.* 254:6068 (1979).
16. Levine, Y. K., *Prog. Biophys. Mol. Biol.* 24:1 (1972).
17. Ruocco, M. J., and Shipley, G. G., *Biochem. Biophys. Acta* 684:59 (1982).
18. Larsson, K., *Chem. Phys. Lipids* 20:225 (1977).
19. Sackmann, E., Ruppel, D., and Gebhardt, C., in *Liquid Crystals of One- and Two-Dimensional Order*, Springer-Verlag, Berlin, 1980, p. 309.
20. Doniach, S., *Chem., Phys.* 70:4587 (1979).

21. Nagle, J. F., and Wilkinson, D. A., *Biophys. J.* 23:159 (1978).
22. Zaccai, G., Büldt, G., Seelig, A., and Seelig, J., *J. Molec. Biol.* 134:639 (1979).
23. Zaccai, G., Blasie, J. K., and Schoenborn, B. P., *Proc. Natl. Acad. Sci., U.S.A.* 72:376 (1975).
24. McIntosh, T. J., *Biochem. Biophys. Acta* 513:43 (1978).
25. McCaughan, L., and Krimm, S., *Biophys. J.* 37:417 (1982).
26. Barton, P. G., and Gunstone, F. D., *J. Biol. Chem.* 250:4470 (1975).
27. Sato, N., Murato, N., Miura, Y., and Ueta, N., *Biochem. Biophys. Acta* 572:19 (1979).
28. Forsyth, J. M., and Frankel, R. D., *Biophys. J.* 37:73a (1982).
29. "Liposomes and Their Uses in Biology and Medicine," D. Papahadjopoulos, Ed., *Ann. NY. Acad. Sciences* 308, 1978.
30. Wennerstrom, H., and Lindman, B., *Phys. Rep.* 52:1 (1979).
31. (a) Israelachvili, J. N., Marcelja, S., and Horn, R. G., *Q. Rev. Biophys.* 13:121 (1980). (b) Israelachvili, J. N., and Mitchell, D. J., *Biochem. Biophys. Acta* 389: 13 (1975).
32. Israelachvili, J. N., Mitchell, D. J., and Ninham, B. W., *J. Chem. Soc., Faraday Trans.* II, 72:1525 (1976).
33. Mason, J. T., and Huang, C., in ref. 29.
34. Tardieu, A., Doctorate Thesis, Université de Paris-Sud (1972).
35. Berden, J. A., Baker, R. W., and Radda, G K., *Biochem. Biophys. Acta* 375:186 (1975).
36. Longmuir, K. J., and Dahlquist, F. W., *Proc. Natl. Acad. Sci.* 73:2716 (1976).
37. Tsai, K. W., and Lenard, J., *Nature* 253:554 (1975).
38. Cullis, P. R., and de Kruijff, B., *Biochem. Biophys. Acta,* 507:207 (1978).
39. Rand, R. P., and Sengupta, S., *Biochem. Biophys. Acta* 255:484 (1972); Cullis, P. R., Verkley, A. J., and Ververgaert, P. H. J. T., *Biochem. Biophys. Acta* 513:11 (1978); Deamer, D. W., Leonard, R., Tardieu, A., and Branton, D., *Biochem. Biophys. Acta* 219:47 (1970).
40. Demel, R. A., Bruckdorfer, K. R., and van Deenen, L. L. M., *Biochem. Biophys. Acta* 255:311 (1972).
41. Watts, A., Harlos, K., Maschke, W., and Marsh, D., *Biochem. Biophys. Acta* 510:63 (1978).
42. Madden, T. D., and Cullis, P. R., *Biochem. Biophys. Acta* 684:149 (1982).
43. de Gennes, P. G., *The Physics of Liquid Crystals*, Oxford University Press, London, 1974.
44. Priestly, E. B., Wojtowicz, P. J., and Sheng, P., *Introduction to Liquid Crystals*, Plenum Press, New York, 1975.
45. Saupe, A., *Z. Naturf.* 19a:161 (1964).
46. Silver, B. L., *Irreducible Tensor Methods*, Academic Press, New York, 1976.
47. Seelig, J., *Q. Rev. Biophys.* 10:353 (1977).
48. Berliner, L. J., in *Spin Labelling, Theory and Applications*, L. J. Berliner, Ed., vol. 1, Academic Press, New York, 1976.
49. Stockton, G. W., and Smith, I. C. P., *Chem. Phys. Lipids* 17:251 (1976).

50. Kinosita, K. Jr., Ikegami, A., and Kawato, S., *Biophys. J.* 37:461 (1982).
51. Lee, A. G., *Prog. Biophys. Molec. Biol.* 29:5 (1975).
52. Pace, R. J., and Chan, S. I., *Biophys. J.* 33:165a (1981).
53. Seelig. A., and Seelig, J., *Biochemistry* 13:4839 (1974).
54. Vogel, H., *Ber. Bunsenges. Phys. Chem.* 85:518 (1981).
55. Gaber, B. P., and Peticolas, W. L., *Biochem. Biophys. Acta* 465:260 (1977).
56. Pink, D. A., Green, T. J., and Chapman, D., *Biochemistry* 19:349 (1980).
57. Snyder, R. G., Cameron, D. G., Casal, H.L., et al., *Biochem. Biophys. Acta* 684:111 (1982).
58. Griffin, R. G., Powers, L., and Pershan, P. S., *Biochemistry* 17:2718 (1978).
59. Herzfeld, J., Griffin, R. G., and Haberkorn, R. A., *Biochemistry* 17:2711 (1978).
60. Kohler, S. J., and Klein, M. P., *Biochemistry* 15:967 (1976).
61. Skarjune, R., and Oldfield, E., *Biochemistry* 18:5903 (1979).
62. Seelig, J., Gally, H.-U., and Wohlgemuth, R., *Biochem. Biophys. Acta* 467:109 (1977).
63. Seelig, J., *Biochem. Biophys. Acta* 515:105 (1978).
64. Cullis, P. R., and de Kruijff, B., *Biochem. Biophys. Acta* 559:399 (1979).
65. Traube, J., *Annalen* 265:27 (1891).
66. Traube, J. *Samml. Chem. Chem.-Tech. Vortr.* 4:255 (1899).
67. Butler, J. A. V., *Trans. Faraday Soc.* 33:229 (1937).
68. Kauzmann, W. J., *Adv. Protein Chem.* 14:1 (1959).
69. Frank, H. S., and Evans, M. W., *J. Chem. Phys.* 13:507 (1945).
70. Tanford, C., *The Hydrophobic Effect: Formation of Micelles and Biological Membranes*, 2nd ed., Wiley, New York, 1980.
71. Ben-Naim, A., *Hydrophobic Interactions*, Plenum Press, New York, 1980.
72. Némethy, G., and Scheraga, H. A., *J. Chem. Phys.* 36:3382 (1962).
73. Goldman, S., *J. Chem. Phys.* 75:4064 (1981).
74. Hildebrand, J., *J. Phys. Chem.* 72:1841 (1968); *Proc. Natl. Acad. Sci.* 76:194 (1979).
75. *Physical Chemistry An Advanced Treatise*, H. Eyring, D. Henderson, and W. Jost, Eds., Academic Press, New York, 1971, Vol. I, p 327.
76. Cheng, W. H., *Biochim. Biophys. Acta* 600:358 (1980).
77. Lee, A. G., *Biochim. Biophys. Acta* 507:433 (1978).
78. Lee, A. G., *Biochim. Biophys. Acta* 472:285 (1977).
79. Ververgaert, P. H. J. Th., Verkleij, A. J., Elbers, P. F., and van Deenen, L. L. M., *Biochim. Biophys. Acta* 311:320 (1973)).
80. Grant, C. W. M., Wu, S. H., and McConnell, H. M., *Biochim. Biophys. Acta* 363:151 (1974).
81. Wu, S. H., and McConnell, H. M., *Biochemistry* 14:847 (1975).
82. Borges, J. L., *Labyrinths*, Penguin Books, London, 1981.
83. Black, S. G., and Dixon, G. S., *Biochemistry* 20:6740 (1981).
84. Klein, R. A., *Q. Rev. Biophys.* 15:667 (1982).
85. Yellin, N., and Levin, I. W., *Biochim. Biophys. Acta* 468:490 (1977).
86. Lee, A. G., *Biochim. Biophys. Acta* 413:11 (1975).

87. Mabrey, S., and Sturtevant, J. M., *Proc. Natl. Acad. Sci.* 73:3862 (1976).
88. Ruckenstein, E., and Nagarajan, R., *J. Phys. Chem.* 79:2622 (1975).
89. Tanford, C., *J. Phys. Chem.* 78:2469 (1974).
90. Israelachvili, J. N., Mitchell, D. J., and Ninham, B. W., *J. Chem. Soc., Faraday Trans. II* 72:1525 (1976).
91. Israelachvili, J. N., Mitchell, D. J., and Ninham, B. W., *Biochim. Biophys. Acta* 470:185 (1977).
92. Evans, E. A., Gershfeld, N. L., Ginsberg, L., and Parsegian, V. A., *Biophys. J.* 37:164a (1982).
93. Chan, D. Y. C., Mitchell, D. J., and Ninham, B. W., *J. Chem. Phys.* 70:2946 (1979).
94. Feynman, R. P., Leighton, R. B., and Sands, M., *The Feynman Lectures on Physics*, Addison-Wesley, New York, 1981, Chapter 6.
95. Reference 94, Chapter 10.
96. Parsegian, V. A., in "Carriers and Channels in Biological Systems," *Ann. New York Acad. Sci.* 264:161 (1975).
97. (a) LeNeveu, D. M., Rand, R. P., and Parsegian, V. A., *Nature (London)* 259:601 (1976).
97. (b) LeNeveu, D. M., Rand, R. P., Gingell, D., and Parsegian, V. A., *Science (Washington, D.C.)* 191;399 (1976).
97. (c) Parsegian, V. A., Fuller, N., and Rand, R. P., *Proc. Natl. Acad. Sci.* 76:2750 (1979).
97. (d) LeNeveu, D. M., Rand, R. P., Gingell, D., and Parsegian, V. A., *Biophys. J.* 18:209 (1977).
98. Chan, D. Y. C., Mitchell, D. J., Ninham, B. W., and Pailthorpe, B. A., *J. Colloid Interface Sci.* 64:194 (1978).
99. Hamaker, H. C., *Physica* 4:1058 (1937).
100. Lifshitz, E. M., *Sov. Phys., JETP* 2:73 (1956).
101. Renne, M. J., and Nijboer, B. R. A., *Chem. Phys. Lett.* 1:317 (1967).
102. Landau, L. D., and Lifshitz, E. M., *Statistical Physics*, Addison-Wesley, Reading, Mass., 1958.
103. London, F., *Trans. Faraday Soc.* 33:8 (1937).
104. McLachlan, A. D., *Proc. Roy. Soc. London* A271:387 (1963); A274:80 (1963).
105. Casimir, H. B. G., and Polder, D., *Phys. Rev.* 73:360 (1948).
106. Israelachvili, J. N., *Proc. Roy. Soc. London* A331:39 (1972).
107. Imura, H., and Okano, K., *J. Chem. Phys.* 58:2763 (1973).
108. McLachlan, A. D., *Disc. Faraday Soc.* 40:239 (1965).
109. Dzyaloshinskii, I.E., Lifshitz, E. M., and Pitaevskii, L. P., *Adv. Phys.* 10:165 (1961).
110. Davies, B., and Ninham, B. W., *J. Chem. Phys.* 56:5797 (1972).
111. Barnes, C., and Davies, B., *J. Chem. Soc. Far. Trans.* II, 71:1667 (1975).
112. Israelachvili, J. N., and Tabor, D., *Proc. Roy. Soc.* A331:19 (1972).
113. Gingell, D., and Parsegian, V. A., *J. Theor. Biol.* 36:41 (1972).
114. Gingell, D., and Parsegian, V. A., *J. Coll. Int. Sci.* 44:456 (1973).
115. Haydon, D. A., and Taylor, J. L., *Nature (London)* 217:739 (1968).
116. Parsegian, V. A., *Ann. Rev. Biophys. Bioeng.* 2:221 (1973).
117. Parsegian, V. A., and Ninham, B. W., *J. Theor. Biol.* 38:101 (1973).

118. LeNeveu, D. M., Rand, R. P., Parsegian, V. A., and Gingell, D., *Biophys. J.* 18:209 (1977).
119. Lis, L. J., McAlister, M., Fuller, N., et al., *Biophys. J.* 37:657 (1982).
120. Marcelja, S., *Biochim. Biophys. Acta* 367:165 (1974).
121. Jehle, H., *Ann. N.Y. Acad. Sci.* 158:240 (1969).
122. Mahanty, J., and Ninham, B. W., *Dispersion Forces*, Academic Press, London, 1976.
123. Israelachvili, J. N., *Q. Rev. Biophys.* 6:341 (1974).
124. Parsegian, V. A., in *Physical Chemistry: Enriching Topics*, J. van Olphen, and J. Mysels, Eds., Theorex, La Jolla, Calif., 1975.
125. Stern, O., *Z. Elektrochem.* 30:508 (1924).
126. See for example: McLaughlin, S. A., *Curr. Top. Membranes Transport* 9:71 (1977).
127. Zakrzewska, K., and Pullman, B., *FEBS Lett.* 131:77 (1981).
128. Ninham, B. W., and Parsegian, V. A., *J. Theor. Biol.* 31:405 (1971).
129. Verwey, E. J. W., and Overbeek, J. Th. G., *Theory of the Stability of Lyophobic Colloids*, Elsevier, Amsterdam, 1948.
130. Parsegian, V. A., and Gingell, D., *Biophys. J.* 12:1192 (1972).
131. Hubbell, W. L., and Cafiso, D. S., *Ann. Rev. Biophys. Bioeng.* 10:217 (1981).
132. Waggoner, A. S., and Grinvald, A., *Ann. N.Y. Acad. Sci.* 303:217 (1977).
133. Lis, L. J., McAlister, M., Fuller, N., et al. Biophys. J. 37:667 (1982).
134. Allis, W. P., and Herlin, M. A., *Thermodynamics and Statistical Mechanics*, McGraw-Hill, New York, 1952, p. 193.
135. Nagle, J. F., *J. Chem. Phys.* 58:252 (1973).
136. Maier, W., and Saupe, A., *Z. Naturforsch., Teil A* 13:564 (1958).
137. Schindler, H., and Seelig, J., *Biochem.* 14:2283 (1975).
138. Caillé, A., Pink, D., de Verteuil, F., and Zuckermann, M. J., *Can. J. Phys.* 58:581 (1980).
139. Caillé, A., Rapini, A., Zuckermann, M. J., et al. *Can. J. Phys.* 56:348 (1978).
140. Nagle, J. F., *Ann. Rev. Phys. Chem.* 31:157 (1980).
141. van der Ploeg, P., and Berendsen, H. J. C., *J. Chem. Phys.* 76:3271 (1982).
142. Post, J. F. M., James, E., and Berendsen, H. J. C., *J. Mag. Res.* 47:251 (1982).
143. Singer, S. J., and Nicholson, G. L., *Science* 175:720 (1972).
144. Hesketh, T. R., Smith, G. A., Houslay, M. D., et al., *Biochem.* 15:4145 (1976).
145. Jost, P. C., Griffith, O. H., Capaldi, R. A., and Van der Kooi, G., *Proc. Natl. Acad. Sci.* 70:480 (1973).
146. Yang, S. Y., Gutowsky, H. S., Huang, J. C., et al., *Biochem.* 18:3257 (1979).
147. Rice, D. M., Meadows, M. D., Scheinman, A. O., et al., *Biochem.* 18:5893 (1979).
148. Pink, D. A., and Chapman, D., *Proc. Natl. Acad. Sci.* 76:1542 (1979).
149. Hoffman, W., Pink, D. A., Restall, C., and Chapman, D., *Eur. J. Biochem.* 114:585 (1981).
150. Pink, D. A., Georgallas, A., and Chapman, D., *Biochem.* 20:7152 (1981).

151. Devaux, P. F., Davoust, J., and Rousselet, A., *Biochem. Soc. Symp.* 46:207 (1981).
152. Marsh, D., Watts, A., Pates, R. D., et al., *Biophys. J.* 37:265 (1982).
153. Yeagle, P. L., *Biophys. J.* 37:227 (1982).
154. Utsumi, H., Tungaal, B. D., and Stoffel, W., *Biochem.* 19:2385 (1980).
155. Yeagle, P. L., Langdon, R. G., and Martin, R. B., *Biochem.* 16:3487 (1977).
156. Taraschi, T., and Mendelsohn, R., *Proc. Natl. Acad. Sci.* 77:2362 (1980).
157. Kleeman, W., Grant, C. W. M., and McConnell, H. M., *J. Supramol. Struct.* 2:609 (1974).
158. Van Dijck, P. W. M., de Kruijff, B., Van Deenen, L. L. M., et al., *Biochim. Biophys. Acta* 455:576 (1976).
159. Boggs, J. M., Wood, D. D., Moscarello, M. A., and Papahadjopoulos, D., *Biochem.* 16:2325 (1977).
160. Jost, P. C., and Griffith, O. H., *Biophys. J.* 37:329a (1982).
161. Melchior, D. L., *Curr. Topics Memb. Trans.* 17:263 (1982).
162. Mabrey, S., Mateo, P. L., and Sturtevant, J.M., *Biochem.* 17:2464 (1978). [Compare Jacob, R., and Oldfield, E., *Biochem.* 18:3280 (1979).]
163. Gally, H. U., Seelig, A., and Seelig, J., *Hoppe-Seyler's Z. Physiol. Chem.* 357:1447 (1976).
164. Jacobs, R., and Oldfield, E., *Biochem.* 18:3280 (1979).
165. Mendelsohn, R., Dluhy, R., Taraschi, T., et al., *Biochem.* 20:6699 (1981).
166. Cornell, B. A., Chapman, D., and Peel, W. E., *Chem. Phys. Lipids* 23:223 (1979).
167. Marcelja, S., *Biochim. Biophys. Acta* 455:1 (1976).
168. Owicki, J. C., and McConnell, H. M., *Proc. Nat. Acad. Sci.* 76:4750 (1979).
169. Landau, L. D., and Lifshchitz, E. M., *Statistical Physics*, Addison-Wesley, Reading, Mass., 1969.
170. Jahnig, F., *Springer Ser. Chem. Phys.* 11:344 (1980).
171. Rubenstein, J. L. R., Smith, B. A., and McConnell, H. M., *Proc. Natl. Acad. Sci.* 75:15 (1979).
172. Freire, E., and Snyder, B., *Biophys. J.* 37:617 (1982).
173. Ikegami, A., Kinosita, K. Jr., Kouyama, T., and Kawato, S., in *Structure, Dynamics and Biogenesis of Biomembranes*, R. Sato, and S-I. Ohnishi, Eds., Japan Scientific Societies Press, Tokyo; Plenum Press, New York, 1982.
174. Gomez-Fernandez, J. C., Goni, F. M., Bach, D., et al., *Biochim. Biophys. Acta* 598:502 (1980).
175. Davoust, J., Bienvenue, A., Fellmann, P., and Devaux, P. F., *Biochim. Biophys. Acta* 596:28 (1980).
176. Kuo, A., and Wade, C. G., *Biochem.* 18:2300 (1979).
177. Webb, W. W., Barak, L. S., Tank, D. W., and Wu, E-S., *Biochem. Soc. Symp.* 46:191 (1981).
178. Roseman, M. A., and Thomson, T. E., *Biochem.* 19:439 (1980).
179. de Kruijff, B., van der Besselaar, A. M. H. P., Cullis, P. R., et al., *Biochem. Biophys. Acta* 514:1 (1978).

180. Zilbersmit, D. B., and Hughes, M. E., *Biochim. Biophys. Acta* 469:99 (1979).
181. Saffman, P. G., and Delbruck, M., *Proc. Natl. Acad. Sci.* 72:3111 (1975).
182. Hughes, B. D., Pailthorpe, B. A., White, L. R., and Sawyer, W. H., *Biophys. J.* 37:673 (1982).
183. Hare, F., *Biophys. J.* 42:205 (1983), and references therein.
184. Kinosita, K., Jr., Kawato, S., and Ikegami, A., *Biophys. J.* 20:289 (1977).
185. Munro, I., Pecht, I., and Stryer, L, *Proc. Natl. Acad. Sci.* 76:56 (1979).
186. Jacobs, R. E., and Oldfield, E., *Prog. NMR Spectroscopy* 14:113 (1981).
187. Axelrod, D., Koppel, D. E., Schlessinger, J., et al., *Biophys. J.* 16:1055 (1976).
188. Peters, R., *Cell Biol. Intern. Rep.* 5:733 (1981).
189. Koppel, D. E., and Sheetz, M. P., *Biophys. J.* 43:175 (1983).
190. Koppel, D. E., Axelrod, D., Schlessinger, J., et al., *Biophys. J.* 16:1315 (1976).
191. Snyder, B., van Osdol, W., and Friere, E., *Biophys. J.* 33:163a (1981).
192. Owicki, J. C., and McConnell, H. M., *Biophys. J.* 30:383 (1980).
193. Hochli, M., and Hackenbrock, C. R., *Proc. Natl. Acad. Sci.* 76:1236 (1979).
194a. Overbeek, J. Th. G., *Prog. Biophys* 6:58 (1956).
194b. Siggel, U., *Bioelect. Bioenerg.* 8:327, 339, 347 (1981).
194c. Fick, A., *Ann. Physik.* 94:59 (1855); *Phil. Mag.* 10:30 (1855).
195. Einstein, A., *Ann. Physik.* 17:549 (1905).
196. Smoluchowski, M. V., *Ann. Physik.* 21:756 (1906).
197. Langevin, P., *C.R. Mebd. Seanc. Acad. Sci., Paris* 146:530 (1908).
198. Eyring, F., Lumry, R., and Woodbury, J. W., *Rec. Chem. Prog.* 10:100 (1949).
199. Planck, M., *Ann. Phys. Chem.* 39:161 (1890).
200. Henderson, P., *Z. Phys. Chem.* 59:118 (1907).
201. Morf, W. E., *Anal. Chem.* 49:810 (1977).
202. Kramers, H. A., *Physica* 7:284 (1940).
203. Goldmann, D. E., *J. Gen. Physiol.* 27:37 (1943).
204. Planck, M., *Ann. Physik Chem.* 35:561 (1890).
205. Hodgkin, A. L., and Katz, B., *J. Physiol. (London)* 108:37 (1949).
206. Attwell, D., and Jack, J. J. B., *Prog. Biophys. Mol. Biol.* 34:81 (1978).
207. Ussing, H. H., *Acta Physiol. Scand.* 19:43 (1949); *Ann. N. Y. Acad. Sci.* 137:543 (1966); in *Membrane Transport in Biology*, Springer-Verlag, Berlin, 1978, Vol. I.
208. Hope, A. B., *Ion Transport and Membranes*, Butterworths, London, 1971.
209. Danielli, J. F., *J. Physiol. London* 96:2P (1939).
210. Davson, H., and Danielli, J. F., *The Permeability of Natural Membranes*, Cambridge University Press, Cambridge, England, 1943, pp. 310–352.
211. Woodbury, J. W., in *Chemical Dynamics: Papers in Honor of Henry Eyring*, J. O. Hirschfelder, Ed., Wiley, New York, 1971.
212. Hodgkin, A. L., and Keynes, R. D., *J. Physiol. London* 116:449 (1952).
213. "Carriers and Channels in Biological Systems," *Ann. N.Y. Acad. Sci.* 264 (1975).

214. Urry, D. W., Goodall, M. C., Glickson, J. D., and Mayers, D. F., *Proc. Natl. Acad. Sci.* 68:1907 (1971).
215. Barr, L., *J. Theor. Biol.* 9:351 (1965).
216. Teorell, T., *Proc. Soc. Exp. Biol. Med.* 33:282 (1935); *Prog. Biophys. Chem.* 3:305 (1953).
217. Meyer, K. H., and Sievers, J. F., *Helv. Chim. Acta* 19:649 (1936).
218. Hille, B., in *Membranes: A Series of Advances*, Eisenmann. Ed., Dekker, New York, 1975.
219. Hille, B. and Schwarz, W., *J. Gen. Physiol.* 72:409 (1978).
220. Levitt, D. G., *Biophys. J.* 37:575 (1982).
221. Lauger, P., *Biochim. Biophys. Acta* 311:423 (1973); 455:493 (1976).
222. Heckmann, K., *Biomembranes* 3:127 (1972).
223. Apel, H. J., Bamberg, E., and Lauger, P., *Biochim. Biophys. Acta* 552:369 (1979).
224. Casper, J., Landuyt-Caufriez, M., Deleers, M., and Ruysschaert, J. M., *Biochim. Biophys. Acta* 554:23 (1979).
225. Attwell, D., *Memb. Trans. Proc.* 3:29 (1979).
226. Lewis, C. A., and Stevens, C. F., *Memb. Trans. Proc.* 3:133 (1979).
227. Moore, C., and Pressman, B. C., *Biochim. Biophys. Res. Comm.* 15:562 (1964).
228. Pressman, B. C., in *Inorganic Biochemistry*, G. L. Eichhorn, Ed., Elsevier, New York, 1973, p. 203.
229. Johnson, S. M., Herrin, J., Liu, S. J., and Paul, I. C., *J. Am. Chem. Soc.* 92:4428 (1970).
230. Kilbourn, B. T., Dunitz, J. D., Pioda, L. A. R., and Simon, W., *J. Mol. Biol.* 30:559 (1967).
231. Pressman, B. C., Harris, E. J., Jagger, W. S., and Johnson, J. H., *Proc. Natl. Acad. Sci.* 58:1949 (1967).
232. Pressman, B. C., and de Guzman, N. T., *Ann. N.Y. Acad. Sci.* 264:373 (1975).
233. Krasne, S., Eisenman, G., and Szabo, G., *Science* 174:412 (1971).
234. Griffith, O. H., Dehlinger, P. J., and Van, S. P., *J. Mem. Biol.* 15:159 (1974).
235. Parsegian, V. A., *Ann. N.Y. Acad. Sci.* 264:161 (1975).
236. Guggenheim, E. A., *Thermodynamics*, 5th rev. ed., North-Holland, Amsterdam, 1967, p. 300.
237. Haydon, D. A., and Myers, V. B., *Biochim. Biophys. Acta* 307:429 (1973).
238. Lauger, P., Benz, R., Stark, G., et al., *Q. Rev. Biophys.* 14:513 (1981).
239. Jordan, P. C., and Stark, G., *Biophys. Chem.* 10:273 (1979).
240. Andersen, O. S., and Fuchs, M., *Biophys. J.* 15:795 (1975).
241. Bruner, L. J., *J. Mem. Biol.* 22:125 (1975).
242. Cafiso, D. S., and Hubbell, W. L., *Biochemistry.* 17:187 (1978).
243. Wulf, J., Benz, R., and Pohl, W. G., *Biochim. Biophys. Acta* 465:429 (1977).
244. Hubbell, W. L., Cafiso, D. S., and Brown, M. F., *Fed. Proc.* 39:1983 (1980).
245. Kotyk, A., and Janacek, K., *Cell Membrane Transport, Principles and Techniques*, 2nd ed. Plenum Press, New York, 1975.

246. Cornish-Bowden, A., *Principles of Enzyme Kinetics*, Butterworths, London, 1976.
247. Hladky, S. B., *Curr. Top. Mem. Trans.* 12:53(1979).
248. Turner, R. J., *Biochim. Biophys. Acta* 689:444 (1982).
249. Deves, R., and Krupka, R. M., *Biochim. Biophys. Acta* 556:533 (1979).
250. Onsager, L., *Phys. Rev.* 37:405 (1931); 38:2265 (1931).
251. Finkelstein, A., and Rosenberg, P. A., *Memb. Trans. Proc.* 3:73 (1979).
252. Prigogine, I., and Lefever, R., *Membranes, Dissipative Structures and Evolution*, G. Nicolis, and R. Lefever, Eds., Wiley, New York, 1975.
253. Vieira, F. L., Caplan, S. R., and Essig, A., *J. Gen. Physiol.* 59:60 (1972).
254. Kedem, O., and Caplan, S. R., *Trans. Faraday Soc.* 61:1897 (1965).
255. Essig, A., and Caplan, S. R., *J. Gen. Physiol.* 8:1434 (1968).
265. Ussing, H. H., in *The Alkali Metal Ions in Biology*, Springer-Verlag, Berlin, 1960.
257. de Groot, S. R., and Mazur, P., *Non-Equilibrium Thermodynamics*, North-Holland, Amsterdam, 1963.
258. Prigogine, I., *Thermodynamics of Irreversible Processes*, Wiley, New York, 1961.
259. Caplan, S. R., *Curr. Top. Bioenerg.* 4:1 (1971).
260. Katchalsky, A., and Curran, P. F., *Non-Equilibrium Thermodynamics in Biophysics*, Harvard University Press, Cambridge, Mass. 1965.
261. Kedem, O., and Katchalsky, A., *Biochim. Biophys. Acta* 27:229 (1958).
262. Hill, A., *Q. Rev. Biophys.* 12:67 (1979).
263. Katchalsky, A., in *Membrane Transport and Metabolism*, A. Kleinzeller, and A. Kotyk, Eds., Academic Press, London, 1961.
264. Staverman, A. J., *Rec. Trav. Chim. Pays-Bas. Belg.* 70:344 (1958).
265. Kedem, O., and Katchalsky, A., *Trans. Faraday. Soc.* 59:1918 (1963).
266. Speigler, K. S., *Trans. Faraday. Soc.* 54:1408 (1958).
267. Kedem, O., and Katchalsky, A., *J. Gen. Physiol.* 45:143 (1961).
268. House, C. R., *Water Transport in Cells and Tissues,* Edward Arnold, London, 1974.
269. Schultz, S. G., *Basic Principles of Membrane Transport,* Cambridge University. Press, Cambridge, England, 1980, p. 65.
270. Essig, A., *Biophys. J.* 8:53 (1968).
271. Hill, T. L., *Free Energy Transduction in Biology,* Academic Press, New York, 1977.
272. Hill, T. L., *J. Chem. Phys.* 76:1122 (1982).
273. Jaynes, E. T., *Ann. Rev. Phys. Chem.* 31:579 (1980).
274. Anderson, P. W., *J. Phys. Soc. Japan* 9:316 (1954); see Abragam, A., *The Principles of Nuclear Magnetism,* Clarendon Press, Oxford, England, 1961.
275. Levine, Y. K., Birdsall, N. J. M., Lee, A. G. and Metcalfe, J. C., *Biochem.* 11:1416 (1972).
276. Lee, A. G., Birdsall, N. J. M., Metcalfe, J. C., et al., *Proc. Roy. Soc. London.*, B, 193:253 (1976).
277. Feller, W., *Probability Theory and Its Applications,* Wiley, New York, 1950.
278. Elson, E. L., and Webb, W. W., *Ann. Rev. Biophys. Bioeng.* 4:311 (1975).

279. Wilkins, M. H. F., Blaurock, A. E., and Engleman, D. M., *Nature (New Biol.)*, 230; 72 (1971).
280. Engleman, D. M., *J. Mol. Biol.* 58:153 (1971).
281. Knutton, S., Finean, J. B., Coleman, R., and Limbrick, A. R., *J. Cell Sci.* 7:357 (1970).
282. Melchior, D. L., and Steim, J. M., *Prog. Surf. Membr. Sci.* 13:211 (1979).
283. Reinert, J. C., and Steim, J. M., *Science* 168:1580 (1970).
284. Davis, J. H., Nichol, C. P., Weeks, G., and Bloom, M., *Biochem.* 18:2103 (1979).
285. Kang, S. Y., Gutowsky, H. S., and Oldfield, E., *Biochem.* 18:3268 (1979).
286. Kang, S. Y., Kinsey, R., Rajan, S., et al., unpublished results quoted in *Prog. N.M.R. Spect.* 14:113 (1980).
287. Gounaris, K., Sen, A., Brain, A.P.P., et al. *Biochim. Biophys. Acta* 728:129 (1983).
288. Pottel, H., van der Meer W., and Herreman, W., *Biochim. Biophys. Acta* 730:181 (1983).
289. Demeduik, P., Cowan, D. L., and Moscatelli, E. A., *Biochim. Biophys. Acta* 730:263 (1983).
290. Kowalska, M. A., and Cierniewski, C. S., *Biochim. Biophys. Acta* 729:275 (1983).
291. Skomurski, J. F., Racine, F. M., and Vary, J. C., *Biochim. Biophys. Acta* 731:428 (1983).
292. Landsberger, F. R., Compans, R. W., Choppin, P. W., and Lenard, J., *Biochem.* 12:4498 (1973).
293. Landsberger, F. R., and Compans, R. W., *Biochem.* 15:2356 (1976).
294. Stoffel, W., Bister, K., Schreiber, C., and Tungaal, B., *Hoppe-Seyler's Z. Physiol. Chem.* 357:905 (1976).
295. Burgess, G. M., Giraud, F., Poggioli, J., and Claret, M., *Biochim. Biophys. Acta* 731:387 (1983).
296. Gordon, L. M., Whetton, A. D., Rawal, S., et al., *Biochim. Biophys. Acta* 729:104 (1983).
297. Hennessey, T. M., and Nelson, D. L., *Biochim. Biophys. Acta* 728:145 (1983).
298. Itoh, S., *Biochim. Biophys. Acta* 548:579 (1979).
299. Barber, J., and Chow, W. S., *FEBS Lett.* 105:5 (1979).
300. Masamoto, K., Itoh, S., and Nishimura, M., *Biochim. Biophys. Acta* 591:142 (1980).
301. Barber, J., and Searle, G. F. W., *FEBS Lett.* 92:5 (1978).
302. Leech, R. M., and Murphy, D. J., in *The Intact Chloroplast,* Vol. 1. *Topics in Photosynthesis,* J. Barber, Ed., Elsevier, Amsterdam, p. 365.
303. Yerkes, C. T., and Babcock, G. T., *Biochim. Biophys. Acta,* 634:19 (1981).
304. Barber, J., and Mills, J., *FEBS Lett.* 68:288 (1976).
305. Gross, E. L., and Hess, S. C., *Biochim. Biophys. Acta* 339: 334 (1974).
306. Rubin, B. T., Chow, W. S., and Barber, J., *Biochim. Biophys. Acta* 634:174 (1981).
307. Sculley, M. J., Duniec, J. T., Thorne, S. W., et al., *Arch. Biochem. Biophys.* 201:339 (1980).

308. Mullet, J. E., and Arntzen, C. J., *Proc. 5th Int. Congr. Photosynth.*, Halkidiki, Greece (1981).
309. Barber, J., *FEBS Lett.* 118:1 (1980).
310. Barber, J., *Biochem. Soc. Trans.* 10:331 (1982).
311. Vanderkooi, G., and Bendler, J. T., in *Structure of Biological Membranes*, S. Abrahamsson and I. Pascher, Eds., Plenum Press, New York, 1977. p. 551.
312. McDonald, A. G., and Cossins, A. R., *Biochim. Biophys. Acta* 730:239 (1983).
313. Scandella, C. J., Devaux, P., and McConnell, H. M., *Proc. Natl. Acad. Sci.* 69:2056 (1972).
314. Cass, A., and Finkelstein, A., *J. Gen. Physiol.* 50:1765 (1967).
315. Leaf, A., and Hays, R. M., *J. Gen. Physiol.* 45:921 (1962).
316. Andersen, B., and Ussing, H. H., *Acta Physiol. Scand.* 39:228 (1957).
317. Schafer, J. A., and Andreoli, T. E., *J. Clin. Invest.* 51:1264 (1972).
318. Hebert, S. C., and Andreoli, T. E., *Biochim. Biophys. Acta* 650:267 (1982).
319. Fant, M. E., Yeakley, J., and Harrison, R. W., *Biochim. Biophys. Acta* 731:415 (1983).
320. Rosenburg, T., and Wilbrandt, W., *J. Gen. Physiol.* 41:289 (1957).
321. Mitchell, P., *Symp. Soc. Exp. Biol.* 8:254 (1954).
322. Levine, M., Oxender, D. L., and Stein, W. D., *Biochim. Biophys. Acta* 109:151 (1965).
323. Hodgkin, A. L., and Katz, B., *J. Physiol.* 108:37 (1949).
324. Rosenberg, S., *Biophys. J.* 9:500 (1969).
325. *Carriers and Channels in Biological Systems*, A. E. Shamoo, Ed., *Ann. N.Y. Acad. Sci.* 264 (1975).
326. Hille, B., *Prog. Biophys. Mol. Biol.* 21:1 (1970).
327. Dionne, V. E., *Memb. Trans. Proc.* 3:123 (1979).
328. Cahalan, M., and Begenisich, T., *J. Gen. Physiol.* 68:111 (1976).
329. Hagiwara, S., Miyazaki, S., Krasne, S., and Ciani, S., *J. Gen. Physiol.* 70:269 (1977).
330. Begenisich, T., and Cahalan, M., *Memb. Trans. Proc.* 3:113 (1979).
331. Hille, B., *J. Gen. Physiol.* 59:637 (1972).
332. Woodhull, A. M., *J. Gen. Physiol.* 61:687 (1973).
333. Cole, K. C., *Membranes, Ions and Impulses*, University of California Press, Berkeley, 1968.
334. Armstrong, C. M., and Bezanilla, F., in reference. 325.
335. Ehrenstein, G., Blumenthal, R., Latorre, R., and Lecar, H., *J. Gen. Physiol.* 63:707 (1974).
336. Vieira, F. L., Caplan, S. R., and Essig, A., *J. Gen. Physiol.* 59:77 (1972).
337. Westerhoff, H. V., and Van Dam, *Curr. Top. Bioenerg.* 9:1 (1979).
338. Gillespie, R. J., Maw, G. A., and Vernon, C. A., *Nature* 171:1147 (1953).
339. Mitchell, P., *Biochem. Soc. Trans.* 4:399 (1976).
340. Jagendorf, A. T., and Uribe, E., *Proc. Natl. Acad. Sci.* 55:170 (1966).
341. Portis, A. R., and Heldt, H. W., *Biochem. Biophys. Acta.* 449:434 (1976).
342. Krause, G. H., *Biochem. Biophys. Acta* 460:500 (1977).

List of Symbols

Symbol	Reference	Meaning
A_{LmR}	(7.19) pg 158	Hamaker Constant
$A_{LmR}(l)$	(7.20) pg 158	Hamaker Coefficient
A	(15.28) pg 319	Affinity
B_1	(6.31) pg 121	Expansion coeffiecient in Margules equation
C_1		Expansion coefficient in Margules equation
C_P	(6.53) pg 128	Specific heat at constant pressure
$C_{BB}(d_{\max})$	Fig 89 Cap pg 229	Connectedness function
c_i	(12.5) pg 261	Concentration of species i
d	Fig 22 Cap pg 40	Repeat distance between bilayers
d_{p-p}	Fig 22 Cap pg 40	Distance between two points of maximum electron density in bilayer
d_w	pg143, equation	Interlayer separation of stacked bilayers
D	(5.7) pg 92	Diffusion constant
$\hat{E}(r)$	(7.30) pg 165	Electric field
e^2qQ/h	below (5.2) pg 81	Quadrupole coupling constant
f	(16.21) pg 333	Partial molar frictional coefficient
F	(11.5) pg 250	Faraday
g	(6.55) pg 129	Aggregation number of micelles
G	(6.59) pg 130	Free energy
$g^{\pm}$	penultimate line pg 93	Gauche conformations
h	(7.82) pg 183	Planck's constant
$\hbar$	(7.9) pg 153	$h/2$
H	(6.15) pg 118	Enthalpy
H_{MF}	(8.12) pg 201	Mean field Hamiltonian
I	(14.3) pg 301	Current
J^{0d}, J^{d0}	(12.51), (12.52)	Unidirectional flows, pg 274
J_i	(12.1) pg 258	Flow of species i
$J(\omega)$	(17.26) pg 351	Spectral density

$\underline{k}$	(13.5) pg 280	Rate constant
k	(13.2) pg 279	Rate constant
K	(7.42) pg 167	Debye constant
$K_{\alpha\beta}$	(8.11) pg 201	Pair interaction energy
L_{ij}	(15.3) pg 313	Expansion coefficient in phenomenological equations
$\mathfrak{L}_{ij}$	(8.11) pg 201	Projection operator
$\langle L \rangle$	(5.12) pg 95	Average chain length
m	(12.3) pg 259	mass
$n_i(x)$	(7.36) pg 166	Ion density
P	(7.56) pg 177	Pressure
P	(17.5) pg 340	Probability
P	(12.14) pg 263	Permeability coefficient
$P_2(\cos\theta)$	(10.14) pg 241	Second Legendre polynomial
Q	(17.30) pg 353	Fluorescence quantum yield
q	(15.46) pg 322	Degree of coupling
$r(t)$	(10.11) pg 241	Fluoresence anisotropy
S	(8.24) pg 203	Entropy
S	(1.1) pg 17	Area taken up by phospholipid at bilayer surface
S_{CD}	(5.6) pg 82	Order parameter of C—D bond
S_{mol}		Order parameter of methylene group
T_1, $T_{1\varrho}$	pg 93 line 10	Nuclear relaxation times
T_t	(Table 1) pg 28	Melting point of lipid chains
t	penultimate line pg 93	Trans conformation
u	(12.16) pg 264	Mobility
u	(9.2) pg 223	Order parameter
$\bar{u}^{\pm}$	(12.33) pg 267	Mean mobilities
$\underline{v}$	(12.3) pg 259	Velocity
$\bar{V}$	(11.5) pg 250	Partial molar volume
$V(z)$	(8.26) pg 204	Potential energy
x	(6.10) pg 117	Mole fraction
z	(7.33) pg 166	Number of charges on ion
Z	(8.17) pg 202	Partition function
Z	(6.37) pg 122	Coordination number
$\alpha(\omega)$	(7.7) pg 152	Polarizability at frequency ω
β	(11.3) pg 249	Partition coefficient
γ	(6.1) pg 112	Activity coefficient
γ	(17.5) pg 340	Conductance
γ	(6.72) pg 134	Surface tension
δ_{ij}	(17.12) pg 342	Kronecker delta
$\Delta\alpha$	(7.82) pg 183	Change in polarizability
$\Delta\tilde{\mu}$		Change in dipole moment
Δp	(18.6) pg 377	Protonmotive force
$\Delta\pi$	(16.4) pg 328	Osmotic pressure
$\Delta\phi_D$	(12.46) pg 272	Diffusion potential
$\Delta\phi_M$		Membrane potential
$\Delta\nu$	(7.82) pg 183	Frequency difference
$\Delta\nu_Q$	(5.2) pg 81	Quadrupole splitting
ΔH_{vH}	(6.49) pg 126	Van't Hoff transition enthalpy

ϵ	(17.30) pg 353	Extinction coefficient
ϵ	(7.2) pg 138	Dielectric constant
ϕ	(15.14) pg 315	Dissipation function
ϕ	(8.28) pg 205	Dihedral angle
ϕ	(8.6) pg 198	Molecular field
ϕ_0	(13.31) pg 290	Surface potential
$\phi(r)$	(7.28) pg 165	Potential
ζ	(7.77) pg 182	Zeta potential
η	(5.7) pg 92	Viscosity
θ	(6.47) pg 125	Extent of transition
λ	(13.1) pg 290	Conductance
μ	(6.1) pg 112	Chemical potential
μ°	(6.1) pg 112	Standard chemical potential
$\tilde{\mu}$	(11.4) pg 250	Electrochemical potential
μ^e	3rd line pg 118	Excess chemical potential
μ_i	(10.6) pg 235	Viscosity
$\pi_{i,j}(t)$	(17.10) pg 342	Conditional probability
$\varrho(x)$	(7.33) pg 166	Point charge density
σ_{ii}	Bottom pg 99	Component of chemical shift tensor

Index